AF386154

The IMA Volumes
in Mathematics
and its Applications

Volume 139

Series Editors
Douglas N. Arnold   Fadil Santosa

Springer
New York
Berlin
Heidelberg
Hong Kong
London
Milan
Paris
Tokyo

Institute for Mathematics and its Applications (IMA)

The **Institute for Mathematics and its Applications** was established by a grant from the National Science Foundation to the University of Minnesota in 1982. The primary mission of the IMA is to foster research of a truly interdisciplinary nature, establishing links between mathematics of the highest caliber and important scientific and technological problems from other disciplines and industry. To this end, the IMA organizes a wide variety of programs, ranging from short intense workshops in areas of exceptional interest and opportunity to extensive thematic programs lasting a year. IMA Volumes are used to communicate results of these programs that we believe are of particular value to the broader scientific community.

The full list of IMA books can be found at the Web site of the Institute for Mathematics and its Applications:

http://www.ima.umn.edu/springer/full-list-volumes.html.

Douglas N. Arnold, Director of the IMA

* * * * * * * * * *

IMA ANNUAL PROGRAMS

1982–1983	Statistical and Continuum Approaches to Phase Transition
1983–1984	Mathematical Models for the Economics of Decentralized Resource Allocation
1984–1985	Continuum Physics and Partial Differential Equations
1985–1986	Stochastic Differential Equations and Their Applications
1986–1987	Scientific Computation
1987–1988	Applied Combinatorics
1988–1989	Nonlinear Waves
1989–1990	Dynamical Systems and Their Applications
1990–1991	Phase Transitions and Free Boundaries
1991–1992	Applied Linear Algebra
1992–1993	Control Theory and its Applications
1993–1994	Emerging Applications of Probability
1994–1995	Waves and Scattering
1995–1996	Mathematical Methods in Material Science
1996–1997	Mathematics of High Performance Computing
1997–1998	Emerging Applications of Dynamical Systems
1998–1999	Mathematics in Biology

Continued at the back

David R. Brillinger   Enders Anthony Robinson
Frederic Paik Schoenberg
Editors

Time Series Analysis and Applications to Geophysical Systems

With 94 Illustrations

Springer

David R. Brillinger
Department of Statistics
University of California,
 Berkeley
367 Evens Hall
Berkeley, CA 94720-3860
USA
E-mail:
 brill@stat.berkeley.edu

Enders Anthony Robinson
Department of Earth and
 Environmental Engineering
Henry Krumb School of
 Mines
Columbia University
918 Seeley Mudd Building
500 120th Street
New York, NY 10027
USA
E-mail: earl1@columbia.edu

Frederic Paik Schoenberg
Department of Statistics
University of California,
 Los Angeles
8142 Math-Science Building
Los Angeles, CA 90095-1554
USA
E-mail: frederic@stat.ucla.edu

Series Editors:
Douglas N. Arnold
Fadil Santosa
Institute for Mathematics and its Applications
University of Minnesota
Minneapolis, MN 55455
USA
http://www.ima.umn.edu

Mathematics Subject Classification (2000): 62M10, 62M15, 62M20, 60G35, 86A15, 86A10, 86A05, 86A22, 86A32, 86A40, 86A60, 62H12, 94A12, 94A13, 60G15, 60G25, 60G60, 62H11, 62M30, 93E10, 93E11, 60K35.

Library of Congress Cataloging-in-Publication Data

On file.

Printed on acid-free paper.

ISBN-13: 978-1-4612-7735-4   e-ISBN-13: 978-1-4612-2962-9
DOI: 10.1007/ 978-1-4612-2962-9

9 8 7 6 5 4 3 2 1   SPIN 10990735

Springer-Verlag is part of *Springer Science+Business Media*

springeronline.com

FOREWORD

This IMA Volume in Mathematics and its Applications

TIME SERIES ANALYSIS AND APPLICATIONS TO GEOPHYSICAL SYSTEMS

contains papers presented at a very successful workshop on the same title. The event which was held on November 12-15, 2001 was an integral part of the IMA 2001-2002 annual program on "Mathematics in the Geosciences." We would like to thank David R. Brillinger (Department of Statistics, University of California, Berkeley), Enders Anthony Robinson (Department of Earth and Environmental Engineering, Columbia University), and Frederic Paik Schoenberg (Department of Statistics, University of California, Los Angeles) for their superb role as workshop organizers and editors of the proceedings. We are also grateful to Robert H. Shumway (Department of Statistics, University of California, Davis) for his help in organizing the four-day event.

We take this opportunity to thank the National Science Foundation for its support of the IMA.

Series Editors

Douglas N. Arnold, Director of the IMA

Fadil Santosa, Deputy Director of the IMA

PREFACE

This volume contains a collection of papers that were presented during the Workshop on Time Series Analysis and Applications to Geophysical Systems at the Institute for Mathematics and its Applications (IMA) at the University of Minnesota from November 12-15, 2001. This was part of the IMA Thematic Year on Mathematics in the Geosciences, and was the last in a series of four Workshops during the Fall Quarter dedicated to Dynamical Systems and Ergodic Theory. The Workshop brought together 28 scientists from around the world and from various scientific backgrounds: many were specialists in the statistical analysis of time series; others were geophysicists, geologists, or climatologists with mainly subject matter expertise. The main goals of this Workshop were to engage discussion between these groups in order to facilitate the application of recent methodological advances in time series analysis to the most important geophysical problems.

Before other matters, we extend our sincerest thanks for making the Workshop a success to Robert Shumway, who not only presented but also served as the local organizer of the Workshop with very little advanced notice, and by all accounts did a superb job. The idea for this workshop, and more generally for the IMA Thematic Year on Mathematics in the Geosciences, was Bill Newman's, and we thank Bill for all his help and for asking us to get involved. We also thank the IMA staff, especially Patricia V. Brick for her enormous help in coordinating and overseeing these proceedings and preparing this book for publication, as well as Alison Givand, Willard Miller, and Douglas N. Arnold for their helpful assistance and supervision. We also thank the authors for their contributions.

Univariate and multivariate time series methods are critical in the analysis and identification of dynamical properties in a wide range of geophysical systems. While traditional approaches are based on the spectral analysis of random processes, more recent developments incorporate ideas from the ergodic theory of dynamical systems. The interaction of these two approaches provides unique opportunities for the application of time series methods to the geosciences.

Much of the path connecting time series and geophysics was paved by John Tukey, who unfortunately died in the year preceding this workshop. Tukey made giant strides in applying concepts such as robust estimation, spectral analysis, and exploratory data analysis to geophysics, and the works in these Proceedings build upon his great contributions. The early interest of Tukey in geophysics goes back to the "Symposium on Autocorrelation Analysis applied to Physical Problems" held at Woods Hole, MA in June 1949. Tukey's paper entitled "The sampling theory of power spectrum estimates" was the high point of this meeting. This paper appears as pages 129-160 in The Collected Works of John W. Tukey, Volume I (1984),

Wadsworth, Belmont, CA. Before Tukey's work, the power spectra computed from empirical autocorrelation functions were too erratic to be of any use in formulating physical hypotheses. Not only did Tukey show correctly how to compute power spectra from empirical data, but he also laid the statistical framework for the analysis of short time series, as opposed to the very long ones envisaged by others.

The works in this volume deal with theoretical and methodological issues as well as real geophysical applications, and are written with both statistical and geophysical audiences in mind. They cover a wide range of important geophysical applications, including the investigation and prediction of climatic variations and the interpretation of seismic signals.

The first four papers deal with the interpretation of seismic signals. Robert H. Shumway, Jessie L. Bonner, and Delaine T. Reiter extend univariate cepstral methods to the problem of deconvolving seismic phases in a multivariate seismic arrays in order to determine the source depth of the seismic event. Genshiro Kitagawa, Tetsuo Takanami, and Norio Matsumoto fit a state space model with time-varying parameters and assess its implications on arrival time estimation, detection of coseismic contamination and spectral changes, and other problems. Enders Robinson suggests a method for obtaining a more refined estimate of a seismic signal in a layered system. Hernando Ombao, Jungeun Heo, and David Stoffer propose fitting piecewise stationary AR models as a way of decoding seismic signals in real time.

The next four papers deal with temperature data. T. Subba Rao and E.P. Tsolaki perform tests and spectral analysis on global climatic data and present nonstationary time series models that explain observed climatic trends and temperature anomalies. Wei Biao Wu similarly investigates global warming trends but from quite a different perspective, namely that of testing whether temperature levels are constant versus isotonic alternatives. T. Subba Rao and Ana Monica Costa Antunes fit space-time ARMA models to data on monthly mean temperatures at various sites in the United Kingdom, and investigate the forecasting performance of the models. Donald B. Percival, James E. Overland, and Harold O. Mofjeld inspect the fit of an autoregressive model, a fractionally differenced model, and a square wave signal plus noise model to a North Pacific climatic index, and conclude that although the three models provide very different predictions, current data are insufficient to discriminate adequately between the models.

The final five papers deal with an assortment of important time series problems and applications. Marc G. Genton and Keith R. Thompson apply a skew-elliptical time series model to hourly sea-level data in Atlantic Canada in order to estimate the risk of flooding. Zhongjie Xie suggests a method for identifying hidden periodicities in spatial time series data and applies the results to data on permeability in Chinese oil fields. T. Ozaki, J.C. Jimenez, H. Peng, and V. Haggan-Ozaki propose using an innovation approach with nonlinear models generally, and radial basis function models in particular, for the description of nonlinear time series data, and illustrate the implications of their results on a variety of different processes including models for the NOx decomposition from thermal power plants. Many geophysical processes appear to be characterized by non-Gaussian noise, though the assumption of Gaussianity is typically assumed in conventional time series methods. Hence the importance of Murray Rosenblatt's discussion of the estimation and prediction of linear non-Gaussian time series models as well as Winston C. Chow and Edward J. Wegman's treatment of, and proposed estimators for, stochastic differential equation models with fractional Gaussian noise.

David R. Brillinger
Department of Statistics
University of California, Berkeley

Enders Anthony Robinson
Department of Earth and Environmental Engineering
Krumb School of Mines
Columbia University

Frederic Paik Schoenberg
Department of Statistics
University of California, Los Angeles

CONTENTS

ASSORTMENT OF IMPORTANT TIME SERIES PROBLEMS AND APPLICATIONS

NONPARAMETRIC DECONVOLUTION OF SEISMIC DEPTH PHASES

ROBERT H. SHUMWAY[*], JESSIE L. BONNER[†], AND DELAINE T. REITER[†]

Abstract. Accurate determination of the source depth of a seismic event is a potentially important goal for better discrimination between deeper earthquakes and more shallow nuclear tests. Earthquakes and explosions generate depth phases such as pP and sP as reflections of the underlying P signal generated by the event. The delay time between the original signal and the pP phase can be used to estimate the depth of the seismic event. Cepstral methods, first used by Tukey and later by others, offer natural nonparametric means for estimating general echo patterns in a single series. Here, we extend the single series methodology to arrays by regarding the ensemble of log spectra as sums of nonstationary smooth functions and a common additive signal whose periods are directly related to the time delays of the seismic phases. Detrending the log spectra reduces the problem to one of detecting a common signal with multiple periodicities in noise. Plotting an approximate cepstral F-statistic over pseudo-time yields a function that can be considered as a deconvolution of the seismic phases. We apply the array methodology to determining focal depths using three component recordings of earthquakes.

Key words. Cepstral F, array processing, signal detection, nuclear monitoring, earthquakes, depth estimation.

1. Introduction. One definitive way of ruling out seismic events as possible nuclear tests is to accurately determine the depth of the event, using the fact that nuclear explosions by their nature must be shallow, whereas earthquakes will be deep. A definitive way to rule out a majority of seismic events as possible nuclear tests is to establish, with high confidence, that the source depth is greater than 15 km.

The use of depth phases or multiple signal arrivals on a seismic record is the most important tool for constraining the depth of a seismic event. Depth phases appear in the seismograms as echoes of the initial P wave that have been reflected at the earth's free surface. Identification of depth phases, such as pP and sP, is dependent upon the amplitude of the arrival at a recording station. Source mechanism, path effects, and reflection coefficients at the earth's surface control the depth phase amplitude. The depths of events are usually estimated by measuring the time delay between the direct (P) and depth (pP and sP) phases. This time delay induces a periodicity in the sample spectrum that is directly related to the the delay of the reflected wave. Relating the delay time to the focal depth of the event can be done by knowing the path propagation properties as in Kennett and Engdahl (1991) . Hence, accurate determination of the delay time

---

[*]Department of Statistics, University of California, Davis, CA 95616.

[†]Weston Geophysical Corporation, 57 Bedford Street, Suite 102 Lexington, MA 02420.

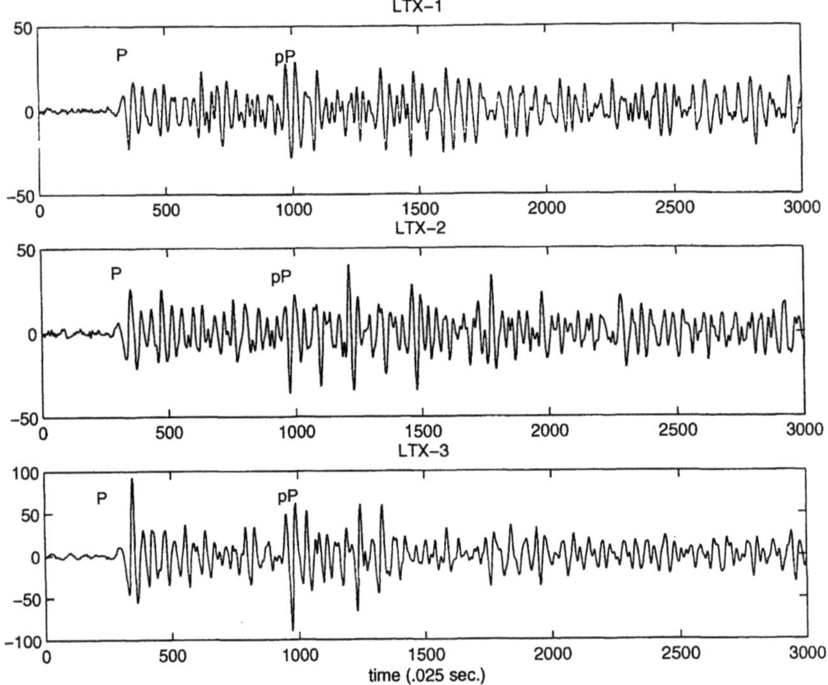

FIG. 1. *A Vertical and two horizontal components for northern Chile earthquake observed at Lajitas Texas. P and pP arrivals are visible after filtering (0.6–4.5 Hz).*

and subsequent depth can serve as a preliminary means for discriminating between earthquakes and explosions.

As an example, consider Figure 1, which shows an event in Northern Chile, as observed on a three component vertical array at Lajitas, Texas. Sampling is at 40 points per second and the data are filtered in a signal pass-band ranging from .6 to 4.5 cycles per second (Hz). The magnitude 5.3 event occurred in May, 2000 and the depth, reported by the U.S. Geological Survey, was 57 km. For a depth of 57 km, the method of Kennett and Engdahl (1991) predicts an arrival of pP at 15.42 seconds or about 620 points at the sampling rate given above. The P and pP arrivals are marked approximately on the three components and there is obvious ambiguity in assigning the delay to be assigned to the second pP phase. The important feature of the second arrival that distinguishes from other apparent arrivals is that it must appear consistently on all three components. If the delay time is denoted by τ and there is a modification of the amplitude of the pP reflection by a multiplier θ, it is natural to express the received signal at each channel as $s(t) + \theta s(t - \tau)$, when $s(t)$ is the underlying signal.

Since, the signal is received in echo form, it is natural to employ signal processing techniques that exploit this feature. The cepstrum was introduced as a technique for echo estimation by Bogert et al. (1962) and has appeared in applications to speech and image processing as well as in seismology where a number of authors (see, for example, Kemerait, 1982, Baumgardt and Ziegler, 1988, Alexander, 1996, Shumway et al., 1998) have utilized it as a technique for modeling multiple arrivals. The idea behind the cepstrum as a tool for analysis is that there will be periodicities induced in the spectrum that are proportional to the delay times of the arrivals. These periodicities are often quite strong over a broad frequency range and are enhanced by looking at the log spectrum. In this paper, we exploit the above properties by thinking of the detrended log spectra at the different channels as the sum of a signal and noise, where the signal is roughly periodic and the same on each channel. This allows application of conventional methods for detecting a signal in a collection of stationarily correlated noise series as in Shumway (1971) and Shumway et al. (1998).

In the next section, we develop a multiplicative signal and noise model that exhibits the log spectrum of the data in terms of an additive model as a function of frequency. In Section 3, the discrete Fourier transform (DFT) of the sample log spectra gives a signal plus noise model in the quefrency or pseudo-time domain that can be handled by the usual analysis of power techniques (see, for example, Shumway and Stoffer, 2000). The F-statistic obtained exhibits the echos at the proper delay times, giving the primary estimated output delay needed for determining depth. In section 4, the test procedure is applied to the Northern Chile earthquake shown in Figure 1.

2. Multiplicative signal models. We suppose here that N observed series $y_j(t), j = 1, 2, \ldots, N$ can be expressed as the convolution of a fixed unknown function $a_j(t)$ with a delayed stochastic unknown signal $s_j(t)$ and a noise process $n_j(t)$, assumed to be a linear process with square summable coefficients. The model for the observed data becomes

$$(1) \qquad y_j(t) = a_j(t) \otimes [s_j(t) + \theta s_j(t - \tau)] \otimes n_j(t),$$

where we assume that the P phase reflection pP is delayed by τ points and scaled by a reflection parameter $|\theta| < 1$. The notation $a(t) \otimes b(t) = \sum_s a(s)b(t - s)$ denotes the convolution of the series $a(t)$ and $b(t)$. It is natural to handle (1) in the frequency domain because the theoretical spectrum of such a process will be of the form

$$
\begin{aligned}
(2) \qquad f_{y_j}(\nu) &= |A_j(\nu)|^2 |1 + \theta e^{-2\pi i \nu \tau}|^2 f_{s_j}(\nu) f_{n_j}(\nu) \\
&= |A_j(\nu)|^2 (1 + \theta^2 + 2\theta \cos 2\pi\nu\tau) f_{s_j}(\nu) f_{n_j}(\nu)
\end{aligned}
$$

where $A_j(\nu)$ is the Fourier transform of $a_j(t)$ and $f_{s_j}(\nu)$ and $f_{n_j}(\nu)$ are the spectra of the signal and noise respectively, with frequency ν measured in cycles per point over the range $-1/2 \leq \nu \leq 1/2$. The above form for the spectrum exhibits it as the product of multiplicative noise, a fixed signal function and a periodic component, with periodicities determined as a known function of the time delays.

We note that the multiplicative model (2) for the spectra is implied by the signal model (1), which assumes that the observed data will be a convolution of the signal and noise. This recognizes signal-generated noise as the major component of the model rather than the usual additive noise. This is mainly a result of noticing that noise within the signal window is generally different and larger than the noise preceding the signal. Multiplicative noise models also dominate a good portion of the time series literature because of the popularity of multiplicative ARMA modelling as developed by Box et al. (1994). In fact, the model above would be close to a seasonal moving average model of order one, with the seasonal period corresponding to the delay τ. A further motivating factor is that taking logarithms leads to a simple additive model for the log spectrum as will be shown below.

The dynamic range spanned by typical spectra tends to show the periodicities as being proportional to the magnitude of the spectral function. Taking logarithms helps stabilize the dynamic range and also leads to an additive model of the form

$$
\begin{aligned}
\log f_{y_j}(\nu) &= \log |A_j(\nu)|^2 + \log f_{s_j}(\nu) \\
&\quad + \log\left(1 + \theta^2 + 2\theta \cos 2\pi\nu\tau\right) + \log f_{n_j}(\nu) \\
&= T_j(\nu) + \log\left(1 + \theta^2 + 2\theta \cos 2\pi\nu\tau\right) + \log f_{n_j}(\nu),
\end{aligned}
\tag{3}
$$

where

$$
T_j(\nu) = \log |A_j(\nu)|^2 + \log f_{s_j}(\nu)
\tag{4}
$$

trend function $T_j(\nu)$ is assumed to be smooth for each channel. In later arguments, we will identify the fixed additive function defined by the sum of the first two terms in (3) with a trend component that is different on each series. The common component in each series has the additive function whose period is proportional to the time delay τ. Hence it seems sensible to consider the Fourier transform of the log spectra as underlying data. For sampled data, consider modeling the $\log |Y_j(\nu)|^2 - T_j(\nu)$, where

$$
Y_j(\nu_\ell) = n^{-1/2} \sum_{t=0}^{n-1} y_j(t) e^{-2\pi i \nu_\ell t}
\tag{5}
$$

is the DFT of the original process and its squared value is the usual periodogram. We may use (3) at frequencies of the form $\nu_\ell = \ell/n, \ell = 0, 1, \ldots, n-1$ cycles per frequency point and think of the detrended version of (3) as a series in psueudo-time ν. Then, compute the sample periodogram again at delays of the form $d_k = k/n, k = 0, 1, \ldots, n-1$, i.e.,

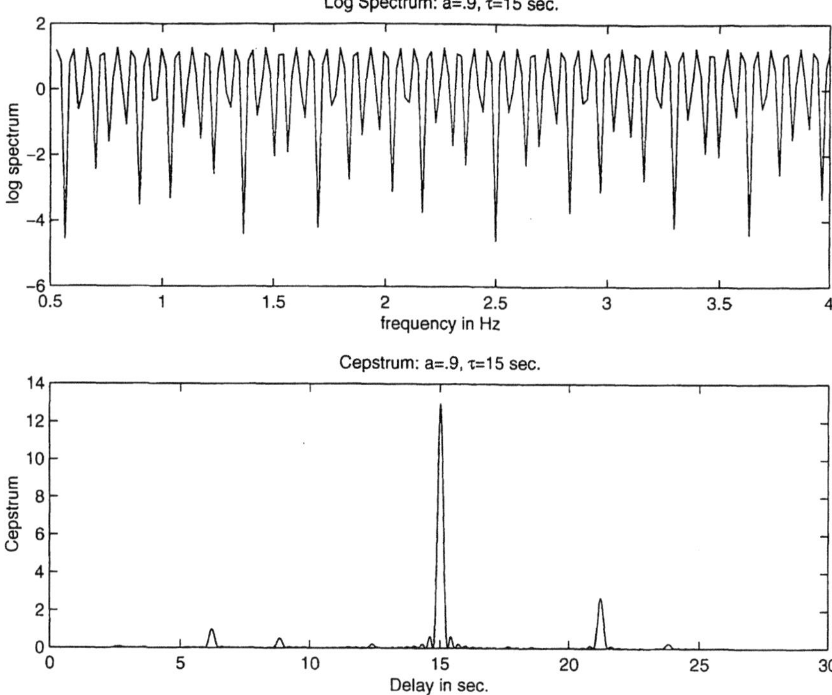

FIG. 2. *Plot of the function* $g(\nu; \tau, \theta)$ *in (7) for* $\theta = .9, \tau = 600$ *i.e, (top panel) and its predicted spectrum or cepstrum (bottom panel) showing pP arrival at about 15 seconds (600 points).*

$$(6) \qquad C_j(d_k) = n^{-1/2} \sum_{\ell=0}^{n-1} \left(\log |Y_j(\nu_\ell)|^2 - T_j(\nu_\ell)\right) e^{-2\pi i d_k \nu_\ell}.$$

The resulting sample cepstra should show peaks at delays corresponding to the periodicities in the spectra.

In order to check how this procedure might work in practice, consider the upper panel in Figure 2 which shows the function

$$(7) \qquad g(\nu; \tau, \theta) = \log \left(1 + \theta^2 + 2\theta \cos 2\pi\nu\tau\right)$$

for $\tau = 600$ points and $\theta = .9$. The Fourier transform of $|g(\nu; \tau, \theta)|^2$ is shown in the lower panel and we note the peak at the correct delay of 15 seconds shows up in the component of (3) that contains the parameters of the reflection. There are, of course, some small peaks due to finite computations.

Before proceeding further it is useful to examine the log spectra of the northern Chile earthquake, as shown in Figure 3 for the bandwidth 1-5 Hz. It is clear that the periodicities noted for the underlying model all have

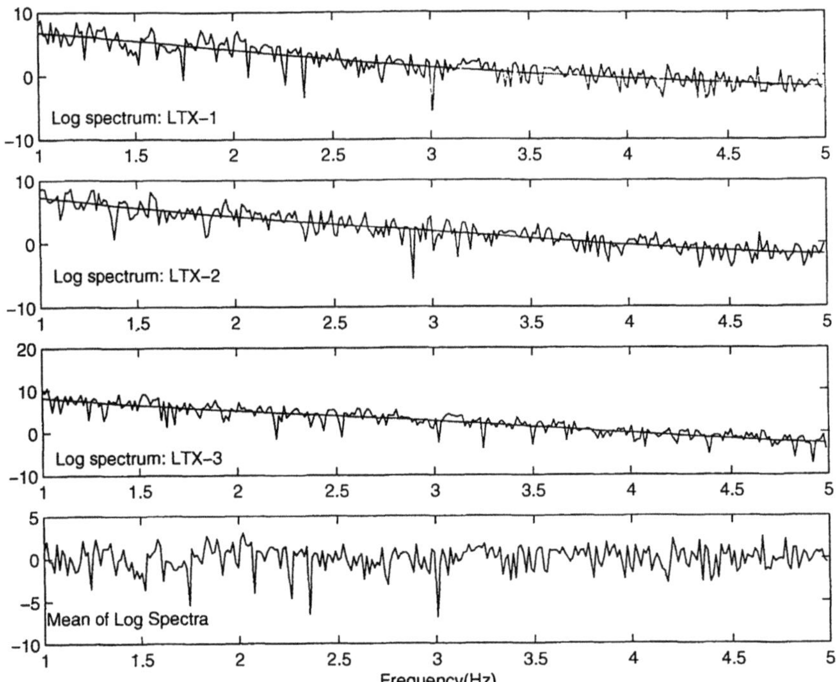

FIG. 3. *Log spectra for the three component data and the mean of the detrended log spectra. The smoother line is the cubic spline with one knot that was used to detrend the log spectrum. The mean series should show common periodicities corresponding to the pP time delay.*

strong trends which was the motivation for the term $\log |A_j(\nu)|^2$ in the model (3). However, there will always be a fairly smooth underlying trend function observed over the frequency band of interest. Hence, a cubic spline with a single knot is usually sufficient for detrending. In this context, we applied a regression spline model of the form

$$(8) \qquad T_j(\nu) = a_{j0} + a_{j1}\nu + a_{j2}\nu^2 + a_{j3}\nu^3 + a_{j4}(\nu - \nu_0)_+^3,$$

where ν_0 is the knot location and $(\nu - \nu_0)_+^3$ is zero for $\nu < \nu_0$. Placing the knot at the middle of the frequency leads to the smooth fitted lines shown in Figure 3. Adjusting each log spectrum for its fitted cubic spline at frequencies of the form $\nu_\ell = \ell/n$, say $\widehat{T_j}(\nu_\ell)$, and adjusting leads to a detrended series of the form

$$(9) \quad \log |Y_j(\nu_\ell)|^2 - \widehat{T_j}(\nu_\ell) = \log \left(1 + \theta^2 + 2\theta \cos 2\pi\nu_\ell\tau\right) + \log f_{n_j}(\nu_\ell).$$

Before proceeding further, it is useful to consider the distribution theory for the residual spectra $f_{n_j}(\nu_\ell)$ in (9). We looked at the residuals from the mean detrended residuals shown in the bottom panel of Figure 3. The distributions were approximately normal except for observations in the left tails that are caused by the expected dips in the series due to failure to fit the theoretical reflection pattern shown in the top panel of Figure 2. It should be noted that we take another Fourier transform in the next section so that there will be a further central limit effect. The autocorrelation functions of all three residuals were essentially zero at all lags. The form of (9) also suggests a nonlinear regression approach with θ and τ as the parameters but Figure 2 implies that the derivatives may be unstable so we opt for simply isolating periodicities via the nonparametric cepstral approach in the next section.

3. Cepstral analysis of power and the F statistic. The residuals in (9) are nearly white and the DFTs (5) will be nearly Gaussian. Hence, the Fourier transform (6), applied to (9) will give a model of the form

$$(10) \qquad C_j(d_k) = S(d_k) + N_j(d_k),$$

where the signal transform $S(d_k)$ have a peak at the time delay d_k corresponding to the periodicity in the function $g(\nu)$ in (7) (see Figure 2) . Noises will be uncorrelated, with variance equal to the cepstral variance $\sigma^2(d_k)$. Since we would like to determine the frequency where the primary signal lives, it is natural to apply the classical approach to detecting a fixed signal in noise, as proposed in Shumway (1971, 1998). Suppose that we ask for the test statistic for testing $S(d_0) = 0$ at a particular delay d_0, and suppose that we assume the variance of the noise to be constant, i.e. $\sigma^2(d)$ over some interval in the neighborhood of d_0. If L(odd) values of $\{C_j(d_0 + d_k), k = -\frac{L-1}{2}, \ldots, 0, \ldots, \frac{L-1}{2}\}$ are observed in some interval, define the total cepstral power as

$$(11) \qquad \text{TCP}(d_0) = \sum_{|k|<L/2} \sum_{j=1}^{N} |C_j(d_0 + d_k)|^2.$$

The total cepstral power, called the stacked cepstrum by Alexander (1996), has been used in the past for detecting time delays. We also write the mean or beam cepstral power as

$$(12) \qquad \text{BCP}(d_0) = N \sum_{|k|<L/2} |\bar{C}(d_0 + d_k)|^2,$$

where the mean is over $j = 1, 2, \ldots N$ channels. The cepstral noise power is

$$\text{NCP}(d_0) = \sum_{|k|<L/2} \sum_{j=1}^{N} |C_j(d_0 + d_k) - \bar{C}(d_0 + d_k)|^2$$

$$(13)$$

$$= \text{TCP}(d_0) - \text{BCP}(d_0).$$

The statistic resulting from the test $C(d_0) = 0$ is given by

$$(14) \qquad F_{2L,2L(N-1)}(d_0) = (N-1)\frac{\text{BCP}(d_0)}{\text{NCP}(d_0)},$$

which is distributed as central F with $2L$ and $2L(N-1)$ degrees of fredom when $\left\{ S(d_0 + k) = 0, k = -\frac{L-1}{2}, \ldots, 0, \ldots, \frac{L-1}{2} \right\}$ and as non-central F otherwise. The non-centrality parameter is proportional to $2NL$ times the integrated signal power over the bandwidth and inversely proportional to the noise spectrum.

We may apply the procedure to the problem of determining the depth of the Northern Chile event shown in Figure 1. As mentioned, earlier the US Geological Survey estimated the depth at 57.2 km, which corresponded to a pP arrival 15.4 seconds after P. For this data the sampling rate is 40 points per second, leading to a folding frequency of 20 Hz. The data in Figure 1 were bandpass filtered to restrict the series to the interval from .6 to 4.5 Hz. The log spectra have already been exhibited in Figure 3 and we note that some periodicity is evident in the mean of the detrended log spectra. Note that the first three channels are $\log(|Y_j(\nu_\ell)|^2, j = 1, 2, 3,$ whereas the last channel is the average of the three residuals on the left side of (9).

Two components of the cepstral power are shown in Figure 4, with the total power $\text{TCP}(\cdot)$ as the dashed line and the beam power $\text{BCP}(\cdot)$ as the solid line. For this analysis, we had 3000 data points and 1500 frequencies from zero to the folding frequency. Taking $L = 51$ would imply a bandwidth of about 1.25 sec in the pseudo-time domain. Since $N = 3$ in this case, the F-statistic shown in the bottom panel has $2(51) = 102$ and $2(51)(3 - 1) = 204$ degrees of freedom. We note the strong peak at 16.3 seconds, the .001 critical value $F_{102,204}(.001) = 1.66$ by a substantial amount. This peak corresponds to a depth of 61.1 km as compared to the quoted U.S. Geological Survey estimate of 57.2 km. If we suppose that upper and lower limits are defined as the narrow part of the peak, we might take 15.2 and 17.2 seconds as rough lower and upper limits; the depths corresponding to these two delays are 56.0 and 65.3 km. The broad limits implied suggest that determination of depth from delays may not be as accurate as assumed when time delays are read off the single series by an analyst.

4. Discussion. We have developed a nonparametric approach to estimating a single delay observed on multiple time series. The method is based on a decomposition of the detrended log periodogram into a smooth function specific to each component and a periodic function common to all components. The periodic function contains the delay information in its period. Transforming again to the pseudo time domain, we obtained a model implying that the transformed log periodogram could be represented as the sum of a deterministic signal and noise. Using known results

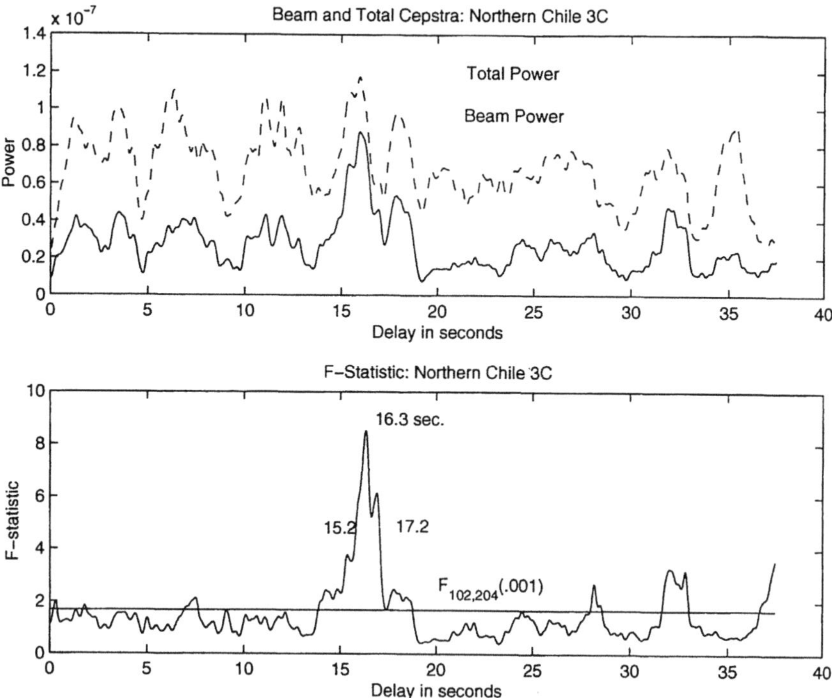

FIG. 4. *Analysis of cepstral power components (top panel) and F-statistics for northern Chile earthquake (bottom panel) showing estimated pP arrival at about 16 seconds. Estimated USGS depth for this event was 57.2 km, which would predict (Kennett and Engdahl, 1991) a travel time of about 15.5 seconds.*

for detecting a deterministic signal in stationary noise gave an F statistic that depended on the sum of the spectra and the spectrum of the mean.

Using the approach to estimate depth was illustrated for one single array recording a northern Chile earthquake. Large arrays and even combinations of arrays should lead to more accurate depth estimates. A large number of depth determinations using data from the Prototype International Data Center, Reviewed Event Bulletin and from the USGS bulletin were made in Reiter et al. (2002). The results show that the method is highly successful for teleseismic distance, slightly less reliable at far regional distances and problematic at distances less than 10 degrees from the source. Other events, data and the software can be found on the website *http://www.weston-geo.com/cepstral_data.html.*

Acknowledgements. The support of the Air Force Technical Applications Center (AFTAC) through the U.S. Department of Defense, Defense Threat Reduction Agency, Contract No. DSWA01-98-C-0142 with Weston Geophysical Corporation is gratefully acknowledged.

REFERENCES

ALEXANDER S.S. (1996). A new method for determining source depth from a single regional station. *Seismic Research Letters*, **67**, p. 63.

BAUMGARDT D.R. AND K.A. ZIEGLER (1988). Spectral evidence for source multiplicity in explosions: Application to regional discrimination of earthquakes and explosions. *Bull. Seismolog. Soc. Amer.*, **78**, 1773–1795.

BOGERT G.M., M.J. HEALY, AND J.W. TUKEY (1962). The quefrency analysis of times series for echoes. In *Proceedings of a Symposium on Time Series Analysis*, ed. M. Rosenblatt. New York: John Wiley.

BOX G.E.P., G.M. JENKINS, AND G.C. REINSEL (1994). *Time Series Analysis, Forecasting, and Control, 3rd ed.*. Englewood Cliffs, N.J.: Prentice Hall.

KEMERAIT R.C. AND A.F. SUTTON (1982). A multidimensional approach to seismic depth estimation. *Geoexploration*, **20**, 113–130.

KENNETT B.L.N. AND E.R. ENGDAHL (1991). Travel times for global earthquake location and phase identification. *Geophys. J. Int.*, **105**, 429–465.

REITER D.T., J.L. BONNER, AND R.H. SHUMWAY (2002). Application of a cepstral F-statistic for improved depth estimation. Final Scientific Report, Contract DSWA01-C-98-0142, Weston Geophysical, 57 Bedford St., Suite 102, Lexington, MA 02420.

SHUMWAY R.H. (1971). On detecting a signal in N stationarily correlated noise series. *Technometrics*, **10**, 523–534.

SHUMWAY R.H., D.R. BAUMGARDT, AND Z.A. DER (1998). A cepstral F statistic for detecting delay-fired seismic signals. *Technometrics*, **40**, 100–110.

SHUMWAY R.H. AND D.S. STOFFER (2000). *Time Series Analysis and Its Applications*. New York: Springer-Verlag.

STATE SPACE APPROACH TO
SIGNAL EXTRACTION PROBLEMS IN SEISMOLOGY*

GENSHIRO KITAGAWA[†], TETSUO TAKANAMI[‡], AND NORIO MATSUMOTO[§]

Abstract. State space methods for extracting signal from noisy seismic data are shown. The method is based on the general state space model, recursive filtering and smoothing algorithms. The self-organizing state space model is used for the estimation of time-varying parameter of the model. In this paper, we show five specific examples of time series modeling for signal extraction problems related to seismology. Namely, we consider 1) the estimation of the arrival time of a seismic signal, 2) the extraction of small seismic signal from noisy data, 3) the detection of the coseismic effect in groundwater level data contaminated by various effects from air pressure etc., 4) the estimation of changing spectral characteristic of seismic record, and 5) spatial-temporal smoothing of OBS data.

Key words. State space model; Monte Carlo filter; Arrival time; Seismic wave; Ground water level; Coseismic effect; Time-varying spectrum.

AMS(MOS) subject classifications. Primary 60G35, 62F15, 86A15.

1. Introduction. By the rapid progress of information technologies, it become possible to obtain various observations from many locations automatically. In Japan, in an attempt to predict big earthquakes, nation-wide seismological network system was established in 1979, and huge amount of various observations have been accumulated. However, in the analysis of such observations causes two problems. Firstly, the seismic signals observed by seismometers are contaminated by various kinds of natural forces, such as microtremors, microseisms, wave, wind, tide, air pressure, precipitation and a variety of human induced sources. Since the noise level is almost a constant independent of the signal, the effect of the background noise becomes more severe for earthquakes with smaller magnitudes. Therefore, to analyze seismic signals with smaller magnitudes, we need to develop a more sophisticated procedure which can handle very noisy data. Secondly, the number of earthquakes increases exponentially as the magnitude decrease and linearly as the number of data increases. Therefore, for the processing of increasingly many earthquakes, it becomes necessary to develop a

---

*This is an expository article based on the previous papers [22],[24],[25]. A part of this study was carried out under the ISM Cooperative Research Program (2001-ISM·CRP-2026).

[†]The Institute of Statistical Mathematics, Minato-ku, Tokyo 106-8569, Japan. The work of the first author was supported in part by Grant-in-Aid for Scientific Research (B)(2) 13558025 and (C)(2) 12680321 from Japan Society for the Promotion of Science.

[‡]Institute of Seismology and Volcanology, Hokkaido University, Sapporo 060-0810, Japan. Concerning Section 7, he is grateful to the coauthors of the paper [25].

[§]Geological Survey of Japan, National Institute of Advanced Industrial Science and Technology, Tsukuba 305-8567, Japan.

computationally efficient statistical method that can automatically detect seismic signal from noisy data.

Statistical modeling procedure mainly consists of three parts, namely models, computational methods and the model evaluation criterion. In this paper, we review the systematic use of state space models for the modeling of seismic signals. For the extraction of the signal and for the decomposition of noisy data, we developed various model and all of them can be expressed in general state space model form. For the estimation of unknown state vector, three types of recursive filtering algorithms, Kalamn filter, non-Gaussian filter and the Monte Carlo filter are used. For the estimation of time-varying parameter of the model, the self-organizing state space model is used. We further use the Akaike Information Criterion (AIC) for the evaluation of the goodness of the model which facilitates establishment of automatic signal extraction procedures.

In this paper, we show five specific examples of time series modeling for signal extraction problems related to seismology. Namely, we consider 1) the estimation of the arrival time of seismic signal, 2) the extraction of small seismic signal from noisy data, 3) the detection of the coseismic effect in groundwater level data contaminated by various effects from air pressure etc., 4) the estimation of changing spectral characteristic of seismic record, and 5) spatial-temporal smoothing of OBS data.

2. Extraction of the signal by state space modeling.

2.1. State Space Models. Given a time series y_n, the ordinary linear Gaussian state space model is defined by

$$(2.1) \qquad x_n = F_n x_{n-1} + G_n v_n$$

$$(2.2) \qquad y_n = H_n x_n + w_n,$$

where x_n is an unknown state vector and $v_n \sim N(0, Q_n)$ and $w_n \sim N(0, R_n)$ are Gaussian white noise sequences. (2.1) and (2.2) are called the system model and the observation model, respectively. The initial state x_0 is assumed to be distributed as $N(x_{0|0}, V_{0|0})$. The matrices F_n, G_n, H_n, Q_n and R_n are assumed to be known here. However in actual data analysis, they usually contain unknown structural parameters or hyper-parameters denoted as θ. The estimation of this θ will be briefly discussed in subsections 2.5 and 2.6.

A nonlinear non-Gaussian state space model is defined by

$$(2.3) \qquad x_n = F_n(x_{n-1}, v_n)$$

$$(2.4) \qquad y_n = H_n(x_n, w_n),$$

where v_n and w_n are the system noise and the observation noise with densities $q_n(v)$ and $r_n(w)$, and $F_n(x, v)$ and $H_n(x, w)$ are possibly nonlinear functions of the state and the noise inputs.

The above nonlinear non-Gaussian state space model specifies the conditional density of the state given the previous state, $p(x_n|x_{n-1})$, and that of the observation given the state, $p(y_n|x_n)$. This is the essential features of the state space model, and it is sometimes convenient to express the model in this general form based on conditional distributions

$$(2.5) \qquad x_n \sim Q_n(\ \cdot\ |x_{n-1})$$

$$(2.6) \qquad y_n \sim R_n(\ \cdot\ |x_n).$$

With this model, it is possible to treat discrete models such as the inhomogeneous Poisson process or hidden Markov model.

2.2. Estimation of the state by the Kalman filter/smoother.
The most important problem in state space modeling is the estimation of the state vector x_n from the observations, $Y_t \equiv \{y_1, \ldots, y_t\}$, since many important problems in time series analysis can be solved by using the estimated state vector. The problem of state estimation can be formulated as the evaluation of the conditional density $p(x_n|Y_t)$. Corresponding to the three distinct cases, $n > t$, $n = t$ and $n < t$, the conditional distribution, $p(x_n|Y_t)$, is called the predictor, the filter and the smoother, respectively.

If the state space model (2.1) and (2.2) is given, the state vector x_n can be estimated by the Kalman filter. Denote the mean and the variance covariance matrix of the state x_n given the observations $y_1, \ldots, y_t$ by $x_{n|t}$ and $V_{n|t}$, respectively. Then the one-step-ahead predictor $x_{n|n-1}$ and the filter $x_{n|n}$ can be obtained recursively by the following Kalman filter [4]

Prediction

$$(2.7) \qquad \begin{aligned} x_{n|n-1} &= F x_{n-1|n-1} \\ V_{n|n-1} &= F V_{n-1|n-1} F^T + G Q_n G^T, \end{aligned}$$

Filter

$$(2.8) \qquad \begin{aligned} K_n &= V_{n|n-1} H^T (H V_{n|n-1} H^T + R)^{-1} \\ x_{n|n} &= x_{n|n-1} + K_n (y_n - H x_{n|n-1}) \\ V_{n|n} &= (I - K_n H) V_{n-1|n-1}. \end{aligned}$$

Here the initial values $x_{0|0}$ and $V_{0|0}$ must be given properly and the filter steps are repeated as long as the observations are obtained.

The final estimates of the state vectors $x_{n|N}$, for $n < N$, are obtained by the following fixed interval smoothing algorithm [4]:

$$(2.9) \qquad A_n = V_{n|n} F^T V_{n|n-1}^{-1}$$

$$(2.10) \qquad x_{n|N} = x_{n|n} + A_n (x_{n+1|N} - x_{n+1|n})$$

$$(2.11) \qquad V_{n|N} = V_{n|n} + A_n (V_{n+1|N} - V_{n+1|n}) A_n^T.$$

2.3. Non-Gaussian filter and smoother. For general state space models, however, the conditional distributions become non-Gaussian and their distributions cannot be completely specified by the mean vectors and the variance covariance matrices. In this case, the following non-Gaussian filter and smoother can yield a precise posterior density [15].

[Non-Gaussian filter]

$$p(x_n|Y_{n-1}) = \int p(x_n|x_{n-1})p(x_{n-1}|Y_{n-1})dx_{n-1}$$

(2.12)

$$p(x_n|Y_n) = \frac{p(y_n|x_n)p(x_n|Y_{n-1})}{p(y_n|Y_{n-1})},$$

where $p(y_n|Y_{n-1})$ is the predictive distribution of y_n and is defined by $\int p(y_n|x_n)p(x_n|Y_{n-1})dx_n.$

[Non-Gaussian smoother]

$$(2.13) \qquad p(x_n|Y_N) = p(x_n|Y_n)\int \frac{p(x_{n+1}|x_n)p(x_{n+1}|Y_N)}{p(x_{n+1}|Y_n)}dx_{n+1}.$$

However, the direct implementation of the formula requires computationally very costly numerical integration and can be applied only to lower dimensional state space models.

2.4. Monte Carlo filtering. To mitigate the computational burden, numerical methods based on Monte Carlo approximation of the distribution have been proposed. In the Monte Carlo filtering [9, 16], we approximate each density function by many particles that can be considered as realizations from that distribution. Specifically, assume that each distribution is expressed by using m (m=10,000, say) particles as follows:

$$\{p_n^{(1)},\ldots,p_n^{(m)}\} \sim p(x_n|Y_{n-1}) \qquad \text{Predictor}$$

$$\{f_n^{(1)},\ldots,f_n^{(m)}\} \sim p(x_n|Y_n) \qquad \text{Filter}$$

$$\{s_{n|N}^{(1)},\ldots,s_{n|N}^{(m)}\} \sim p(x_n|Y_N) \qquad \text{Smoother}$$

$$\{v_n^{(1)},\ldots,v_n^{(m)}\} \sim p(v_n) \qquad \text{System noise}$$

Namely, $p(x_n|Y_{n-1})$ is approximated by the probability function

$$(2.14) \qquad \Pr(x_n = p_n^{(j)}|Y_{n-1}) = \frac{1}{m}, \qquad \text{for } j = 1,\cdots,m.$$

Then it can be shown that a set of particles approximating the one step ahead predictor $p(x_n|Y_{n-1})$ and the filter $p(x_n|Y_n)$ can be obtained recursively as follows.

[Monte Carlo Filter]

1. *Generate a random number $f_0^{(j)} \sim p_0(x)$ for $j = 1, \ldots, m$.*
2. *Repeat the following steps for $n = 1, \ldots, N$.*
   (a) *Generate a random number $v_n^{(j)} \sim q(v)$, for $j = 1, \ldots, m$.*
   (b) *Compute $p_n^{(j)} = F(f_{n-1}^{(j)}, v_n^{(j)})$, for $j = 1, \ldots, m$.*
   (c) *Compute $\alpha_n^{(j)} = p(y_n | p_n^{(j)})$ for $j = 1, \ldots, m$.*
   (d) *Generate $f_n^{(j)}$, $j = 1, \ldots, m$ by resampling of $p_n^{(1)}, \ldots, p_n^{(m)}$.*
       with the weights proportional to $\alpha_n^{(1)}, \ldots, \alpha_n^{(j)}$.

In this filtering algorithm, the function $F(f, v)$ and the noise density $q(v)$ are known. If they contain some unknown parameters, they can be estimated by the method shown in subsection 2.5 or 2.6. For the observation models, (2.2), (2,4) and (2.6), the $p(y_n | p_n^{(j)})$ in step (c) are respectively given by $\varphi(y_n - H_n p_n^{(j)}; 0, R_n), r_n(S(y_n; p_n^{(j)}) | \frac{\partial G}{\partial y_n} |)$ and $Q_n(y_n | p_n^{(j)})$, where $\varphi(x; 0, R_m)$ is the Gaussian density with mean 0 and the variance covariance matrix R_n and $S(y; p_n^{(j)})$ is the inverse function of $y = H(x, p_n^{(j)})$ given $p_n^{(j)}$.

The above algorithm for Monte Carlo filtering can be extended to smoothing by a simple modification. The details of the derivation of the algorithm is shown in [16].

An algorithm for smoothing is obtained by replacing the Step 2 (d) of the algorithm for filtering by

(d-S) *Generate $\{(s_{1|n}^{(j)}, \cdots, s_{n-1|n}^{(j)}, s_{n|n}^{(j)})^T, \ j = 1, \ldots, m\}$ by the resampling of $\{(s_{1|n-1}^{(j)}, \cdots, s_{n-1|n-1}^{(j)}, p_n^{(j)})^T, \ j = 1, \ldots, m\}$.*

In this modification, the particles of the past state $s_{1|n-1}^{(j)}, \cdots s_{n-1|n-1}^{(j)}$, are preserved and $\{(s_{1|n-1}^{(j)}, \cdots, s_{n-1|n-1}^{(j)}, p_n^{(j)})^T, \ j = 1, \ldots, m\}$ is resampled with the same weights, $\alpha_n^{(j)}$, as the one obtained by step 2 (d).

This algorithm realizes fixed interval smoothing for nonlinear non-Gaussian state space model. However, in practice, since the number of particles is finite, the repetition of the resampling (d-S) will gradually decrease the number of different particles in $\{s_{i|n}^{(1)}, \ldots, s_{i|n}^{(m)}\}$ and eventually the accuracy of the distribution will be lost.

To mitigate this problem, the step (d-S) needs to be replaced by:

(d-L) *For fixed L, generate $\{(s_{n-L|n}^{(j)}, \cdots, s_{n-1|n}^{(j)}, s_{n|n}^{(j)})^T, \ j = 1, \ldots, m\}$ by the resampling of $\{(s_{n-L|n-1}^{(j)}, \cdots, s_{n-1|n-1}^{(j)}, p_n^{(j)})^T, \ j = 1, \ldots, m\}$ with $f_n^{(j)} = s_{n|n}^{(j)}$.*

This algorithm realizes the L-lag fixed lag smoother [16]. The increase of lag, L, will improve the accuracy of the $p(x_n | Y_{n+L})$ as an approximation to $p(x_n|Y_N)$, while it is very likely to decrease the accuracy of $\{s_{n|N}^{(1)}, \cdots, s_{n|N}^{(m)}\}$ as representatives of $p(x_n|Y_{n+L})$. Since $p(x_n|Y_{n+L})$ usually converges rather quickly to $p(x_n|Y_N)$, it is recommended to take L not so large.

2.5. Estimation of the model parameters. In actual estimation, the parameters of the model θ is unknown. The log-likelihood of the state space model (2.5) and (2.6) is given by

$$(2.15) \qquad \ell(\theta_m) = \sum_{n=1}^{N} \log p(y_n|Y_{n-1})$$

where $p(y_n|Y_{n-1})$ is obtained by $p(y_n|Y_{n-1}) = \int p(y_n|x_n)p(x_n|Y_{n-1})dx_n$. In particular, for linear Gaussian state space model, the log-likelihood is obtained by

$$(2.16) \qquad \ell(\theta_m) = -\frac{N}{2}\log 2\pi - \frac{1}{2}\sum_{n=1}^{N}\log r_n - \frac{1}{2}\sum_{n=1}^{N}\frac{\varepsilon_n^2}{r_n},$$

where $\varepsilon_n = y_n - Hx_{n|n-1}$ and $r_n = Hv_{n|n-1}H^T + \sigma^2$ [13].

2.6. Self-organizing state space model. If many actual problems, the parameters of the model gradually changes with time. In this case, we consider a Bayesian estimation by augmenting the state vector as

$$(2.17) \qquad z_n = \begin{bmatrix} x_n \\ \theta_n \end{bmatrix}.$$

The state space model for this augmented state vector z_n is easily obtained from the original state space model and the model for the time evolution of the parameter, e.g.,

$$(2.18) \qquad \theta_n = \theta_{n-1} + u_n.$$

Assume that we obtain the posterior distribution $p(z_n|Y_N)$ given the entire observations $Y_N = \{y_1, \cdots, y_N\}$. Since the original state vector x_n and the parameter vector θ_n are included in the augmented state vector z_n, it immediately yields the marginal posterior densities of the parameter and of the original state.

3. Estimation of the arrival time of seismic signal. We shall show a method of estimating arrival time of seismic signal by locally stationary AR modeling [18, 28]. When an earthquake signal arrives, the characteristics of the record of seismograms, such as the variances and the spectrum, change abruptly. For simplicity, it is assumed that each of the seismogram before and after the arrival of the seismic wave is stationary and can be expressed by an autoregressive model as follows [33, 34]:

Background noise model: (for $n = 1, \ldots, k$)

$$(3.1) \qquad y_n = \sum_{i=1}^{m} a_i y_{n-i} + v_n, \qquad v_n \sim N(0, \tau^2),$$

Seismic signal model: (for $n = k+1, \ldots, N$)

$$(3.2) \qquad y_n = \sum_{i=1}^{\ell} b_i y_{n-i} + w_n, \qquad w_n \sim N(0, \sigma^2),$$

where the change point $k + 1$, the autoregressive orders m and ℓ, the autoregressive coefficients $a_1, \ldots, a_m$, $b_1, \ldots, b_\ell$, the innovation variances τ^2 and σ^2 are unknown parameters. The vector consisting of the unknown parameters is denoted by $\theta_{m\ell} = (a_1, \ldots, a_m, \tau^2, b_1, \ldots, b_\ell, \sigma^2)^T$.

Given the observations, $y_1, \ldots, y_N$, the log-likelihood can be approximated by

$$\ell(k, m, \ell, \theta_{m\ell}) = \ell_B(k, m, a_1, \ldots, a_m, \tau^2) + \ell_S(k, \ell, b_1, \ldots, b_\ell, \sigma^2)$$

$$(3.3) \qquad = -\frac{k - M}{2} \log 2\pi\tau^2 - \frac{1}{2\tau^2} \sum_{n=M+1}^{k} \left(y_n - \sum_{j=1}^{m} a_j y_{n-j} \right)^2$$

$$-\frac{N - k}{2} \log 2\pi\sigma^2 - \frac{1}{2\sigma^2} \sum_{n=k+1}^{N} \left(y_n - \sum_{j=1}^{\ell} b_j y_{n-j} \right)^2,$$

where ℓ_B and ℓ_S denote the log-likelihoods of the background noise model and the seismic signal model, respectively, and M is the possible maximum AR order of the background noise model, $m \leq M \ll k$.

The estimates $\hat{\theta}_{m\ell} = (\hat{a}_1, \ldots, \hat{a}_m, \hat{b}_1, \ldots, \hat{b}_\ell, \hat{\tau}^2, \hat{\sigma}^2)^T$ are obtained by maximizing this log-likelihood function. Namely, $\hat{a}_1, \ldots, \hat{a}_m$ and $\hat{\tau}^2$ are obtained by maximizing ℓ_B and $\hat{b}_1, \ldots, \hat{b}_\ell$ and $\hat{\sigma}^2$ by maximizing ℓ_S. In actual computations, these parameters can be obtained by the least squares method.

For automatic determination of the change point by the minimum AIC procedure, we have to fit and compare $(K + 1)(M + 1)^2$ models. Here K is the number of possible change points. By a simple minded method, the amount of necessary computations is of the order of NM^3K. As shown in the next subsection, estimation of the parameters and search for the minimum AIC model can be realized by the state space representation and the Kalman filter. Another implementation based on the least squares method via the Householder transformation is shown in [24, 34]. By these methods, the necessary amount of computations is reduced to the order of NM^2. Note that if $M = 10$ and $K = 1,000$, the amount of necessary computation is reduced to about $1/10,000$ of the original method.

3.1. State space method. For the computation of AIC's of the background models, we consider the following state space representation

$$(3.4) \quad x_n = \begin{bmatrix} a_{n1} \\ \vdots \\ a_{nm} \end{bmatrix}, \quad F = I_m, \quad H_n = [y_{n-1}, \ldots, y_{n-m}], \quad v_n = 0.$$

Note that in this representation, the system noise is zero, so that the in the prediction step we have $x_{n|n-1} = x_{n-1|n-1}$ and $V_{n|n-1} = V_{n-1|n-1}$. The AIC of the highest order background noise model $\mathrm{AIC}_k^B(M)$ is obtained by

$$(3.5) \quad \begin{aligned} \mathrm{AIC}_k^B(M) &= (k - M)\log 2\pi\tau_k^2(M) + 2(M+1) \\ \sigma_n^2(M) &= \frac{n-1}{n}\tau_{n-1}^2(M) + \frac{1}{n}\varepsilon_n^2. \end{aligned}$$

AIC's of the lower order AR models $(m = M-1, \ldots, 1)$ can be obtained by

$$(3.6) \quad \begin{aligned} a_j^{m-1} &= \frac{a_j^m + a_m^m a_{m-j}^j}{1 - (a_m^m)^2} \\ \sigma_n^2(m-1) &= \sigma_n^2(m)\left(1 - \left(a_m^2\right)^2\right)^{-1} \\ \mathrm{AIC}_k^B(m) &= (k-M)\log 2\pi\tau_k^2(m) + 2(m+1). \end{aligned}$$

Similarly, the AIC's of the seismic signal models AIC_k^S, $(k = K, K-1, \ldots, 0)$, can be computed by backward recursion. Then the AIC of the locally stationary AR models is given by

$$(3.7) \quad \mathrm{AIC}_k = \mathrm{AIC}_k^B + \mathrm{AIC}_k^S \qquad \text{for } k = 0, 1, \ldots, K,$$

and the arrival time $(k^* + 1)$ is determined by finding the minimum of AIC_k, $k \in \{L, \ldots, L+K\}$.

3.2. Examples. The left plots of Figure 1 show three components of the seismogram [32] observed at Hokkaido, Japan, y_k, $k = 2800, \ldots, 3200$, where the P-wave (primary wave, wave velocity $V_p = 5.2 \sim 6.1$ km/sec in this area) arrived in the middle of the plot. The sampling interval is about $\Delta T = 0.02$ second. The right plots of Figure 1 show the change of AIC_k for $k = 2800, \ldots, 3200$ when locally stationary AR models are fitted to the data y_j, $j = 2400, \ldots, 3600$. From this figure, it can be seen that the AIC has a clear minimum at around $k = 3000$, depending on the component. Corresponding to the high signal to noise ratio, the AIC's have steep minima and the determination of the arrival time is easy. Note that the difference of AIC is over 2000.

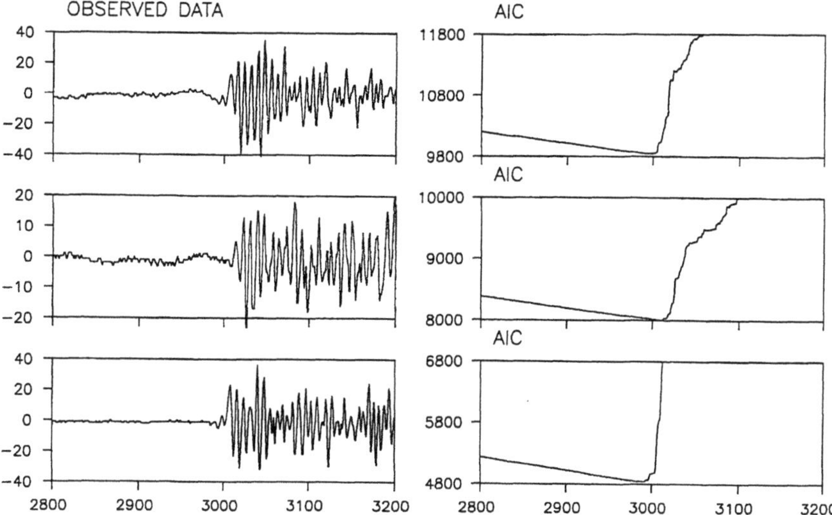

FIG. 1. *Estimation of the arrival times of P-waves. Observed time series (left plots) and the AIC values (right plots). EW-component (top), NS-component (middle) and UD-component (bottom).*

Figure 2 shows the latter part of the seismogram shown in Figure 1 where the S-wave (secondary wave, $V_s = 3.0 \sim 3.6$ *km/sec* in this area) arrived. In this case, the variance and the wave form do not change so significantly as the case of the arrival of P-wave. Corresponding to this, change of AIC are not so large (about 150, 80 and 30, respectively). Further in UD-component, there are two significant local minima at around 3700 and 3800. These suggest the difficulty in estimating precise estimates of arrival times of S-waves.

3.3. Prior distribution of the arrival time. The epicenters of the earthquakes are usually estimated by the constrained least squares method. By this method, inclusion of an inappropriate observation site may result in very inaccurate estimation. This problem may occur when the posterior probability has several significant local maxima and especially when the correct arrival time fails to attain the global maximum, since the least squares method assume the unimodality and the symmetry of the posterior probability of the arrival time. Therefore, a possible remedy is to use the posterior distribution of the arrival times directly in the estimation of the epicenter of the earthquake.

In [2], it was shown that $\exp\{-AIC/2\}$ can be considered as the likelihood of the model whose parameters are estimated by the maximum likelihood method. Therefore, for a prior probability $p(k)$, say $p(k) = 1/(K+1)$, the posterior probability of the arrival time is given by

$$(3.8) \qquad p(k|x) \propto p(k)\exp\{-AIC_k/2\}.$$

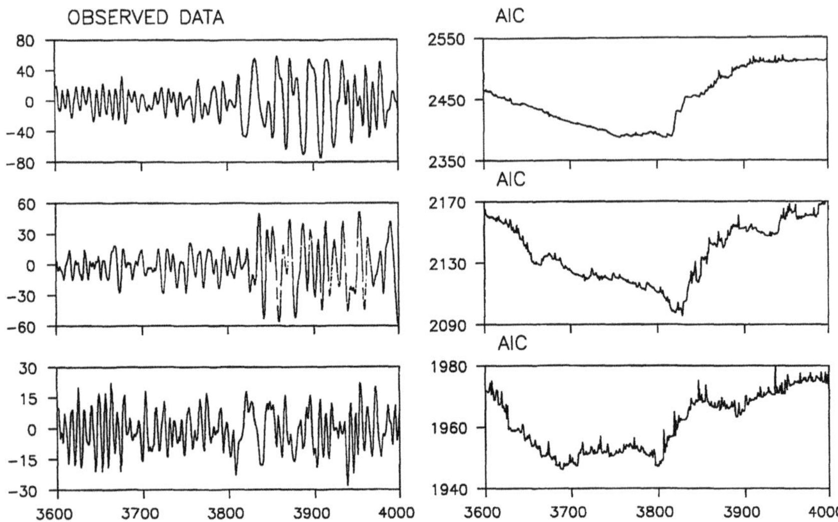

FIG. 2. *Estimation of the arrival times of S-waves. Observed time series (left plots) and the AIC values (right plots). EW-component (top), NS-component (middle) and UD-component (bottom).*

Figure 3 shows the posterior probabilities of the arrival times of P-wave (left) and S-wave (right). Left plot shows that even when the AIC looks to have a clear minimum, there are actually two minima and the posterior probability distribution has two sharp peaks. On the other hand, in the case of S-wave, two peaks of the posterior probability are about 100 points (2 seconds) apart. It is expected that by using these posterior probability in estimating the epicenter, it may be possible to develop a much more robust estimate.

3.4. Multivariate model for the estimation of the arrival time. The locally stationary AR modeling for the estimation of the arrival time of seismic signal can be easily generalized to multivariate time series [34]. By the modeling in 3-dimensional space, it is possible to take into account the characteristics of the wave in 3-D space. This is useful for the estimation of the arrival time of the S-wave or when the signal to noise ratio is very small.

4. Extraction of seismic signal from noisy data. The earth's surface is under continuous disturbances due to a variety of natural forces and human induced sources. Therefore, if the amplitude of the earthquake signal is very small, it will be quite difficult to distinguish it from the background noise. In this section, we consider a method of extracting small seismic signals from relatively large background noise [17], [24].

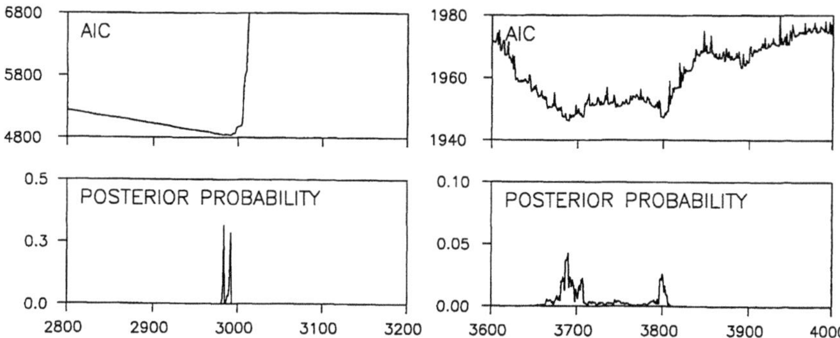

FIG. 3. *AIC's and the posterior probabilities of the arrival times of P-wave (left) and S-wave (right). Top: AIC values. Bottom: Posterior probabilities of the arrival time.*

4.1. The models for the extraction of seismic signal.

For the extraction of the small seismic signal from background noise, we consider the model

$$(4.1) \qquad y_n = r_n + s_n + \varepsilon_n,$$

where r_n, s_n and ε_n denote the background noise, the signal and the observation noise, respectively. To separate these three components, it is assumed that both r_n and s_n are respectively expressed by the autoregressive models

$$(4.2) \qquad r_n = \sum_{i=1}^{m} a_i r_{n-i} + u_n, \qquad s_n = \sum_{i=1}^{\ell} b_i s_{n-i} + v_n,$$

where the AR orders m and ℓ and the AR coefficients a_i and b_i are unknown and u_n, v_n and ε_n are white noise sequences with $u_n \sim N(0, \tau_1^2)$, $v_n \sim N(0, \tau_{2n}^2)$ and $\varepsilon_n \sim N(0, \sigma^2)$ [23]. The models in (4.1) and (4.2) can be combined in the state space model form

$$(4.3) \qquad \begin{aligned} x_n &= F x_{n-1} + G w_n \\ y_n &= H x_n + \varepsilon_n, \end{aligned}$$

where $x_n = (r_n, \ldots, r_{n-m+1}, s_n, \ldots, s_{n-\ell+1})^T$, and $w_n = (u_n, v_n)^T$ is a two dimensional system noise. F, G and H are respectively $(m + \ell) \times (m + \ell)$, $(m + \ell) \times 2$ and $1 \times (m + \ell)$ matrices defined by

$$
(4.4) \quad F = \left[\begin{array}{cccc|ccccc}
a_1 & a_2 & \cdots & a_m & & & & & \\
1 & & & & & & & & \\
& \ddots & & & & & 0 & & \\
& & 1 & & & & & & \\
\hline
& & & & b_1 & b_2 & \cdots & b_\ell & \\
& & & & 1 & & & & \\
& 0 & & & & \ddots & & & \\
& & & & & & & 1 &
\end{array}\right], \quad
G = \left[\begin{array}{cc}
1 & 0 \\
0 & 0 \\
\vdots & \vdots \\
0 & 0 \\
0 & 1 \\
0 & 0 \\
\vdots & \vdots \\
0 & 0
\end{array}\right]
$$

$$
H = [\, 1 \quad 0 \quad \cdots \quad 0 \mid 1 \quad 0 \quad \cdots \quad 0 \,].
$$

The variance of the system noise w_n is given by

$$
(4.5) \qquad Q_n = \left[\begin{array}{cc} \tau_1^2 & 0 \\ 0 & \tau_{2n}^2 \end{array}\right].
$$

4.2. Estimation of the time varying variance.

Note that the variance of the autoregressive model τ_{2n}^2 corresponds to the amplitude of the seismic signal and is actually time varying. This variance parameter plays the role of a signal to noise ratio, and the estimation of this parameter is the key problem for the extraction of the seismic signal. Recently, a self-organizing state space model was successfully applied to the estimation of the time-varying variance [17].

In this method, the original state vector x_n is augmented with the time-varying parameter θ_n as

$$
(4.6) \qquad z_n = \left[\begin{array}{c} x_n \\ \theta_n \end{array}\right],
$$

where the parameter θ_n is defined by

$$
(4.7) \qquad \theta_n = \log_{10} \tau_{2n}^2.
$$

The logarithm of the variance is used to assure the positivity of τ_{2n}^2. We further assume that this parameter θ_n changes according to the random walk model

$$
(4.8) \qquad \log_{10} \tau_{2,n}^2 = \log_{10} \tau_{2,n-1}^2 + \eta_n,
$$

where η_n is the Gaussian white noise with $\eta_n \sim N(0, \xi^2)$.

The state space model for this augmented state is easily obtained from the original state space model for x_n and (4.8). Then by applying the nonlinear non-Gaussian smoother based on the Monte Carlo method [16], we can estimate the state z_n. Since the augmented state z_n contains x_n and θ_n, this means that the marginal posterior density of x_n and θ_n can be obtained simultaneously and that it is not necessary to estimate the parameter θ_n numerically by repeating the filtering many times.

4.3. Example. Top plot of Figure 5 shows the record of the N-S component of a foreshock of an earthquake [32]. Figure 4 shows the spectra of the estimated AR models for the background noise and the seismic signal. Although the variance of the seismic signal is time-varying, in this computation, it is assumed to be a constant and estimated from the data $y_n, n = 1201, \ldots, 2000$. The background noise has high power in low frequency range and has a peak at around 1 Hz. On the other hand the seismic signal has broad spectrum and has higher power in 2–20 Hz.

The observed signal is miniscule relative to the background noise. The background noise model was estimated using the first 500 observations. The AR(5) model was the best AIC model. Using this AR model for the background noise, the AR(5) model was fitted to the data from $n = 1201$ to $n = 2000$ where the seismic signal apparently exists.

The second and the third plots in Figures 5 show the extracted background noise and the seismic signal, respectively. It is clear that the background noise exists even after the arrival of the seismic signal. The arrival of the seismic signal becomes apparent by this decomposition.

The bottom plot shows the estimated variance function $\log_{10} \tau_{2n}^2$. It can be seen that the τ_{2n}^2 is approximately 10^{-4} for the background noise, increases to 10^0 due to the arrival of P-wave and S-wave and then gradually decreases to the original noise level.

4.4. Possible extensions of the method. The method described in this section can be directly generalized to decompose the observed time series into background noise, P-wave and S-wave. However, this decomposition is sometimes very delicate and requires very careful modeling.

The seismograms are records of seismic waves in 3-dimensional space and actually 3- components, namely East-West, North-South and Up-Down components, are observed. The signal extraction method shown in this section can be generalized to 3-dimensional case by using multivariate AR models.

A more precise extraction of the seismic signal will be possible by using the explicit modeling of the characteristics of the P-wave and S-wave. Namely, P-wave is a compression wave and it moves along the wave direction. Therefore it can be approximated by a one-dimensional model,

$$(4.9) \qquad p_n = \sum_{j=1}^{m} a_j p_{n-j} + u_n.$$

On the other hand, S-wave moves on a plane perpendicular to the wave direction and thus can be expressed by 2-dimensional model,

$$(4.10) \qquad \begin{bmatrix} q_n \\ r_n \end{bmatrix} = \sum_{j=1}^{\ell} \begin{bmatrix} b_{j11} & b_{j12} \\ b_{j21} & b_{j22} \end{bmatrix} \begin{bmatrix} q_{n-j} \\ r_{n-j} \end{bmatrix} + \begin{bmatrix} v_{n1} \\ v_{n2} \end{bmatrix}.$$

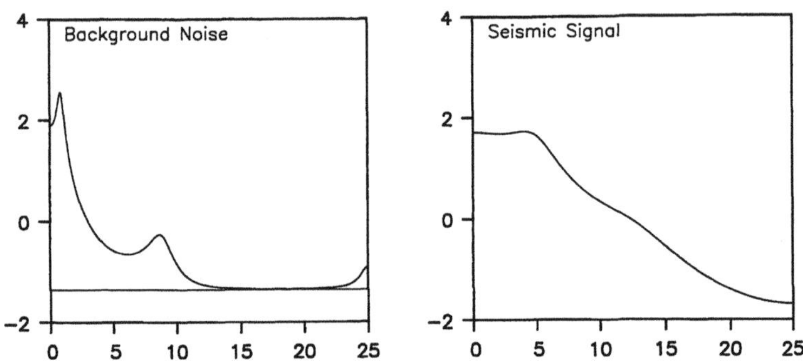

FIG. 4. *AR spectra of estimated component models in logarithmic scale. Left: Spectra of background and observation noises. Right: Spectrum of seismic signal.*

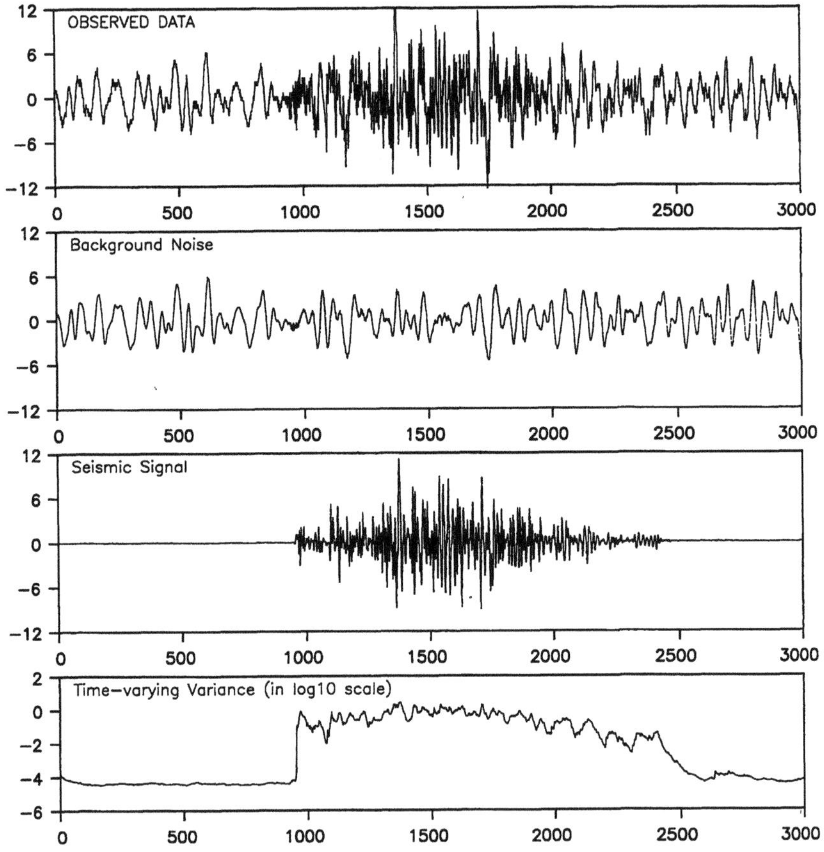

FIG. 5. *Decomposition of a seismogram. From top to bottom, Seismogram record, estimated background noise, extracted micro earthquake signal, and estimated time varying log-variance* $\log \tau_{2n}^2$.

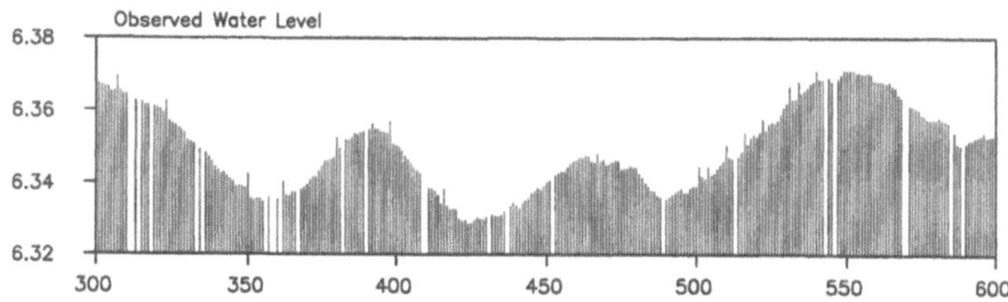

FIG. 6. *A small portion of the observations of groundwater level data.*

Therefore, the observed three-variate time series can be expressed as

$$(4.11) \qquad \begin{bmatrix} x_n \\ y_n \\ z_n \end{bmatrix} = \begin{bmatrix} \alpha_{1n} & \beta_{1n} & \gamma_{1n} \\ \alpha_{2n} & \beta_{2n} & \gamma_{2n} \\ \alpha_{3n} & \beta_{3n} & \gamma_{3n} \end{bmatrix} \begin{bmatrix} p_n \\ q_n \\ r_n \end{bmatrix} + \begin{bmatrix} w_n^x \\ w_n^y \\ w_n^z \end{bmatrix}.$$

In this approach, the crucial problem is the estimation of time-varying wave direction, α_{jn}, β_{jn} and γ_{jn}. They can be estimated by the principle component analysis of the 3D data.

5. Extraction of seismic effect in groundwater level data.

5.1. Ground water level data. In an attempt to predict big earthquakes anticipated in Tokai area, Japan, various sensors have been set since 1979. The groundwater level has been measured in many observation well by Geological Survey of Japan. However, they are strongly contaminated by various effects such as barometric air pressure, earth tide and rain, and it is difficult to detect preseismic or even coseismic effects. [Kearey], [Reoloffs] Further, the groundwater level data contains huge amount of missing and outlying observations. Figure 6 shows a small portion of the observations of groundwater level data, where gaps indicate the missing observations and the upper spikes indicate the outlying observations due to malfunction of the measuring device. Therefore, the development of the method of treating missing observations and correcting outliers is necessary. In this subsection, we show a method based on simple non-Gaussian state space model: [22]

$$(5.1) \qquad \begin{aligned} t_n &= t_{n-1} + w_n, \\ y_n &= t_n + \varepsilon_n, \end{aligned}$$

where t_n is the "signal" and w_n is a Gaussian white noise with variance τ^2. For the observation noise ε_n, we consider three models (see Figure 7)

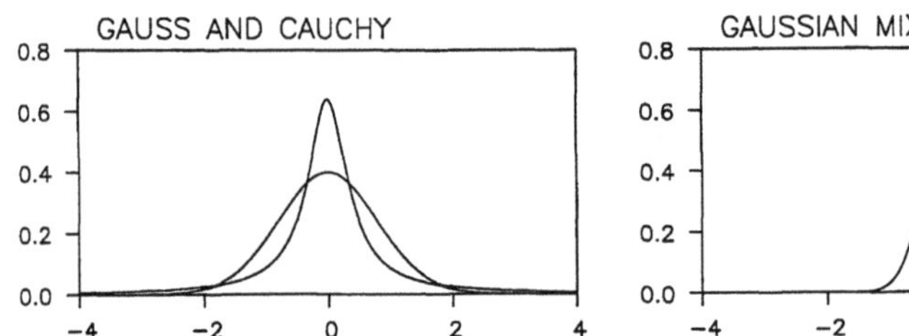

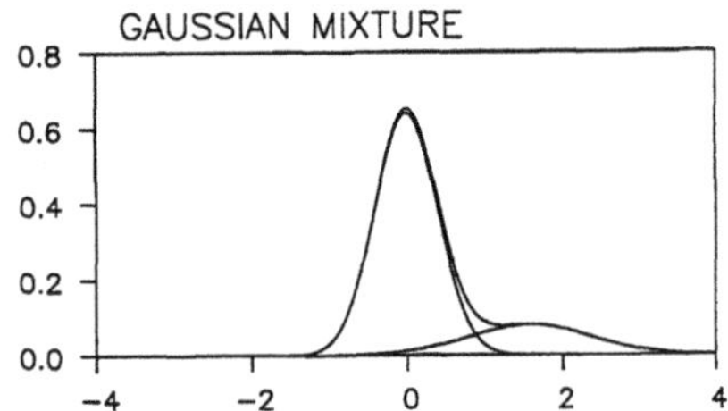

FIG. 7. *Models for outliers. Lest: Gaussian and Cauchy models. Right: Gaussian-mixture model and Gaussian components.*

$$
\text{Gauss} \quad r(w) = \frac{1}{\sigma}\varphi\left(\frac{w}{\sigma}\right)
$$

(5.2)
$$
\text{Cauchy} \quad r(w) = \frac{\sigma}{\pi(w^2 + \sigma^2)}
$$

$$
\text{Mixture} \quad r(w) = \frac{1-\alpha}{\sigma}\varphi\left(\frac{w}{\sigma}\right) + \frac{\alpha}{\eta}\varphi\left(\frac{w-\mu}{\eta}\right),
$$

where $\varphi(x)$ is the standard Gaussian density function, α is the mixture weight, and μ and η^2 are the mean and the variance of the outliers. The Gaussian distribution can be considered as noise distribution in normal situation. On the other hand, the Cauchy distribution is a typical example of heavy-tailed distribution and the Gaussian-mixture distribution can express asymmetric and/or heavy-tailed distributions. Such heavy-tailed densities allow the occurrence of large deviations with a low probability.

Table 1 shows the AIC values for three models for observation noise. The Gaussian mixture model fits best. The Cauchy model fits worse than the Gaussian model because it ignores the fact that the outliers appear only on the positive side. The entire data set was cleaned with this Gaussian mixture noise model and is used for the analysis in the next subsection.

TABLE 1
AIC's of three models for observation noise [22].

Model	AIC
Gauss	−8741
Cauchy	−8655
Gauss Mixture	−8936

5.2. Extraction of coseismic effect. The top plot of Figure 8 shows the cleaned hourly groundwater level data. Even after interpolating the missing observations and correcting the outliers, the groundwater level data is very noisy, and is difficult to pick out the effect of earthquake. Obvi-

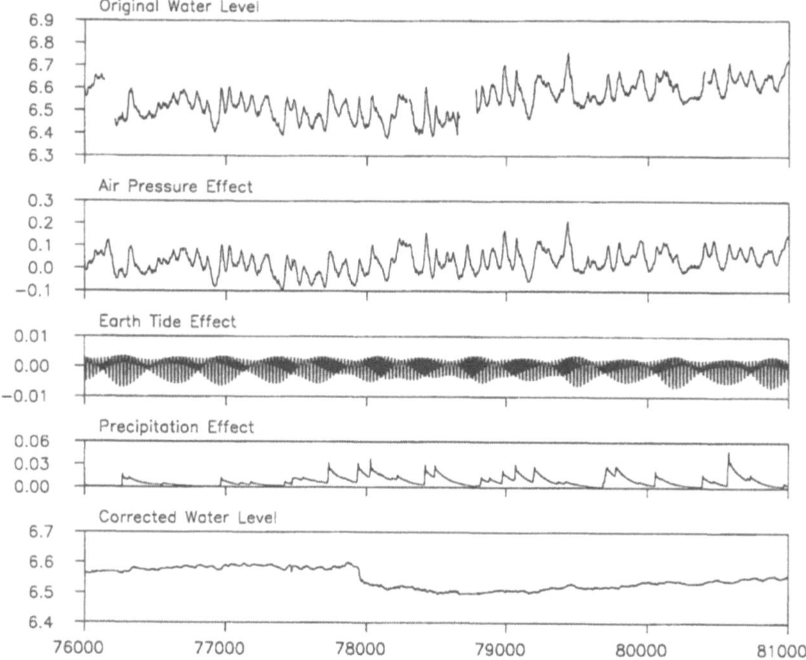

FIG. 8. *Decomposition of a part of the water level data into the air pressure effect, the earth tide effect, the precipitation effects and the corrected water level by the Kalman smoother.*

ously, this is because the data is affected by many other covariates such as barometric air pressure, earth tide and precipitation.

To remove the effects of the covariates and extract the seismic effect in the groundwater level data, we considered the following model

$$(5.3) \qquad y_n = t_n + P_n + E_n + R_n + \varepsilon_n,$$

where t_n, P_n, E_n, R_n and ε_n are the trend, the barometric pressure effect, the earth tide effect, the rainfall effect and the observation noise components, respectively. We assumed that these components follow the models

$$(5.4) \qquad t_n = t_{n-1} + w_n, \qquad P_n = \sum_{i=0}^{m} a_i p_{n-i}$$

$$E_n = \sum_{i=0}^{\ell} b_i et_{n-i}, \qquad R_n = \sum_{i=1}^{k} c_i R_{n-i} + \sum_{i=1}^{k} d_i r_{n-i} + v_n.$$

Here p_n, et_n and r_n are the observed barometric pressure, the theoretical earth tide and the observed precipitation at time n, respectively. An

ARMAX type model [6] was used for precipitation effect, because it was seen that rain effect continues for very long time. It is expected that, after removing the effects of covariates, P_n, E_n and R_n, the trend component will capture the effect of the earthquakes.

The model (5.3) with the component models (5.4) can be expressed by a state space model

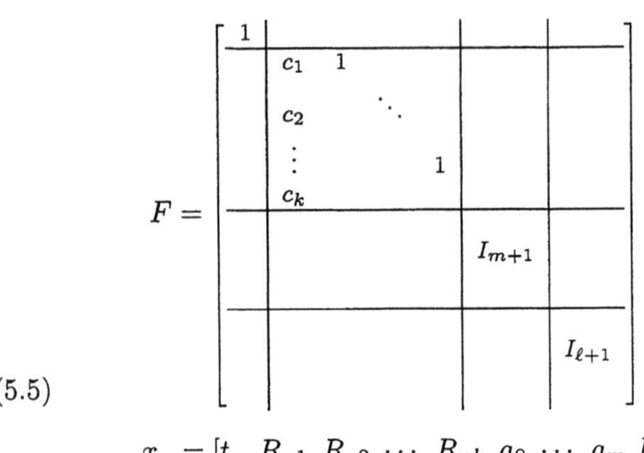

$$(5.5)$$

$$x_n = [t_n, R_{n1}, R_{n2}, \cdots, R_{nk}, a_0, \cdots, a_m, b_0, \cdots, b_\ell]^T$$

$$r_n = [0, d_1, d_2, \cdots, d_k, 0, \cdots, 0, 0, \cdots, 0]^T$$

$$v_n = [u_n, v_{n1}, v_{n2}, \cdots, v_{nk}, 0, \cdots, 0, 0, \cdots, 0]^T.$$

Since the state vector contains the regression coefficients a_j and b_j, they can be estimated by the Kalman filter. On the other hand, $c_1, \ldots, c_k$ and $d_1, \ldots, d_k$ need to be estimated by numerically maximizing the likelihood function.

5.3. Example of the analysis and the implementation of the method. Figure 8 shows decomposition of a part of the smoothed groundwater data into trend, air pressure effect, earth tide effect and precipitation effects by the Kalman smoother. The orders obtained by minimizing the AIC were $m = 25$, $\ell = 2$ and $k = 5$. The second plot shows the estimated air pressure effect. It is quite similar to the groundwater level data indicating that the most of the variation in the observed groundwater level is due to the change of air pressure. Actually most of the range of about $40cm$ variations in the top plot can be considered as the effect of barometric air pressure. The third plot shows the estimated earth tide effect. It clearly reflects almost cyclic tidal effect but the amplitude is small (about 1 cm). The fourth plot shows the precipitation effect. It can be seen that the effect of the precipitation is at largest 4 cm and the effect remains for about three weeks. The bottom plot shows the corrected water level data. It becomes very smooth and the sudden drop of the water level is clearly detected.

From the analysis of data for about 20 years, we obtained the following important observations: (1) The drop of water level can be seen for most of

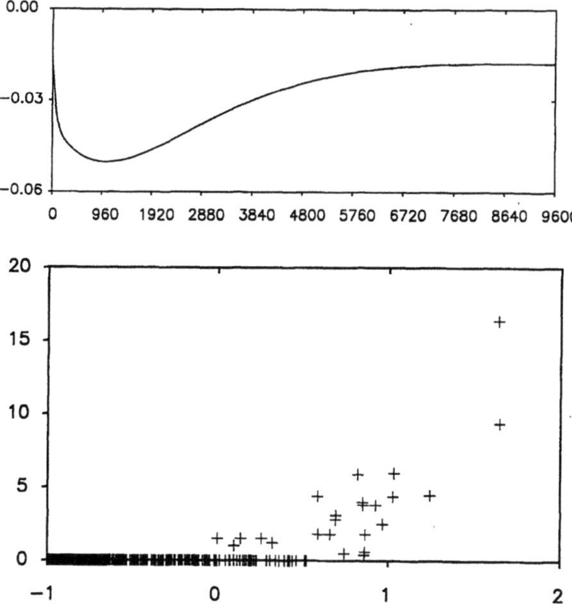

FIG. 9.   *Left: Response function of ground water level to the earthquake.*
Right: Amount of change of water level against modified magnitude.

the earthquakes with magnitude larger than $M > 2.45 \log_{10} D$, where D is
the hypocentral distance. (2) The amount of the drop can be explained as
a function of $M - 2.45 \log_{10} D$. (3) Exclusive of the coseismic effect drop,
the trend regularly increases at the rate of about $6cm$ per year. This might
be considered as a possible indication of the increase of the stress in this
area and is an important subject of future research.

The left plot of Figure 9 shows the estimated response function of
the water level to the earthquake. It is interesting that the effect of an
earthquake remains even after one year. The right plot shows there is a
linear relation between the modified magnitude ($M' = M - 2.45 \log D$, D
is the distance) and the amount of the change of water level for $M' > 0$.

The method outlined in this section was implemented by the Geological
Survey of Japan and is now used in daily routine work (see the Web site
http://gxwell.aist.go.jp/GSJ_E/).

6. Estimation of changing spectrum.

6.1. Time-varying coefficient AR model. When a seismic signal
arrives, characteristics of time series such as the variance and the power
spectrum change significantly. For stationary time series, its characteristics
and power spectrum can be reasonably approximated by an autoregressive
model. Therefore, if the characteristics of the series change, the parameters
of the autoregressive model, i.e., the autoregressive coefficients, change

with time. Motivated by this, we consider an AR model with time-varying coefficients,

$$(6.1) \qquad y_n = \sum_{j=1}^{m} a_{jn} y_{n-j} + w_n,$$

where the autoregressive coefficients a_{jn} is changing with time [14], [19].

Once the autoregressive model with time-varying coefficients is estimated, the instantaneous spectrum is defined by

$$(6.2) \qquad p_n(f) = \frac{\sigma^2}{|1 - \sum_{j=1}^{m} a_{jn} e^{-2\pi i j f}|^2}.$$

For the estimation of the time-varying coefficients, we use the following smoothness prior model for each a_{jn}:

$$(6.3) \qquad \Delta^2 a_{jn} = v_{jn}, \quad v_{jn} \sim N(0, \tau^2),$$

where Δ is the difference operator with respect to n defined by $\Delta a_{jn} = a_{jn} - a_{j,n-1}$, and v_{jn} is the white noise sequence. The time-varying autoregressive model (6.1) and the model for the time-varying coefficients (6.3) can be expressed in state space form with

$$(6.4) \qquad F = \begin{bmatrix} 2 & & & -1 & & \\ & \ddots & & & \ddots & \\ & & 2 & & & -1 \\ \hline 1 & & & & & \\ & \ddots & & & 0 & \\ & & 1 & & & \end{bmatrix}, \quad G = \begin{bmatrix} 1 & & \\ & \ddots & \\ & & 1 \\ \hline & & \\ & 0 & \\ & & \end{bmatrix},$$

$$H = [\, y_{n-1}, \ldots, y_{n-m} \mid 0 \,, \ldots, \, 0 \,]$$

$$x_n = [a_{1n}, \cdots, a_{mn} | a_{1,n-1}, \cdots, a_{m,n-1}], \quad v_n = [v_{1n}, \cdots, v_{mn}]^T.$$

6.2. Estimation of the time-varying parameters. By using this state space representation, the time varying coefficients a_{jn} can be estimated by the Kalman filter and smoother. Two variances, τ^2 and σ^2 can be estimated by maximizing the log-likelihood of the model. Then the ratio of the variances, τ^2/σ^2, plays the crucial role of the trade-off parameter.

When seismic signal arrives, not only the spectrum but also the amplitude of the series changes significantly. Hence in (6.2), the variance σ^2 is actually time-varying. Therefore, in the actual analysis we first estimate the time-varying variance function [19], [20].

Further, when earthquake signal arrives the spectrum of the series changes suddenly and may not be well represented by the random walk model shown in (6.3). For seismic signal, it is rather easy to determine the arrival time by the method shown in Section 3. By specifying a larger value for the system noise variance at the arrival times, it is possible to detect both the sudden and gradual changes of the spectrum.

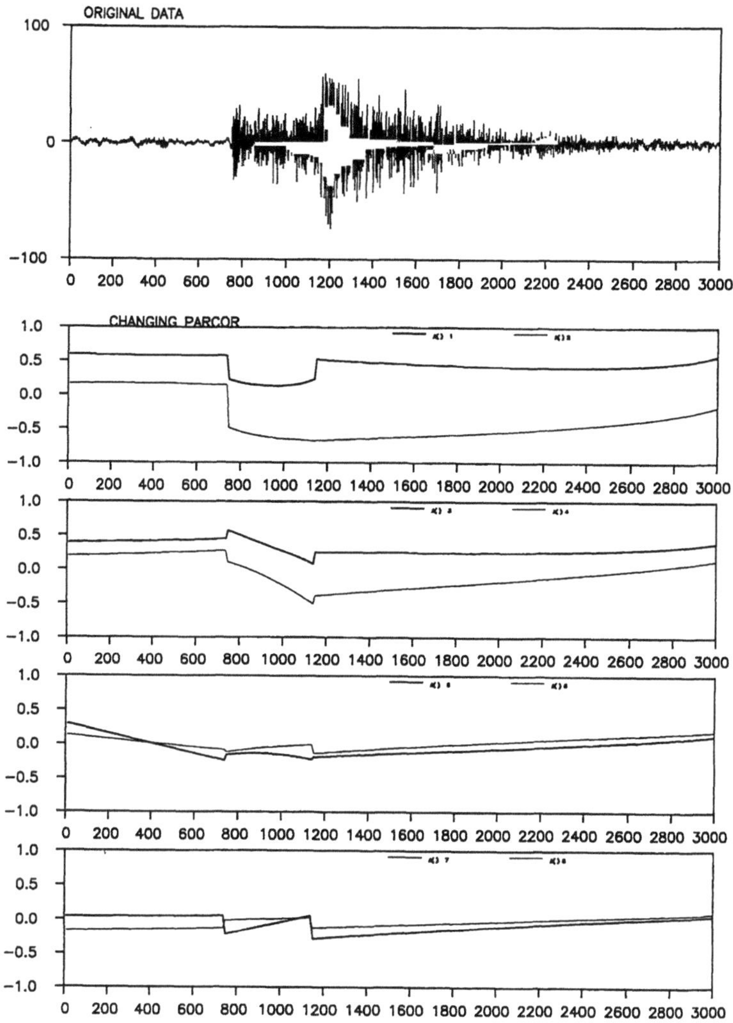

FIG. 10. *Top:A seismogram observed at Hokkaido, Japan and estimated time-varying AR coefficients, the second to the forth plots; time varying partial autocorrelation coefficients $b_{jn}, j = 1, 3, 5, 7$ (thick lines) and $j = 2, 4, 6, 8$ (thin lines).*

6.3. Examples. The top plot of Figure 10 shows a part of the seismogram observed at Moyori, Hokkaido, Japan. The sampling interval is about $\Delta T = 0.02$ second. According to the method shown in Section 2, the P-wave and the S-wave arrived at $n = 744$ and 1150, respectively. The variance of the series increases when these waves arrive.

Other plots in Figure 10 show the changes of the estimated time-varying partial autocorrelation coefficients, $b_{jn}, j = 1, \ldots, 8$, when the AR order m is set to 8. Note that b_{jn} are obtained by the inverse Levinson

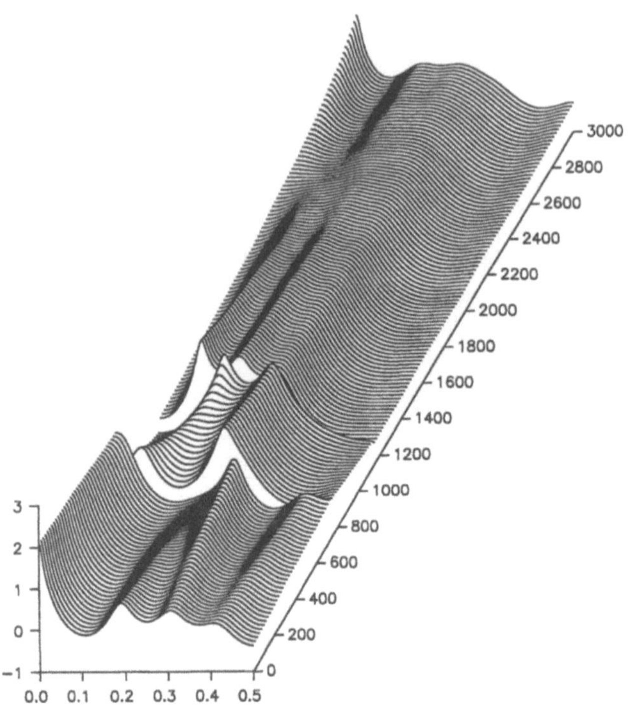

FIG. 11. *The time-varying spectrum.*

recursion formula. Since a large value of variance of system noise is used for arrival times of two waves, abrupt changes of the coefficients at two points are detected. The AR coefficients change very smoothly except for these two points.

Figures 11 shows the time-varying spectrum obtained from the model shown in Figure 10. Sudden changes of the spectrum at two time points and the gradual change of spectrum in S-wave is clearly seen. It can be seen that the low frequency spectrum ($f = 0$) dominates in background noise ($1 \leq n \leq 750$), high frequency wave ($f \doteq 0.25$, about 12Hz) dominates in P-wave ($750 \leq n \leq 1150$), medium frequency wave ($f \doteq 0.10$, about 5Hz) dominates in the S-wave. However the peak of the spectrum gradually shifts to the right (high frequency side) and finally goes back to the background noise.

6.4. Bayesian method and multivariate version. In the time-varying coefficient AR modeling, selection of an appropriate order m is necessary. In the program TVCAR in the TIMSAC-84 package (Akaike et al. 1985), the selection of the order is avoided by introducing smoothness priors for the smoothness of the time change of the coefficients and the

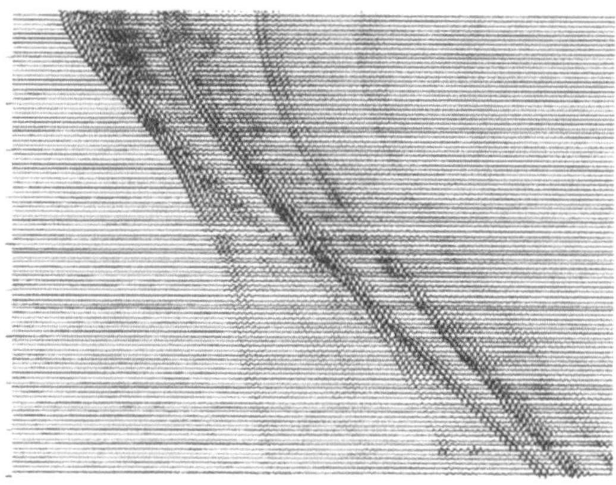

FIG. 12. *Observed data obtained by OBS.*

smoothness of the AR operators. The details of the method is explained in Kitagawa and Gersch [20].

The time-varying AR model shown in this section can be generalized to the multivariate case. Jiang and Kitagawa [12] developed a method of fitting time-varying multivariate AR model based on the simultaneous response model representation. Gersch and Stone developed a method of estimating time-varying multivariate AR model via PARCOR VAR modeling [7].

7. Analysis of OBS data. In this section, we show the analysis of the ocean bottom seismographs (OBS) to explore the underground velocity structure. 1560 4-channel time series with 7500 observations were obtained (Figure 12). The details of the analysis can be seen in [25].

7.1. Time-lag structure of the data and spatial smoothing. The actual time series observed at OBS contains signals of a direct wave and its multiples, reflection waves, refraction waves and observation noise. Just beneath the air-gun, the direct wave (compression wave with velocity about 1.48km/sec.) that travels through the water arrives first and dominates in the time series. However, since the velocity of the waves in the solid structure is larger than that in the water (2-8 km/sec.), in our present case, a reflection and refraction waves arrive before the direct wave and its multiples for the offset distance larger than approximately 1.4 km and 14 km, respectively.

As an example, assume the following three-horizontal-layer structure: the depth and the velocity of the water layer: $h_0$km and $v_0$km/sec., the width and the velocities of three layers: h_1, h_2, $h_3$km, v_1, v_2, $v_3$km/sec, respectively.

The wave path is identified by Wave($i_1 \cdots i_k$), ($i_j = 0, 1, 2, 3$), where Wave(0) denote the direct water wave that travels directly from the air-

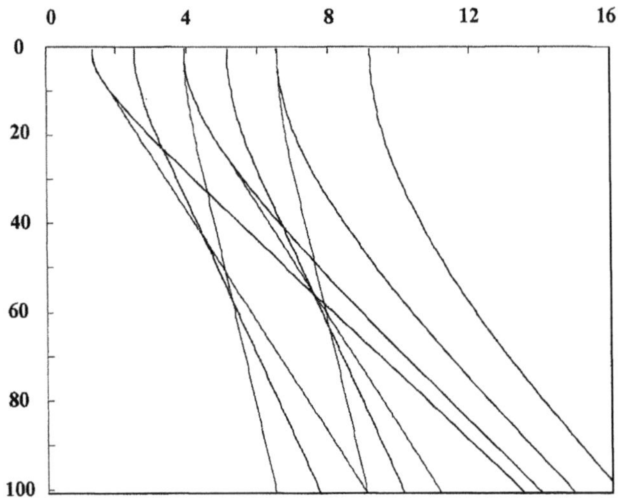

FIG. 13. *Left: Examples of wave types in two-layered half space structure: Wave(0), Wave(000) and Wave(0121). Right: Travel times of various waves. Vertical axis: offset distance D (km), horizontal axis: reduced travel time t − D/6(sec.). From left to right in horizontal axis, Wave (01), (0), (0121), (0001), (000), (012321), (000121), (00000), (00012321).*

gun to the OBS, Wave(01) denotes the wave that travels on the seafloor, Wave(000121) denotes the wave that reflects once at the sea bottoms and the sea surface, then penetrates into the first layer, traveling the interface between the first and the second layers, then goes up to the OBS through the first layer.

Table 2 shows the travel times of various waves. At each OBS these waves arrive successively [35]. Figure 13 shows the arrival times, t, versus the offset distances, D, for some typical wave paths. The parameters of the 3-layer structure are assumed to be $h_0 = 2.1$km, $h_1 = 2$km, $h_2 = 3$km, $h_3 = 5$km, $v_0 = 1.5$km/sec, $v_1 = 2.5$km/sec, $v_2 = 3.5$km/sec, $v_3 = 7.0$km/sec. It can be seen that the order of the arrival times changes in a complex way with the horizontal distance D, even for such simplest horizontally layered structure.

7.2. Difference of the arrival time. At each OBS, many time series were observed, with the location of the explosion shifted by 200 m. Therefore by considering the time-lag structure of the signal, it is expected that we can detect the information that was difficult to obtain from a single time series.

Table 3 shows the difference of the arrival times between two consecutive time series (moveout), computed for each wave type and for some offset distance, D. The moveout of the waves that travel on the surface between

TABLE 2
Wave types and arrival times.

Wave type	Arrival time
Wave (0^{2k-1})	$v_0^{-1}\sqrt{(2k-1)^2 h_0^2 + D^2}$
Wave $(0^{2k-1}1)$	$(2k-1)v_0^{-1}\sqrt{h_0^2 + d_{01}^2} + v_1^{-1}(D - (2k-1)d_{01})$
Wave $(0^{2k-1}121)$	$(2k-1)v_0^{-1}\sqrt{h_0^2 + d_{02}^2} + 2v_1^{-1}\sqrt{h_1^2 + d_{12}^2} + v_2^{-1}d_2$
Wave (012321)	$v_0^{-1}\sqrt{h_0^2 + d_{03}^2} + 2v_1^{-1}\sqrt{h_1^2 + d_{13}^2} + 2v_2^{-1}\sqrt{h_2^2 + d_{23}^2} + v_3^{-1}d_3$

$$d_{ij} = v_i h_i / \sqrt{v_j^2 - v_i^2}, \; d_2 = D - (2k-1)d_{02} - 2d_{12}, \; d_3 = D - d_{03} - 2d_{13} - 2d_{23}.$$

TABLE 3
Wave types and delay of arrival times for various offset distance.

Wave type	Offset Distance (km)				
	0	5	10	15	20
Wave(0)	0.6	16.5	16.5	16.7	16.7
Wave(0^3)	0.2	14.5	14.5	15.8	16.0
Wave(0^5)	0.1	8.0	12.0	14.2	15.0
Wave(01)	—	10.5	10.0	10.1	9.9
Wave(0121)	0.2	7.5	7.1	7.2	7.1
Wave(012321)	0.1	3.8	3.6	3.6	3.5

two layers, such as Wave(01), (0121), (012321), are constants independent on the offset distance D. The moveout becomes small for deeper layer or faster wave. On the other hand, for the direct wave and its multiples that path through the water, such as Wave(0), (000), (00000), the amount of the delay-time gradually increases with the increase of the offset distance D, and converges to approximately 17 for distance $D > 5$km. This indicates that for $D > 5$, the arrival time is approximately a linear function of the distance D.

7.3. Spatial-temporal model. We consider the spatial-temporal smoothing by combining the time series model and the spatial model. The basic observation model is a multi-variate analogue of the decomposition model in (4.1):

$$(7.1) \qquad\qquad y_{n,j} = r_{n,j} + s_{n,j} + \varepsilon_{n,j}$$

where $r_{n,j}$, $s_{n,j}$ and $\varepsilon_{n,j}$ denote the direct wave or its multiples, reflection/refraction wave and the observation noise component in channel j.

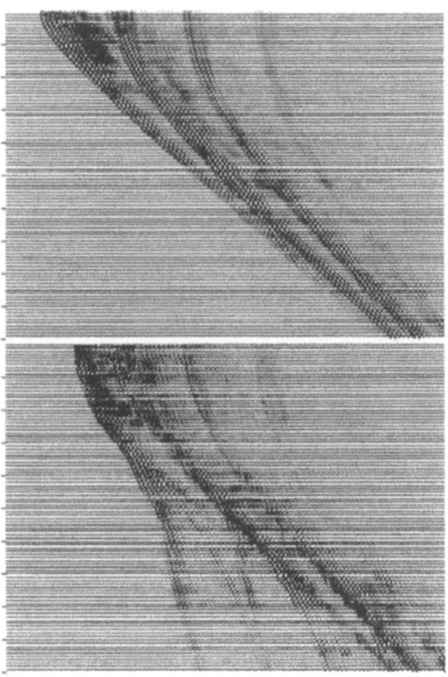

FIG. 14. *Extracted direct wave and its multiples (left) and reflection/refraction waves (right). CH972–1071, data length n=2000, $\Delta t = 1/125$ second.*

As in subsection 3.2, the direct water wave, and the reflection and refraction wave components are assumed to follow the AR models

$$(7.2) \quad r_{n,j} = \sum_{i=1}^{m} a_{i,j} r_{n-i,j} + v^r_{n,j}, \qquad s_{n,j} = \sum_{i=1}^{\ell} b_{i,j} s_{n-i,j} + v^s_{n,j},$$

respectively.

On the other hand, by considering the delay structure discussed in the previous subsection, we also use the following spatial models

$$(7.3) \quad r_{n,j} = r_{n-k,j-1} + u^r_{n,j}, \qquad s_{n,j} = s_{n-h,j-1} + u^s_{n,j}.$$

Here the moveouts k and h are actually functions of the wave type and the distance D. For the direct water wave and other reflection/refraction waves, they are given by $k_j = \Delta T_j(\text{Wave}(0))$ or $\Delta T_j(\text{Wave}(000))$ etc., and $h_j = \Delta T_j(\text{Wave}(X))$, respectively.

Figure 14 shows the results of the decomposition of the data shown in Figure 12. The left plot shows the extracted direct wave and its multiples. Waves(0^k), $k = 1, 3, 5, 7$ are clearly detected. The right plot shows the extracted reflection waves and the refraction waves. Several waves presumably the Wave(0^k12321), (0^k121), $k = 1, 3, 5$ are enhanced by this decomposition.

8. Conclusion. We can develop various procedures for signal extraction problems, by systematic use of state space models. Non-Gaussian filter and Monte Carlo filter enables computationally efficient estimation of the unknown signal even for nonlinear non-Gaussian or general state space model. Time-varying parameter of the model can be estimated by using the self-organizing state space model. The information criterion AIC plays an important role in automatic model selection, and computationally efficient recursive filtering and smoothing algorithms.

In this paper, we showed five examples of time series modeling for signal extraction problems related to seismology. Namely, we considered the estimation of the arrival time of seismic signal, extraction of small seismic signal from noisy data, detection of the seismic signal in groundwater level data, estimation of changing spectral characteristic of seismic record, and a spatial-temporal smoothing of OBS data.

Acknowledgment. The authors are grateful to the careful reading and useful comments on the previous version of the manuscript.

REFERENCES

[1] H. AKAIKE, *Information theory and an extension of the maximum likelihood principle*, in Second International Symposium on Information Theory, Akademiai Kiado, Budapest, 1973, 267–281. (Reproduced in *Selected Papers of Hirotugu Akaike*, Parzen E., Tanabe K., and Kitagawa G. (eds.), Springer-Verlag, New York, 1998).

[2] H. AKAIKE, *A Bayesian extension of the minimum AIC procedure of autoregressive model fitting*, Biometrika, **66**, 1979, 237–242.

[3] H. AKAIKE AND G. KITAGAWA, *The Practice of Time Series Analysis*, Springer-Verlag, New York, 1998.

[4] B.D.O. ANDERSON AND J.B. MOORE, *Optimal Filtering*, New Jersey, Prentice-Hall, 1979.

[5] E. BERG, L. AMUNDSEN, A. MORTON, R. MJELDE, H. SHIMAMURA, H. SHIOBARA, T. KANAZAWA, S. KODAIRA, AND J.P. FJEKKANGER, *Three dimensional OBS-data processing for lithology and fluid prediction in the mid-Norway margin, NE Atlantic, Earth, Planet and Space*, **53**, No. 2, 2001, 75–90.

[6] G.E.P. BOX AND G.M. JENKINS, *Time Series Analysis: Forecasting and Control*, (2nd ed.), Holden-Day, San Francisco, 1976.

[7] W. GERSCH AND D. STONE, *Multi-variate autoregressive time series modeling: One scalar autoregressive model at-a-time, Communications in Statistics. Theory and Methods*, **24**, 1995, 2715–2733.

[8] G.H. GOLUB, *Numerical methods for solving linear least squares problems, Numerische Mathematik*, No. 7, 1965, 206–219.

[9] N.J. GORDON, D.J. SALMOND, AND A.F.M. SMITH, *Novel approach to nonlinear /non-Gaussian Bayesian state estimation, IEE Proceedings-F*, **140**, No. 2, 1993, 107–113.

[10] B. GUTENBERG AND C.F. RICHTER, *Seismicity of the Earth, Geol. Soc. Am., Spec. Pap.*, **34**, 1941, p. 133.

[11] A.C. HARVEY, E. RUIZ, AND N. SHEPARD, *Multivariate stochastic variance model, Review of Economic Studies*, **61**, 1994, 247–264.

[12] X-Q JIANG AND G. KITAGAWA, *A time varying vector autoregressive modeling of nonstationary time series, Signal Processing*, **33**, 1993, 315–331.

[13] R.H. JONES, *Maximum likelihood fitting of ARMA models to time series with missing observations*, Technometrics, **22**, 1980, 389–395.

[14] G. KITAGAWA, *Changing spectrum estimation*, Journal of Sound and Vibration, **89**, No. 4, 1983, 433–445.

[15] G. KITAGAWA, *Non-Gaussian state-space modeling of nonstationary time series*, Journal of the American Statistical Association, **82**, 1987, 1032–1063.

[16] G. KITAGAWA, *Monte Carlo filter and smoother for non-Gaussian nonlinear state space models*, Journal of Computational and Graphical Statistics, **5**, 1996, 1–25.

[17] G. KITAGAWA, *Self-organizing State Space Model*, Journal of the American Statistical Association, **93**, No. 443, 1998, 1203–1215.

[18] G. KITAGAWA AND H. AKAIKE, *Procedure for the modeling of non-stationary time series*, Annals of the Institute of Statistical Mathematics, **30**, 1978, 351–363.

[19] G. KITAGAWA AND W. GERSCH *A smoothness priors-time varying AR coefficient modeling of nonstationary covariance time series*, IEEE Transactions on Automatic Control, **30-ac**, 1985, 48–56.

[20] G. KITAGAWA AND W. GERSCH, *Smoothness Priors Analysis of Time Series*, Lecture Notes in Statistics, No. 116, Springer-Verlag, New York, 1996.

[21] G. KITAGAWA AND T. HIGUCHI, *Automatic transaction of signal via statistical modeling*, The proceedings of The First Int. Conf. on Discovery Science, Springer-Verlag Lecture Notes in Artificial Intelligence Series, 1998, 375–386.

[22] G. KITAGAWA AND N. MATSUMOTO, *Detection of coseismic changes of underground water Level*, Journal of the American Statistical Association, **91**, No. 434, 1996, 521–528.

[23] G. KITAGAWA AND T. TAKANAMI, *Extraction of signal by a time series model and screening out micro earthquakes*, Signal Processing, **8**, 1985, 303–314.

[24] G. KITAGAWA, T. TAKANAMI, AND N. MATSUMOTO, *Signal Extraction Problems in Seismology*, Intenational Statistical Review, **69**, No. 1, 2001, 129–152.

[25] G. KITAGAWA, T. TAKANAMI, Y. MURAI, H. SHIMAMURA, AND A. KUWANO, *Extraction of Signal from High Dimensional Time Series: — Analysis of Ocean Bottom Seismograph Data — Lecture Notes in Computer Science*, 2002, to appear.

[26] A. KUWANO, *Crustal structure of the passive continental margin, west off Svalbard Islands, deduced from ocean bottom seismographic studies*, Master's Theses, Hokkaido University, 2000.

[27] N. MATSUMOTO, *Detection of groundwater level change related to earthquakes*, in *The Practice of Time Series Analysis*, Akaike, H. and Kitagawa, G. eds., Springer-Verlag, New York, 1999, 341–352.

[28] T. OZAKI AND H. TONG, *On the fitting of nonstationary autoregressive models in time series analysis*, Proceedings of 8th Hawaii International Conference on System Science, Western Periodical Company, 1975, 224–226.

[29] E.A. ROELOFFS, *Hydrologic precursors to earthquakes: a review*, , Pure & Appl. Geophys, **126**, 1988, 177–206.

[30] H. SHIMAMURA, *OBS technical description*, Cruise Report, Inst. of Solid Earth Physics Report, Univ. of Bergen, eds. Sellevoll, M.A., **72**, 1988.

[31] P.L. STOFFA (ed.), *Tau-p, A Plane Wave Approach to the Analysis of Seismic Data*, Kluwer, 1989.

[32] T. TAKANAMI, *ISM data 43-3-01: Seismograms of foreshocks of 1982 Urakawa-Oki earthquake*, Annals of the Institute of Statistical Mathematics, **43**, No. 3, 1991, p. 605.

[33] T. TAKANAMI, *High precision estimation of seismic wave arrival times*, in *The Practice Time Series Analysis*, Akaike H. and Kitagawa G. eds., Springer-Verlag, New York, 1999, 79–94.

[34] T. TAKANAMI AND G. KITAGAWA, *Estimation of the arrival times of seismic waves by multivariate time series model*, Annals of the Institute of Statistical Mathematics, **43**, No. 3, 1991, 407–433.

[35] W.M. TELFORD, L.P. GELDART, AND R.E. SHERIFF, *Applied Geophysics, Second edition*, Cambridge University Press, Cambridge, 1990.

[36] T. YOKOTA, S. ZHOU, M. MIZOUE, AND I. NAKAMURA, *An automatic measurement of arrival time of seismic waves and its application to an on-line processing system, Bulletin of Earthquake Research Institute*, **55**, 1981, 449–484 (in Japanese with English abstract).

IMPROVED SIGNAL TRANSMISSION THROUGH RANDOMIZATION

ENDERS A. ROBINSON*

Abstract. The transmission of energy and information is basic to science and engineering. A signal is transmitted from source to receiver by means of waves passing through a medium. A homogeneous medium transmits the direct wave only, and thus provides the best transmission. Transmission performance is less for a heterogeneous medium. Mathematically a continuously varying heterogeneous medium is difficult to handle, but it can be approximated by a finely divided layered system. A layered system is characterized by the sequence of Fresnel reflection coefficients of the successive interfaces between layers. A layered system not only transmits the direct wave, but also transmits internal multiple reflections. The multiples degrade the transmission performance. Ideally the multiples should be kept small, so that most of the transmitted energy occurs in the direct wave. Transmission performance improves as the reflection coefficients become smaller in magnitude. Transmission performance can also be improved in another significant way. That way is randomization. High performance is achieved when, in addition to being small in magnitude, the reflection coefficients are a realization of random white stochastic process. Transmission though a layered system with small white reflection coefficients closely approximates the ideal transmission though a homogeneous medium.

Key words. Layered system, lattice model, acoustic tube model, thin-film model, energy transmission, impedance matching, small random white reflection coefficients, Durbin-Levinson recursion, Schur polynomial.

1. Introduction. The seismic method in petroleum exploration represents an instrument for remote detection. It has much in common with other disciplines that use non-invasive techniques to find the structure of an inaccessible body. Petroleum is found in sedimentary geologic basins. Sedimentary rocks are formed from sediment that has accumulated in layers. The seismic exploration technique is the most important method for the discovery of new deposits of oil and natural gas deep within the earth. Echolocation is the determination, as by a bat, of the position of an object by the emission of sound waves, which are reflected back to the sender as echoes. Seismic waves are elastic waves that propagate through the earth. Seismic exploration is an echolocation technique that makes use of seismic waves to detect the interfaces between the subsurface sedimentary layers. The exploration geophysicist generates a seismic source signal and sends it into the earth. The source and the receivers are either on the surface of the earth or at shallow depths. The subsurface geologic structures of interest can be as deep as four or five miles. The subsurface rock layers transmit and reflect seismic waves. Because the layering in the sedimentary rocks, the signal encounters many interfaces. Reflections occur each time that a wave strikes an interface. Hence many multiple reflections are generated,

---

*Department of Earth and Environmental Engineering, Columbia University, New York, NY 10027.

and these multiples degrade the quality of the transmission. In fact, multiple reflections represent the most common and particularly troublesome type of interference that can occur on seismic records. The energy content of these multiples can be quite large with respect to the desired primary reflections, and thus the removal of the multiple reflections is important. Because all the energy of the multiples can never be completely removed, it is better to say multiple suppression instead of multiple removal. The question then is what kind of layered system is best able to suppress the unwanted multiple reflections.

Impedance matching, as used in the design of electrical transmission lines, is a well-known method for the suppression of the multiple reflections. A transmission line is made up of sections. Each section has a different impedance. The reflection coefficient of the interface between two sections depends upon the impedance contrast. If there is no impedance contrast, then the reflection coefficient is zero. By matching the impedances of adjacent sections as well as possible, the magnitude of the reflection coefficients are reduced in magnitude. The smaller reflection coefficients reduce the multiple energy, thereby improving the transmission performance. In the same way, a geological prospect with small reflection coefficients produces smaller multiples than one with large reflection coefficients. Prospects with small reflection coefficients were amenable to seismic interpretation in the early days before the digital revolution. However there is another consideration, which is just as important for high performance signal transmission. That consideration is randomness. A white random process as used in this paper is a second-order stationary stochastic process for which any pair of observations are uncorrelated. Signal transmission is improved if the reflection coefficients are observations from a white random process. In summary, high performance signal transmission occurs in a layered system with small white reflection coefficients.

2. The layered earth model. The first step in seismic analysis is the construction of a model that can be used to explain the propagation of seismic waves. Three-dimensional models are the most valuable. The mathematics of theoretical 3D models is much too involved to obtain closed solutions except in simple cases. As a result, most 3D models are determined empirically. However, the most pronounced variations in the earth layering are along the vertical scale. As a result, a theoretical one-dimensional vertical model can often be used to advantage (Brekhovskikh, 1960; Ewing et al., 1957). The foremost 1D model, namely, the so-called layered-earth model, is mathematically identical to the lattice model for electric transmission lines (Mitra and Kaiser, 1993). The model is also mathematically identical both to the acoustic tube model used in speech processing (Gray and Markel, 1973) and to the thin-film model used in optics (Heavens, 1991). The layered-model makes use of discrete closely spaced horizontal layers to represent an inhomogeneous medium.

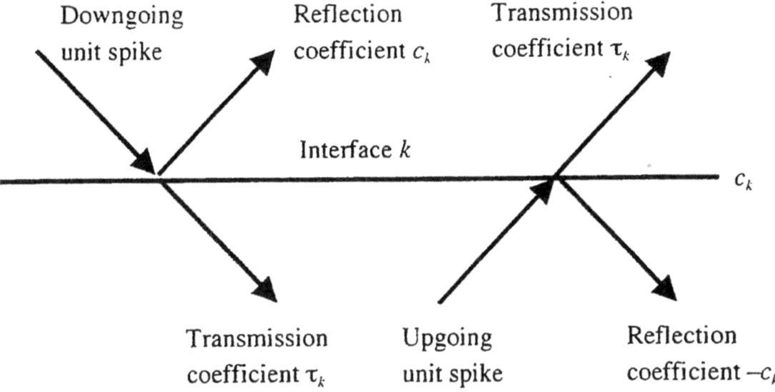

FIG. 1. *The Fresnel reflection and transmission coefficients. (Left) for a downgoing incident wave. (Right) for an upgoing incident wave.*

In the end, a computer will use an advanced mathematical model in the form of numerical algorithms to process the seismic data. The geophysicist will see the input data and the output results. In many cases this may be enough, but often it is important for the geophysicist to have a good concept of what the computer is doing. Simple models are useful for this purpose because their action can be understood in relatively straight-forward mathematical terms. For example, a curve can be approximated by a spline function. However, for basic understanding, a simple model using just straight-line segments can be useful. In computer processing, complicated models for wave propagation in an inhomogeneous medium are used. However, there is still a need for the simple layered-earth model in order to obtain a visualization of what is going on in the earth. The layered-earth model gives a basic understanding of the relationship between the earth structure (as represented by the reflection coefficients) and the transmission of seismic waves through the earth (as represented by the transmission response). To show how this relationship is obtained is the purpose of the paper. One must realize that the approach given here is primarily used so as to give insight as to nature of the propagation of waves traveling through an inhomogeneous medium.

Let us look at a single horizontal interface between two layers, for example, interface k. Fig. 1 illustrates the Fresnel coefficients. Time in the figure is shown by the horizontal coordinate. The physical ray paths are at normal incidence to the interface. However, the ray paths shown in the figure have a horizontal displacement that indicates the passage of time as they travel. Suppose that an incident downgoing spike of unit amplitude strikes the interface. We assume that all amplitudes are measured in units of square root of energy. As we know from classical physics, some of the energy is transmitted through the interface and some is reflected back from the interface. The Fresnel reflection coefficient c_k is defined as the amplitude of the resulting upgoing reflected spike, and the Fresnel transmission coefficient τ_k is defined as the amplitude of the resulting downgoing transmitted spike. This relationship is illustrated in Fig. 1 (left). An incident upgoing unit spike striking the interface from below gives rise to a downgoing reflected spike of amplitude $-c_k$ and an upgoing transmitted spike of amplitude τ_k. See Fig. 1 (right). The relationship between the Fresnel reflection and transmission coefficients is given by

$$\tau_k = +\sqrt{1 - c_k^2} \, .$$

3. The case of an arbitrary number of horizontal interfaces. In the layered-model, the interfaces between consecutive layers are numbered from 1 to N. The top interface 1 represents surface of the earth. The bottom interface is interface N. The reflection coefficient c_1 characterizes interface 1, the reflection coefficient c_2 characterizes interface 2, and so on. The reflection coefficient c_N characterizes interface N. The model is constructed so the two-way travel time in each layer is the same. This common unit of time is chosen as the discrete time unit for the digitized signals. If two adjacent layers have the same impedance, then the interface between the layers has a zero reflection coefficient. As a result, these two layers make up a single layer with double the thickness. By setting reflection coefficients equal to zero, layers of any thickness can be approximately obtained in the model.

The sequence $\{c_1, \ c_2, \ \cdots, \ c_N\}$ of reflection coefficients is called the reflectivity sequence, or simply the reflectivity. The transmission factor for the entire system is defined as the product of the individual transmission coefficients. In other words, the transmission factor is

$$\sigma_N = \tau_1 \, \tau_2 \, \cdots \, \tau_N = \sqrt{(1 - c_1^2)\,(1 - c_2^2) \, \cdots \, (1 - c_N^2)} \, .$$

This factor always lies between zero and one.

Let the source be a downgoing unit spike incident on the upper interface. The transmission response is the downgoing signal escaping from the lower interface. The transmission response is made up of the direct arrival together the later arrivals of multiple reflections. The direct arrival is denoted by the coefficient t_0, the first multiple arrival is denoted by the coefficient t_1, the second multiple arrival is denoted by the coefficient t_2, and so on. The generating function of the transmission response $\{t_0, \ t_1, \ t_2, \ \cdots\}$ is the power series in the dummy variable s given by

$$T_N(s) = t_0 + t_1 s + t_2 s^2 + t_3 s^3 + \cdots \, .$$

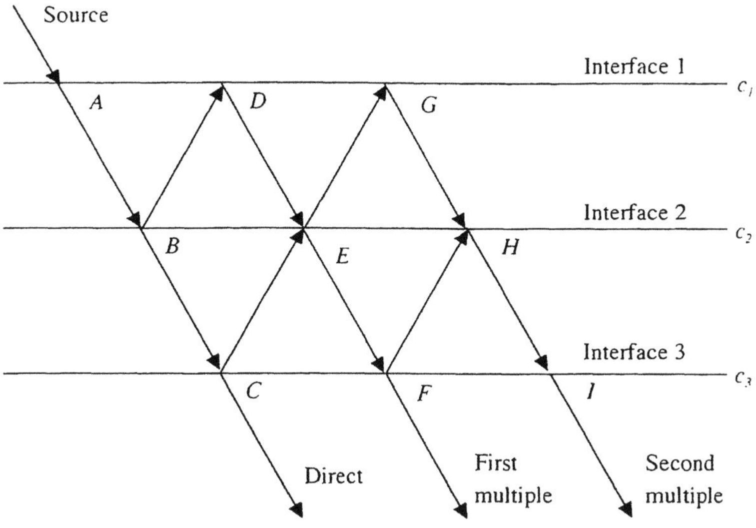

FIG. 2. *A three-interface system.*

The dummy variable s represents the unit time-delay operator. The direct arrival t_0 is the first term. The first multiple arrival t_1 is delayed one time unit (as indicated by the factor s) from the direct arrival. The second multiple arrival t_2 is delayed two time units (as indicated by the factor s^2) from the direct arrival, and so on. An example is given in the next section.

4. The case of three horizontal interfaces. Fig. 2 depicts a three-interface system with equal travel times in each layer. In this case the reflectivity function is $\{c_1,\ c_2,\ c_3\}$. The direct arrival t_0 follows the path ABC, and so is the product $\tau_1\ \tau_2\ \tau_3$ of the transmission coefficients of the three interfaces. Thus the direct arrival t_0 is equal to the transmission factor; that is, $t_0 = \sigma_3$.

The first multiple arrival t_1 is the result of waves travelling two paths, namely ABDEF and ABCEF. The pulse following path ABDEF suffers a transmission coefficient τ_1 at A, a reflection coefficient c_2 at B, a reflection coefficient $-c_1$ at D, a transmission coefficient τ_2 at E and a transmission coefficient τ_3 at F. Thus this path contributes $\tau_1\ c_2(-c_1)\ \tau_2\ \tau_3$. Similarly path ABCEF contributes $\tau_1\ \tau_2\ c_3(-c_2)\ \tau_3$. Thus the first multiple arrival, which is the sum of these two contributions, is given by

$$t_1 = -\sigma_3\ (c_1\ c_2 + c_2\ c_3).$$

The second multiple arrival is made up of four paths, as seen in Fig. 2. The first three paths ABDEGHI, ABCEFHI, and ABDEFHI are first-order reverberation paths and the remaining path ABCEGHI is a second-order reverberation path. The pulse following path ABDEGHI suffers a transmission coefficient τ_1 at A, a reflection coefficient c_2 at B, a reflection coefficient $-c_1$ at D, a reflection coefficient c_2 at E, a reflection coefficient $-c_1$ at G, a transmission coefficient τ_2 at H, and a transmission coefficient τ_3 at I. Thus this path contributes $\tau_1 \, c_2(-c_1) \, c_2(-c_1) \, \tau_2 \, \tau_3$. In the same way we can obtain the other three contributions. The sum of the four contributions gives the second multiple arrival as

$$t_2 \;=\; \sigma_3 \, c_1^2 \, c_2^2 \;+\; \sigma_3 \, c_2^2 \, c_3^2 \;+\; \sigma_3 \, c_1 \, c_2^2 \, c_3 \;-\; \sigma_3 \, (1 - c_2^2) \, c_1 \, c_3 \;.$$

Rearranging terms, the second multiple arrival becomes

$$t_2 \;=\; -\sigma_3 \, (c_1 \, c_3 \;-\; c_1^2 \, c_2^2 \;-\; 2 \, c_1 \, c_2^2 \, c_3 \;-\; c_2^2 \, c_3^2) \;.$$

We can continue this process indefinitely and find expressions for all the higher multiple arrivals t_3, t_4, t_5, ... However the derivation of the transmission response by this method is unnecessarily cumbersome. A concise expression for the transmission response is given in the next section.

5. The fundamental polynomials. The sequence of fundamental polynomials $P_k(s)$ and auxiliary polynomials $Q_k(s)$ for $k = 1, 2, \ldots, N$ are generated by recursion (Robinson, 1982). Although $P_k(s)$ is actually a polynomial of degree $k - 1$, it is treated as if it were a polynomial of degree k with last coefficient 0. The polynomial $Q_k(s)$ is of degree k, but its first coefficient is zero. Define the delayed-reverse polynomials $P_k^R(s)$ and $Q_k^R(s)$ as

$$P_k^R(s) = s^k P_k(s^{-1})$$
$$Q_k^R(s) = s^k Q_k(s^{-1}) \;.$$

The recursion can be written as

$$P_k(s) = P_{k-1}(s) - c_k s Q_{k-1}^R(s)$$
$$Q_k(s) = Q_{k-1}(s) - c_k s P_{k-1}^R(s)$$

with the initial conditions $P_0(s) = P_0^R(s) = 1$ and $Q_0(s) = Q_0^R(s) = 0$. The coefficients of the first four polynomials are

$$P_1 : \;\; \{\, 1, 0 \,\}$$
$$Q_1 : \;\; \{\, 0, -c_1 \,\}$$
$$P_2 : \;\; \{\, 1, c_1 c_2, 0 \,\}$$
$$Q_2 : \;\; \{\, 0, -c_1, -c_2 \,\}$$
$$P_3 : \;\; \{\, 1, c_1 c_2 + c_2 c_3, c_1 c_3, 0 \,\}$$
$$Q_3 : \;\; \{\, 0, -c_1, -c_2 - c_1 c_2 c_3, -c_3 \,\}$$
$$P_4 : \;\; \{\, 1, c_1 c_2 + c_2 c_3 + c_3 c_4, c_1 c_3 + c_2 c_4 + c_1 c_2 c_3 c_4, c_1 c_4, 0 \,\}$$
$$Q_4 : \;\; \{\, 0, -c_1, -c_2 - c_1 c_2 c_3 - c_1 c_3 c_4, -c_3 - c_1 c_2 c_4 - c_2 c_3 c_4, -c_4 \,\} \;.$$

It is seen that the coefficients of the fundamental polynomials P_k involve only even products of the reflection coefficients, whereas the coefficients of the auxiliary polynomials Q_k involve only odd products of the reflection coefficients.

For N interfaces, the fundamental polynomial $P_N(s)$ provides a compact expression for the generating function of the transmission response (Robinson, 1982), namely

$$T_N(s) = t_0 + t_1 s + t_2 s^2 + t_3 s^3 + \cdots = \frac{\sigma_N}{P_N(s)} \ .$$

For example, in the case of three interfaces, the generating function is

$$\begin{aligned}
T_3(s) &= \frac{\sigma_3}{1 + (c_1 c_2 + c_2 c_3)s + c_1 c_3 s^2} \\
&= t_0 + t_1 s + t_2 s^2 + t_3 s^3 + \cdots \\
&= \sigma_3 - \sigma_3(c_1 c_2 + c_2 c_3)s - \sigma_3(c_1 c_3 - c_1^2 c_2^2 - 2c_1 c_2^2 c_3 - c_2^2 c_3^2)s^2 + \cdots
\end{aligned}$$

which agrees with the results of the previous section. Ideally, we would like the transmission response to approximate a spike $\{t_0, \ 0, \ 0, \ \cdots\}$. Such a situation would occur if all the multiple arrivals were so small that they could be neglected. Thus the task is to find the type of reflectivity sequence that would produce negligible multiple arrivals. If the fundamental polynomial is close to a spike, then the transmission response will also be close to a spike. In the next section, we examine the classical method to make this situation possible.

6. The case of small reflection coefficients. The classical method used to provide good performance in electrical transmission lines is impedance matching. The impedances of two connecting circuits are matched if they are complex conjugates of each other. Impedance matching is important because the better the matching, the better is the transfer of power. The layered model can handle waves travelling at various angles to the interfaces, as is done in the study of the optics of thin films (Heavens, 1991). For mathematical simplicity, the treatment given here is restricted to vertically travelling waves. In such a case, the impedances as well as the reflection coefficients are all real numbers. If the impedances of two adjacent layers are close in value, the reflection coefficient is small. A reflection coefficient can be positive or negative, but its magnitude must be less than one. Generally, the magnitudes of the reflection coefficients $\{c_1, \ c_2, \ \cdots, \ c_N\}$ encountered in seismic prospecting are much less than one in value. Whenever the reflection coefficients cluster around the mean value of zero, they are considered small. In such a case, the transmission factor σ_N is nearly equal to one. An essential mathematical simplification occurs in the case of small reflection coefficients. For small reflection coefficients, the higher order products in the fundamental polynomial become so small that they can be neglected. If higher-order products are neglected, the coefficients of the fundamental polynomial take on a simplified form. For example, for $N = 4$ we can neglect the higher order product $c_1 c_2 c_2 c_4$ in the fundamental polynomial $P_4(s)$. As a result the coefficients become approximately

$$P_4 : \quad \{\, 1,\ c_1 c_2 + c_2 c_3 + c_3 c_4,\ c_1 c_3 + c_2 c_4,\ c_1 c_4, 0 \,\} \ .$$

The unnormalized autocovariance coefficients g_i of the reflectivity $\{c_1,\ c_2,\ c_3,\ c_4\}$ are defined as

$$g_0 = c_1^2 + c_2^2 + c_3^2 + c_4^2$$
$$g_1 = g_{-1} = c_1 c_2 + c_2 c_3 + c_3 c_4$$
$$g_2 = g_{-2} = c_1 c_3 + c_2 c_4$$
$$g_3 = g_{-3} = c_1 c_4 \ .$$

Hence, for small reflection coefficients, the coefficients of the fundamental polynomial $P_4(s)$ are approximately given by $(1, g_1, g_2, g_3, 0)$. Thus the generating function of the transmission response is approximately given by

$$T_4(s) \approx \frac{\sigma_3}{1 + g_1\, s\, +\, g_2\, s^2\, +\, g_3\, s^3}$$

which, on expansion, becomes

$$T_4(s) \approx \sigma_4 [\, 1 - g_1\, s\, -\, (g_2 - g_1^2)\, s^2\, -\, (g_3 - 2 g_1\, g_2 + g_1^3)\, s^3 - \cdots] \ .$$

The reflection coefficients should be small enough to make the autocovariance coefficients also small. As a result, higher order products of the autocovariance coefficients can be neglected, and hence the above equation gives the approximation

$$\{t_0,\ t_1,\ t_2,\ t_3,\ \cdots \} \approx \{\sigma_4, -\sigma_4\, g_1, -\sigma_4\, g_2, -\sigma_4\, g_3,\ 0,\ 0,\ \cdots \} \ .$$

In this approximation the multiple arrivals are proportional to the autocovariance coefficients, and hence the multiples are small. The same result applies to the general case of N layers. Impedance matching works. However, it is often the case in practice that there is a limit as to how small the reflection coefficients can be made. The next section goes one step further in order to gain increased performance in signal transmission.

7. The case of small random reflection coefficients. A multiple reflection represents seismic energy that has been reflected more than once. As seen in the previous section, small reflection coefficients produce small multiple arrivals. However one further step can be taken in order to reduce the magnitudes of the multiple arrival even more. That step is randomization. As we have seen the multiples involve an intricate pattern of behavior. This pattern can be disturbed if the reflection coefficients are randomized. If the reflection coefficient sequence is generated by a random white process, then its autocovariance coefficients (except the coefficient for lag zero) are approximately zero. Lithology refers to the structure and the composition of rock formations. As far as seismic transmission is concerned, the most important characteristic is that the rock formation has reflection coefficients that are small in magnitude and that are a realization of random white stochastic process. Such a rock formation is called a small white lithologic section. Thus, for a small white lithologic section, the fundamental polynomial reduces approximately to

$$P_N(s) \approx 1 + g_1 s + g_2 s^2 + \cdots + g_{N-1} s^{N-1} \approx 1 .$$

Thus the transmission response of a small white lithologic section is approximately the spike

$$\{t_0,\ t_1,\ t_2,\ t_3,\ \cdots \} \approx \{\sigma_N,\ 0,\ 0,\ 0,\ \cdots\} .$$

Thus to a good approximation, small white reflection coefficients produce no significant multiple reflections on the transmission response. In other words, a lithologic section with small white reflection coefficients passes a signal in transmission with no change in shape, but attenuated by the scale factor σ_N. The randomization of small reflection coefficients produces high-performance signal transmission.

The following table gives the reflectivity and the transmission response for each of two cases: A. Small cyclic reflectivity and B. small white reflectivity. In both cases there are 20 interfaces. Both have the same transmission factor 0.904, which means that the reflection coefficients are moderately small. The transmission response is infinitely long, but only the first twenty values are shown in the table. The direct arrival in the transmission response is at time zero, and the multiple arrivals are at the later times. In the case of small cyclic reflectivity, the transmission response has somewhat large multiples so the transmission though the layered system is relatively poor. In the case of small random reflectivity, the transmission response has small multiples so the transmission though the layered system is excellent.

8. Conclusion and suggestions for further work. In any remote detection problem, the first question that comes up is how well does the medium transmit a signal. An ideal system would transmit a signal with no change in the shape of the signal. In other words, a good transmitting medium would be a medium whose impulse transmission response approximates a spike. This means that the later arrivals in the transmission response due to internal multiple reflections should be as small as possible. A layered system with small random white reflection coefficients provides such a transmission response. In summary, randomization improves signal transmission.

TABLE 1

A. Small cyclic reflectivity and B. small white reflectivity, and their transmission responses.

Interface index	A. Small cyclic reflectivity	B. Small white reflectivity	Discrete time index for signals	Transmission response for A.	Transmission response for B.
1	0.1	−0.1	0	1.00	1.00
2	−0.1	−0.1	1	0.19	0.01
3	0.1	0.1	2	−0.16	0.00
4	−0.1	0.1	3	0.13	−0.03
5	0.1	−0.1	4	−0.10	−0.02
6	−0.1	0.1	5	0.08	0.03
7	0.1	0.1	6	−0.06	−0.08
8	−0.1	−0.1	7	0.04	−0.01
9	0.1	0.1	8	−0.03	−0.02
10	−0.1	0.1	9	0.01	−0.01
11	0.1	0.1	10	0.00	0.04
12	−0.1	0.1	11	−0.01	0.01
13	0.1	0.1	12	0.01	−0.01
14	−0.1	−0.1	13	−0.02	0.01
15	0.1	0.1	14	0.02	0.00
16	−0.1	−0.1	15	−0.02	−0.02
17	0.1	0.1	16	0.02	0.01
18	−0.1	0.1	17	−0.02	0.03
19	0.1	0.1	18	0.01	0.00
20	−0.1	−0.1	19	−0.01	−0.01

In this paper, randomness is discussed only in terms of second-order statistics, namely the autocovariance sequence. However, randomness as evidenced by higher order statistics could be invoked to see how such randomness increases the efficacy of signal transmission. The well-known Schur polynomials A are generated by the Durbin-Levinson recursion (Whittle, 1963, page 37; Gardner, 1990, page 302) in the fitting of autoregressive processes of successively increasing orders. It is worth noting that the Schur polynomial A is equal to the sum of P and Q. In this sense, the polynomials P and Q are more basic. However, except for their use in layered systems, as given in this paper, the writer had never seen the polynomials P and Q used in mathematics.

REFERENCES

BREKHOVSKIKH L.M. (1960), *Waves in Layered Media*, Academic Press, NY.

EWING M., W. JARDETZKY, AND F. PRESS (1957) *Elastic Waves in Layered Media*, McGraw Hill, NY.

GARDNER W.A. (1990), *Introduction to Random Processes*, Second Edition, McGraw Hill, NY.

GRAY A. AND J. MARKEL (1973), Digital lattice and ladder filter synthesis, *IEEE Trans Audio Electroacoust.*, AU-21, 491–500.

HEAVENS O.S. (1991), *Optical Properties of Thin Solid Films*, Dover, NY.

MITRA S. AND J.F. KAISER (1993), *Handbook for Digital Signal Processing*, John Wiley, NY.

ROBINSON E.A. (1982), Spectral Approach to Geophysical Inversion by Lorentz, Fourier, and Radon Transforms, *Proceedings of the IEEE*, Vol. **70**, pp. 1039–1054.

WHITTLE P. (1963), *Prediction and Regulation*, The English Universities Press, London.

ONLINE ANALYSIS OF SEISMIC SIGNALS

HERNANDO OMBAO[*], JUNGEUN HEO[†], AND DAVID STOFFER[‡]

Abstract. Seismic signals can be modeled as non-stationary time series. Methods for analyzing non-stationary time series that have been recently developed are proposed in Adak [1], West, et al. [25] and Ombao, et al. [12]. These methods require that the entire series be observed completely prior to analyses. In some situations, it is desirable to commence analysis even while the time series is being recorded. In this paper, we develop a statistical method for analyzing seismic signals while it is being recorded or observed. The basic idea is to model the seismic signal as a piecewise stationary autoregressive process. When a block of time series becomes available, an AR model is fit, the AR parameters estimated and the Bayesian information criterion (BIC) value is computed. Adjacent blocks are combined to form one big block if the BIC for the combined block is less than the sum of the BIC for each of the split adjacent blocks. Otherwise, adjacent blocks are kept as separate. In the event that adjacent blocks are combined as a single block, we interpret the observations at those two blocks as likely to have been generated by one AR process. When the adjacent blocks are separate, the observations at the two blocks were likely to have been generated by different AR processes. In this situation, the method has detected a change in the spectral and distributional parameters of the time series.

Simulation results suggest that the proposed method is able to detect changes in the time series as they occur. Moreover, the proposed method tends to report changes only when they actually occur. The methodology will be useful for seismologists who need to monitor vigilantly changes in seismic activities. Our procedure is inspired by Takanami [23] which uses the Akaike Information Criterion (AIC). We report simulation results that compare the online BIC method with the Takanami method and discuss the advantages and disadvantages of the two online methods. Finally, we apply the online BIC method to a seismic waves dataset.

Key words. Non-stationary time series, Autoregressive models, Akaike information criterion, Bayesian information criterion, Time-frequency analysis, Seismic signals.

AMS(MOS) subject classifications. Primary 62M10, Secondary 86A32.

1. Introduction. Many time series datasets can be modeled as realizations of non-stationary processes. For example seismic waves (Figure 1), which are ground vibrations, have distributional properties that may change over time and space during a seismic activity. In particular, their amplitude increases during the arrival of the seismic P and S waves. In this paper, we will propose a statistical method that can detect changes in the distributional properties of a time series and report the detected changes as they occur. It is ideal that the method report changes only

[*]Department of Statistics, University of Illinois, Champaign, IL 61822. The work of H. Ombao was supported in part by NIMH 62298 and NSF DMS-0102511.

[†]Department of Statistics and Department of Psychiatry, University of Pittsburgh, Pittsburgh, PA 15260. The work of J. Heo was supported in part by NIMH 55123.

[‡]Department of Statistics, University of Pittsburgh, Pittsburgh, PA 15260. The work of D. Stoffer was supported in part by NSF DMS-0102511.

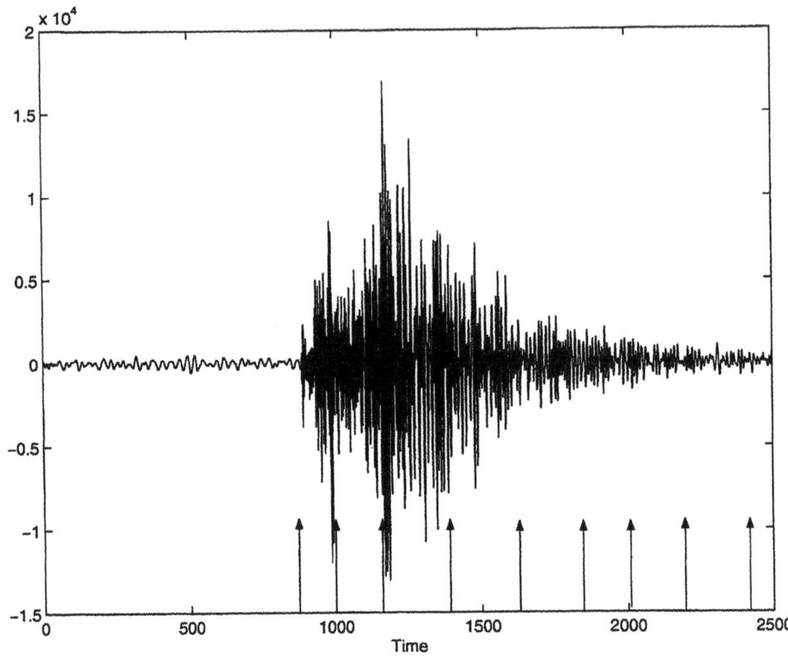

FIG. 1. *Seismic waves. Total number of time points is $T = 2500$.*

when they actually occur. Our method will also include a procedure for estimating the time-varying spectra of a non-stationary time series.

Adak [1] developed a time-varying spectral method that uses a windowed Fourier transform. The Adak method essentially divides the time series sequentially in a dyadic manner (i.e., the time series is divided into two half blocks, then four quarter blocks, etc.) and then estimates the spectrum at each of the blocks. If the spectral estimates at adjacent children blocks (blocks that result from splitting a bigger block) are sufficiently close according to some distance measure, then they are kept as one block. Some distance measures proposed included the Kolmogorov-Smirnov and Cramér-von Misses. West, et al. [25] proposed to model non-stationary time series using time-varying coefficient autoregressive models with coefficients modeled by some random process. Recently, Ombao, et al. [12] developed an automatic method that simultaneously segments and estimates the spectra and coherence of a bivariate time series. The SLEX method is based on the SLEX (**S**mooth **L**ocal **C**omplex **EX**ponential) vectors which are like Fourier complex exponentials but are localized both in time and frequency domains. Hence they are ideal at representing non-stationary time series, i.e., the time series whose spectral properties evolve over time. The SLEX model of a non-stationary random process is discussed in Ombao, et al. [14]. The Adak; West; and SLEX methods all require that the entire time series be observed prior to any data analysis. In many applications, it is desirable to do online analysis of time series. In other words, some situations require us to analyze the time series dataset even while the time series is being recorded or observed. For example, neurologists would like to analyze the brain waves of epileptic patients who are being monitored continuously. A timely detection of changes in the spectrum or changes in wave amplitude could prompt the patient of these changes some of which might be indicative of an onset of epileptic seizure. As another example, seismologists can benefit from an automatic method that monitors and detects changes in seismic activities.

Takanami [23], following earlier work in Takanami and Kitagawa [21, 22] developed an online procedure for estimating the arrival of seismic waves. The procedure fits a piecewise stationary autoregressive model to the time series. Ozaki and Tong [15], Kitagawa and Akaike [9] and others have developed the idea of a locally stationary autoregressive model and have applied this to many seismic recordings. When blocks of time series observations become available, the autoregressive (AR) model is fit separately to the blocks. The Takanami method uses the Akaike information criterion (AIC) [2] as a criterion for combining or splitting adjacent blocks. The basic procedure is to combine adjacent blocks if the AIC for the stationary autoregressive model that is fit to the combined block is smaller than the sum of the AIC's for a piecewise stationary AR model (i.e., a stationary AR model is fit separately to each of the blocks.) When the method combines adjacent blocks into a single block, the interpretation is that the observations at those two blocks were likely to have been generated by the same AR process. When the method keeps the adjacent blocks separate, then we say that the observations at the two blocks were likely to have been generated by different AR processes. Our simulation results suggest that the Takanami method is sensitive to changes in the time series, i.e., it is able to detect actual changes. However, it has a tendency to declare adjacent blocks as different even when there is no actual change in the time series.

In this paper, we propose a procedure that is inspired by Takanami [23]. Our proposed method, however, differs from the Takanami approach in the following respects. The proposed method treats the order of the AR model as constant across all blocks whereas the Takanami method allows the order of the AR model to change from block to block. The benefits of using a model that has a fixed order outweigh the flexibility of the Takanami approach. Foremost, fixing the order of the model allows the AR coefficients to be tracked over time and changes in the AR coefficients can be interpreted in a meaningful manner. Moreover, there is a well developed body in the literature on time varying coefficient AR models whose order is constant over time (see Kitagawa and Gersch [10] and Dahlhaus [5]). The second major difference is the criterion used in making a decision whether to combine or split adjacent blocks. Takanami uses the AIC while our method uses the Bayesian information criterion (BIC). The results of our simulation study suggest that when the block size used is "sufficiently" large, then both the online BIC and the Takanami methods are sensitive to true changes in the time series. The Takanami method, however, suffers from having high rate of "false positives" or "false alarms". It tended to declare a change even when there was none actually present. When the block size is small, the Takanami method is more sensitive to actual changes than the online BIC method. However, the online BIC method gives a lower "false positive" rate.

The rest of this paper is organized as follows. We describe our algorithm in Section 2, present simulation results in Section 3, and analyze seismic waves datasets in Section 4.

2. The online method. In analyzing two adjacent blocks of time series, we fit essentially two models, namely, (i.) stationary model that is fit to the combined blocks (ii.) piecewise AR stationary model (AR process is fit separately to the two blocks). We then need to choose between the two models by using an objective criterion for model selection. Our proposed online method uses the Bayesian information criterion for model selection.

2.1. The Bayesian information criterion. A Bayesian type approach to model selection was proposed in Schwarz [19]. This approach assigns a prior probability to each candidate model from a well-defined family of models and a prior distribution to the parameters conditioned on the model. The approach first computes the posterior probability of each model, conditioned on the observations, and then selects the model with the largest posterior probability. In other words, after observing the data, the best model is that which most likely generated the given observations. Schwartz developed the BIC for linear models with observations that are independent and have identical distributions from the regular exponential family. The BIC has been extended to other situations. Haughton [7] derived the BIC for curved exponential family; and Cavanaugh and Neath [4] derived the BIC for general likelihood that satisfies fairly non-restrictive regularity conditions.

We sketch the derivation of the BIC. First, we define the family of models to be the set $\Omega = \{M_1, \ldots, M_L\}$. Let the parameters for the model M_k be denoted as $\boldsymbol{\theta}_k$, $k = 1, \ldots, L$. Denote the observed dataset to be $Y = (Y_1, \ldots, Y_T)$. Let $\mathcal{L}(\boldsymbol{\theta}_k|Y)$ be the likelihood for Y based on the model M_k; π_k be a discrete prior of model M_k being correct; and $g(\boldsymbol{\theta}_k|M_k)$ be the prior density for the parameter vector $\boldsymbol{\theta}_k$ conditional on M_k being correct. Applying Bayes' theorem, the posterior probability that M_k is the correct model for a given observation vector Y is

$$(2.1) \qquad P(M_k|Y) = \frac{\pi_k \int \mathcal{L}(\boldsymbol{\theta}_k|Y)\, g(\boldsymbol{\theta}_k|M_k)\, d\boldsymbol{\theta}_k}{\sum_{k=1}^{L} \pi_k \int \mathcal{L}(\boldsymbol{\theta}_k|Y) g(\boldsymbol{\theta}_k|M_k) d\boldsymbol{\theta}_k}.$$

The Bayesian model selection rule is to find the model $M_{k^*} \in \Omega$ that maximizes Equation (2.1). This is equivalent to minimizing $-2 \ln P(M_k \mid Y)$. Thus, if we assume $g(\boldsymbol{\theta}_k | M_k) \propto 1$ then we obtain the Bayesian information criterion of the form

$$(2.2) \qquad BIC = -2 \ln \mathcal{L}(\hat{\boldsymbol{\theta}}_k | Y) + \dim(\boldsymbol{\theta}_k) \ln(T)$$

where $\hat{\boldsymbol{\theta}}_k$ is the MLE and $\dim(\boldsymbol{\theta}_k)$ is the dimension of the parameter vector $\boldsymbol{\theta}_k$. For a detailed derivation, see Neath and Cavanaugh [11] where they also proposed correction terms that could significantly improve the performance of the BIC for small sample sizes. In this paper, we do not implement the small sample corrections since they will not be necessary for datasets that we will analyze.

2.2. BIC for AR(p) models. We now derive the BIC for the autoregressive model. For completeness, we define the time series $Y_1, \ldots, Y_T$ to be generated by an autoregressive model of order p with parameters $a_1, \ldots, a_p$ if we can represent the time series as

$$Y_t = a_1 Y_{t-1} + \cdots + a_p Y_{t-p} + \epsilon_t, \qquad t = 1, \ldots, T$$

where the random innovation ϵ_t is i.i.d. with zero mean and variance σ^2. We denote this model as AR(p) $(a_1, \ldots, a_p)$. The AR(p) model is stationary if the coefficients $(a_1, \ldots, a_p)$ satisfy the condition that the roots of the equation $v(z) = z^p + a_1 z^{p-1} + \cdots a_p$ lie inside the unit circle (see Priestley [17]). In the derivation that follows, we will use a Gaussian AR(p) model, i.e., we assume that ϵ_t has a Gaussian distribution. Let the vector of coefficients be defined as $\boldsymbol{\theta} = (a_1, \ldots, a_p)'$. Then the likelihood can be written as

$$\mathcal{L}(\boldsymbol{\theta}, \sigma^2) = f(y_1, \ldots, y_p) f(y_{p+1} | y_p, \ldots, y_1) \ldots f(y_T | y_{T-1}, \ldots, y_1).$$

Since $Y_t \mid y_{t-1}, \ldots, y_{t-p} \sim \mathrm{N}(\mu_t, \sigma^2)$ where $\mu_t = a_1 y_{t-1} + \ldots + a_p y_{t-p}$, we may then write the likelihood as

$$\mathcal{L}(\boldsymbol{\theta}, \sigma^2) = f(y_1, \ldots, y_p) \prod_{t=p+1}^{T} f_\epsilon \left[\, (y_t - a_1 y_{t-1} - \cdots - a_p y_{t-p}) \, \right]$$

where $f_\epsilon(u) = 1/\sqrt{2\pi\sigma^2} \cdot \exp\left[-u^2/(2\sigma^2)\right]$ is the density of ϵ_t.

The estimation procedure for the full likelihood can be quite complicated even for the AR(1) model (see Shumway and Stoffer [20]). For illustration purposes, let us consider the least complicated case by choosing $p = 1$. Under the AR(1) model, the density $f(y_1)$ is $\mathrm{N}[0, \sigma^2/(1 - a_1^2)]$ and the likelihood becomes

$$\mathcal{L}(a_1, \sigma^2) = (2\pi\sigma^2)^{-T/2}(1 - a_1^2)^{1/2}\exp\left[-Q(a_1, \sigma^2)/(2\sigma^2)\right]$$

where $Q(a_1) = (1-a_1^2)(y_1^2) + \sum_{t=2}^{T}[y_t - a_1 y_{t-1}]^2$ is called the unconditional least squares. Denote the maximum likelihood estimate of σ^2 to be $\hat{\sigma}^2 = Q(\hat{a}_1)/T$ where $\hat{a}_1$ is the MLE of a_1. Next, we take negative logs, ignore constants and use $\hat{\sigma}^2$, then we have the estimator $\hat{a}_1$ to be the minimizer of the log-likelihood

$$\ell(a_1) = \ln(Q(a_1)/T) - [\ln(1 - a_1^2)]/T.$$

Clearly, minimizing $\ell(a_1)$ is quite complicated and has to be accomplished numerically. A way to simplify the estimation is to use the conditional likelihood and exploit the property of AR models that they are linear models conditioned on their initial values. When we condition on $y_1, \ldots, y_p$, the conditional likelihood becomes

$$\mathcal{L}(\boldsymbol{\theta}, \sigma^2 | y_1, \ldots, y_p) = (2\pi\sigma^2)^{-(T-p)/2} \exp[-Q_c(\boldsymbol{\theta})/(2\sigma^2)]$$

where $Q_c(\boldsymbol{\theta}) = \sum_{t=p+1}^{T} [y_t - (a_1 y_{t-1} + \cdots + a_p y_{t-p})]^2$ is the conditional sum of squares. The conditional MLE of σ^2 is $\hat{\sigma}^2 = Q_c(\hat{\boldsymbol{\theta}})/(T-p)$ where $\hat{\boldsymbol{\theta}}$ is the minimizer of $Q_c(\boldsymbol{\theta})$.

Substituting the conditional MLE of $(\boldsymbol{\theta}, \sigma^2)$ into the conditional likelihood and taking the logs, we have the log-likelihood proportional to

$$- [(T - p)/2] (\ln 2\pi + 1) - [(T - p)/2] \ln(\hat{\sigma}^2).$$

We obtain the BIC of AR(p) to be

$$BIC = (T - p)(\ln 2\pi + 1) + (T - p)\ln(\hat{\sigma}^2) + (p + 1)\ln(T).$$

We are now ready to apply the BIC to the online analysis of non-stationary time series.

2.3. Split or combine? Consider two adjacent blocks of time series, namely left and right blocks, which we denote by $B^{\text{left}} = (Y_1, \ldots, Y_N)$ and $B^{\text{right}} = (Y_{N+1}, \ldots, Y_{2N})$ respectively. We fit two models and select the one that gives a smaller BIC value. The first model is the stationary AR(p) $(a_1, \ldots, a_p)$, that we fit to the combined block $B^{\text{com}} = B^{\text{left}} \cup B^{\text{right}} = (Y_1, \ldots, Y_{2N})$. Denote the random innovations on the first model to be ϵ_t i.i.d. $N(0, \sigma^2)$. The second model is piecewise stationary AR(p), i.e., an AR(p) model that is fit separately to each of the left and right blocks:

$$Y_t = \begin{cases} Y_t^{(1)} & \text{if } 1 \leq t \leq N \\ Y_t^{(2)} & \text{if } N + 1 \leq t \leq 2N \end{cases}$$

where $Y_t^{(1)}$ is AR(p) $(a_1^{(1)}, \ldots, a_p^{(1)})$ and $Y_t^{(2)}$ is $(a_1^{(2)}, \ldots, a_p^{(2)})$.

Denote the random innovations on the left block to be $w_t^{(1)}$ i.i.d. $N(0, \tau_1^2)$ and on the right block to be $w_t^{(2)}$ i.i.d. $N(0, \tau_2^2)$. The BIC for the combined block (model 1) is

$$BIC_c = (2N - p)(\ln 2\pi + 1) +$$
$$(2N - p)\ln(\widehat{\sigma}^2) + (p + 1)\ln(2N).$$

The BIC for the split block (model 2) is $BIC_s = BIC(\text{left}) + BIC(\text{right})$ where

$$BIC(\text{left}) = (N - p)(\ln 2\pi + 1) + (N - p)\ln(\widehat{\tau_1}^2) + (p + 1)\ln(N)$$
$$BIC(\text{right}) = (N - p)(\ln 2\pi + 1) + (N - p)\ln(\widehat{\tau_2}^2) + (p + 1)\ln(N).$$

The decision rule is to choose the "combined blocks" (stationary model) if $BIC_c \leq BIC_s$. Otherwise, choose the "split blocks" (piecewise stationary model).

2.4. The algorithm for determining change points. In our algorithm, we analyze the time series as blocks of N observations become available. We denote $B^{(m)}$ as the $m - th$ block of N observations. Thus, $B^{(m)}$ consists of observations $(Y_{(m-1)N+1}, \ldots, Y_{mN})$.

Step 0. INITIALIZE.
　　Set B^{left} = empty; B^{right} = empty; B^{com} = empty;
　　Set $BIC(\text{left}) = 0$; $BIC(\text{right}) = 0$; $BIC_s = 0$; $BIC_c = 0$.
Step 1. When the block $B^{(1)}$ is complete then
　　Set $B^{\text{left}} = B^{(1)}$;
　　Fit an AR(p) model;
　　Compute $BIC(\text{left})$.
Step 2. When the block $B^{(2)}$ is complete then
　　Set $B^{\text{right}} = B^{(2)}$;
　　Fit an AR(p) model;
　　Compute $BIC(\text{right})$.
Step 3. Compute $BIC_s = BIC(\text{left}) + BIC(\text{right})$.
Step 4. Set $B^{\text{com}} = B^{\text{left}} \cup B^{\text{right}} = B^{(1)} \cup B^{(2)}$ and
　　Fit an AR(p) model to B^{com};
　　Compute BIC_c.
Step 5. Decision. Let $\Delta = BIC_c - BIC_s$.
　　If $\Delta \leq 0$ then
　　　　Combine blocks B^{left} and B^{right};
　　　　Set $B^{\text{left}} = B^{\text{com}}$;
　　　　Set $BIC(\text{left}) = BIC_c$;
　　　　Proceed to Step 6.
　　If $\Delta > 0$ then
　　　　Split the blocks;
　　　　Set $B^{\text{left}} = B^{\text{right}}$;

Set $BIC(\text{left}) = BIC(\text{right})$;

The spectral estimate on B^{left} is

$$\widehat{f}(\omega) = \widehat{\sigma}^2/|1 - \widehat{a}_1\exp(2\pi\omega) - \cdots - \widehat{a}_p\exp(2\pi p\omega)|^2$$

where $\widehat{\sigma}^2$ is the conditional MLE of the variance of the innovations and $\widehat{a}_1, \ldots, \widehat{a}_p$ are the conditional MLE of the coefficients of the AR(p) model when fitted to the observations on block B^{left}. Proceed to Step 6.

Step 6. When block $B^{(3)}$ is complete then (proceed as in Step 2):

Set $B^{\text{right}} = B^{(3)}$

Fit AR(p) model

Compute $BIC(\text{right})$.

Step 7. Compute $BIC_s = BIC(\text{left}) + BIC(\text{right})$ as in Step 3.

Step 8. Form the combined block as in Step 4:

Form $B^{\text{com}} = B^{\text{left}} \bigcup B^{\text{right}}$

Fit an AR(p) model

Compute BIC_c.

Step 9. Decision as in Step 5.

Step 10. CONTINUE as a new block is completed.

2.5. Remarks.

(i.) The online BIC method requires the user to supply the AR model order p and the block size N. In determining the value of p, one may use similar past data. In analyzing seismic waves, one can use datasets recorded at a nearby location at the same time period. We outline our procedure for determining the optimal order using past data. Without any loss of generality, suppose that we have a time series of length T that can be divided into b blocks each with length N, i.e., $T = Nb$. Denote the b blocks to be $B^{(1)}, \ldots, B^{(b)}$. Let $\mathcal{P} = \{1, 2, \ldots\}$ be the set of AR order under consideration. For each $p \in \mathcal{P}$ we fit an AR(p) model and compute the BIC at each of the blocks which we denote to be $BIC_1(p), \ldots, BIC_b(p)$. We then compute the "total" BIC for order p: $BIC(p) = \sum_{\ell=1}^{b} BIC_\ell(p)$. The optimal order p^* is the minimizer $BIC(p)$ over all $p \in \mathcal{P}$.

(ii.) The proposed BIC method assumes that the AR order p is fixed over time. Fixing the order allows for the AR coefficients to be compared meaningfully. In practice, we allow the order p to be large enough so that when the order at some time blocks is, say p' where $p' < p$ then $a_{p'+1} = \ldots = a_p = 0$. If this is indeed the case then $\widehat{a}_{p'+1} \approx 0, \ldots, \widehat{a}_p \approx 0$. Davis, et al. [6] also discusses this point.

(iii.) The choice of the block length N should involve opinion from a scientific expert. For example, seismologists have a good idea about what would be an appropriate time resolution of seismic waves that are recorded during an earthquake or explosion. In addition to expert advice, the choice of N should also be guided by the following statistical principles. We need N to be sufficiently small so that the assumption of stationarity within each block is valid. Otherwise, the method will not be able to capture the changes in the time series. However, we should exercise care so that the length of the blocks is not smaller than what is necessary. This will help control variance inflation of the AR parameter estimates.

(iv.) The proposed method can be made more general by fitting piecewise stationary ARMA models rather than AR models. The exact form of the BIC, however, is going to be more complicated and will need to be adjusted accordingly. There are many established advantages for using AR models. Foremost, when conditioned on initial observations, they are linear and hence allow for a simpler approach to estimation. In addition, AR models have been proven to be sufficient and quite useful in various scientific disciplines. As examples, see Wada, et al. [24] and Sato, et al. [18] for applications of AR models in physiology; Inouye, et al. [8] in electroencephalography; and Pagani, et al. [16] in cardiology and psychobiology.

(v.) Finally, the online BIC method also provides the estimate of the time-varying spectrum by fitting the AR spectral estimate at each of the stationary blocks.

3. Simulation study. In this small simulation study we wanted (i.) to determine how often the online BIC method is able to detect actual changes; (ii.) to determine how often the online BIC method incorrectly reports a change when there is none; and finally (iii.) to compare the performance of the online BIC method with the Takanami method.

3.1. Description of the numerical experiments. For study 1–3, we generated 1000 time series datasets of length $T = 1000$ from a piecewise stationary AR(4) model:

$$(3.1) \qquad Y_t = \begin{cases} Y_t^{(1)}, & \text{if } 1 \leq t \leq 200 \\ Y_t^{(2)}, & \text{if } 201 \leq t \leq 600 \\ Y_t^{(3)}, & \text{if } 601 \leq t \leq 1000 \end{cases}$$

where

$$Y^{(1)} \sim AR(4)\,(a_1^{(1)} = 1.35, a_2^{(1)} = -0.70, a_3^{(1)} = 0.40, a_4^{(1)} = -0.31)$$
$$Y^{(2)} \sim AR(4)\,(a_1^{(2)} = 1.45, a_2^{(2)} = -0.60, a_3^{(2)} = 0.30, a_4^{(2)} = -0.25)$$
$$Y^{(3)} \sim AR(4)\,(a_1^{(3)} = 1.37, a_2^{(3)} = -0.55, a_3^{(3)} = 0.35, a_4^{(3)} = -0.35).$$

A typical realization from this piecewise stationary AR process is given in Figure 2.

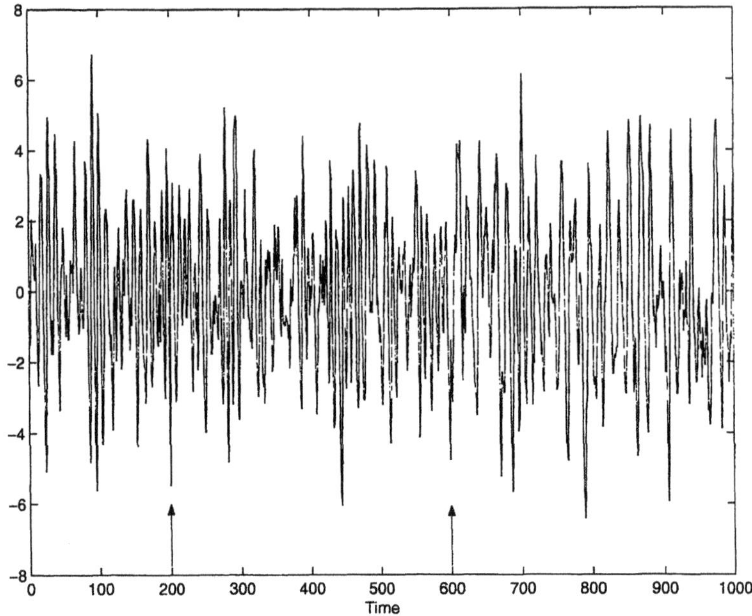

FIG. 2. *A typical time series realization from the piecewise AR(4) process used in the simulation study. The arrow indicates the change points at $t = 200, 600$.*

We then applied the online BIC method and the Takanami method to each datasets to determine the change points. In study 1, we used in our analysis the model order $p = 2$ and block size $N = 200$; in study 2, $p = 4$ and $N = 200$ and in study 3, $p = 6$ and $N = 200$. We report a summary of the percentages of data sets that had change points at $t = 200, 400, 600, 800$.

In study 4, we generated 1000 time series data sets from a process that is similar to that in study 1 to 3 except that the changepoints occur at time points that are not multiples of 100:

$$(3.2) \qquad Y_t = \begin{cases} Y_t^{(1)}, & \text{if } 1 \le t \le 207 \\ Y_t^{(2)}, & \text{if } 208 \le t \le 615 \\ Y_t^{(3)}, & \text{if } 616 \le t \le 1000 \end{cases}$$

where $Y_t^{(1)}, Y_t^{(2)}$ and $Y_t^{(3)}$ are the same as those in study 1–3. We applied the online BIC method and the Takanami method to each time series dataset using model order $p = 4$ and block length $N = 200$.

In study 5, we investigated the effect of using a block size of $N = 100$ which is smaller than the one used in previous studies. The model order $p = 4$ was used.

3.2. Results. The simulation results provide empirical evidence that the online BIC method is quite sensitive to actual changes in the time series. In study 1–3, the detection rate of both the Takanami and the online BIC methods at actual change points is close to 100%. The two methods, however, differ significantly the rates of giving false positives. The Takanami method showed a greater tendency to declare a change even there is no actual change. This is not particularly desirable because when stationary blocks are split needlessly then the sample size at the split blocks are reduced and the variance estimates of the parameters become inflated. Moreover, in practice, we would like a method that is sensitive to changes when they are present and at the same time does not give too many "false alarms".

In study 1 and 3, we used AR model order that are not correct. The correct model order to use was $p = 4$. The results from these studies suggest that both the online BIC method and Takanami method are robust to mild departure from the true model order. This fact is particularly reassuring given that the model order is usually not known and has to be estimated. As long as the estimated model order is "close" to the true order, both the online BIC and the Takanami methods are still expected to perform satisfactorily.

In study 4, where the change points do not coincide exactly with the block size N used in the analysis, we noted that both the Takanami and the online BIC methods were able to detect changes at decent rates. The online BIC method was slightly better than the Takanami method. However, as we have observed in study 1–3, the online BIC method produced a lower "false positive" rate than the Takanami method.

In study 5, we used a smaller block length of $N = 100$. The results suggest that the Takanami method is more sensitive to the changes in the time series but it gave a higher false positive rate than the online BIC method.

3.3. Discussion. When the combined block $B^{\mathrm{com}} = [Y_1, \ldots, Y_{2N}]$ is truly stationary, then we would want our proposed procedure to choose the combined blocks over the split blocks. In other words, it is desirable that our procedure should declare "no change" between blocks $[Y_1, \ldots, Y_N]$ and $[Y_{N+1}, \ldots, Y_{2N}]$.

Define $\Delta = BIC_c - BIC_s$. Following the notation in Section 2.3, we calculate Δ to be

$$\Delta = p\left[\ln 4\pi + 1 + \ln(\widehat{\sigma}^2) - \ln(N)\right] + (N - p)\left[\ln\left(\frac{\widehat{\sigma}^2}{\widehat{\tau}_1^2}\frac{\widehat{\sigma}^2}{\widehat{\tau}_2^2}\right)\right] - \ln(N/2).$$

We say that our procedure will make the correct decision of choosing combined blocks if $\Delta < 0$.

At this point, we propose a conjecture on how the false positive rate of the online BIC method can be controlled. When $[Y_1, \ldots, Y_{2N}]$ is stationary,

TABLE 1

Study 1–4: The true process is piecewise stationary AR(4) process with actual change points at $t = 200$ and $t = 600$. In each of study 1 to 3, 1000 time series datasets were generated. To each time series dataset, we applied both the Takanami and online BIC methods. The change points determined for each dataset were then recorded and report a summary of the percentages of datasets that had change points at $t = 200, 400, 600, 800$. In study 1, we used $p = 2, N = 200$; in study 2, $p = 4, N = 200$; in study 3, $p = 6$, $N = 200$; in study 4, $p = 4$ and $N = 200$. Interpretation of the results: In study 1, the Takanami method declared a change at $t = 200$ in 100% of the datasets; at $t = 400$ in 25%; at $t = 600$ in 100% and at $t = 800$ in 36% of the datasets. On the other hand, the online BIC method declared a change at $t = 200$ in 100%; at $t = 400$ in 2%; at $t = 600$ in 100%; at $t = 800$ in 1% of the datasets. The bold time points indicate the correct change points for study 1 to 3 and the best approximation for the change points in study 4.

Study		Change points at $t =$	**200**	400	**600**	800
1	$p = 2$	Takanami	100	25	100	36
		online BIC	100	2	100	1
2	$p = 4$	Takanami	98	18	99	20
		online BIC	98	2	98	4
3	$p = 6$	Takanami	98	20	100	33
		online BIC	96	1	99	2
4	$p = 4$	Takanami	74	48	74	38
		online BIC	79	12	87	9

TABLE 2

Study 5: The true process is piecewise stationary AR(4) process with actual change points at $t = 200$ and $t = 600$. We generated 1000 time series datasets each of length $T = 1000$. To each time series dataset, we applied the Takanami and online BIC methods using $p = 4$ and $N = 100$. The change points determined for each dataset were then recorded and a summary of the percentages of datasets that had change points at $t = 200, 300, 400, 500, 600, 700, 800$ and 900 are reported.

Selected Change points at $t =$	**200**	300	400	500	**600**	700	900
Takanami	80	31	29	24	97	42	33
online BIC	61	14	19	14	89	7	14

then $\tau_1^2 = \tau_2^2 = \sigma^2$. Under stationarity, when these error variances are known, we have

$$\Delta = p[\ln(4\pi) + 1 + \ln(\sigma^2)] - \ln(N^{p+1}/2).$$

Hence, for a sufficiently large N and small σ^2, we see that $\Delta < 0$. The proposed online BIC method is protective against false positives. In other words, it is expected to declare "no change" between adjacent blocks when there is in fact no change in the parameters in the AR model. It combines adjacent blocks when they are in fact generated by the same AR process.

We now derive an analogue of Δ for the Takanami method. The AIC for the combined block model is $AIC_c = (2N - p)\ln(\widehat{\sigma}^2) + 2(p + 1)$. The AIC for the split block model is $AIC_s = (N-p)[\ln(\widehat{\tau_1}^2)+\ln(\widehat{\tau_2}^2)]+4(p+1)$. Let $\delta = AIC_c - AIC_s$. We calculate δ to be

$$\delta = p\ln(\widehat{\sigma}^2) + (N - p)\left[\ln\left(\frac{\widehat{\sigma}^2}{\widehat{\tau_1}^2}\frac{\widehat{\sigma}^2}{\widehat{\tau_2}^2}\right)\right] - 2(p + 1).$$

Similarly, the Takanami method will make the correct decision of choosing the combined blocks if $\delta < 0$.

Again, when $[Y_1, \ldots, Y_{2N}]$ is stationary, then $\tau_1^2 = \tau_2^2 = \sigma^2$. When these error variances are known, we have

$$\delta = p\ln(\sigma^2) - 2(p + 1).$$

Note that, unlike Δ in the online BIC method, the δ in the Takanami method is independent of the block size N. Consequently, regardless of the length of the time series, the rate of false positives for the Takanami method cannot be controlled even when the block size is allowed to increase. The ability of the Takanami method to make the correct decision of not splitting a stationary block depends only on the error variance and the AR model order.

The discussion above is based on the assumption that the error variances are known. More generally, for any given block of stationary time series, let us define

$$\gamma \doteq \Delta - \delta$$
$$= p(\ln 2\pi + 3) + (p + 1)\ln 2 + 2 - \ln(N^{p+1}).$$

When $\gamma < 0$ then $\Delta < \delta$ which then implies that the online BIC method is more likely than the Takanami method to make the correct decision, i.e., it is more likely to choose the combined blocks over the split blocks. On the other hand, when $\gamma > 0$, the Takanami method is more likely to make the correct decision to combine stationary blocks.

The plot of γ as a function of N for $p = 5$ is given in Figure 3. Note that when $N < 160$, $\gamma > 0$. Hence, that the Takanami method is more likely than the online BIC method to choose the combined blocks over split blocks. On the other hand, when $N > 160$, which is a moderately large, the online BIC method is more likely to make the correct decision of not splitting a stationary block.

The plot of γ for $p = 10$ is given in Figure 4. When $N < 190$ the Takanami method is more likely the online BIC method to choose combined blocks. When $N > 190$, the online BIC method is more likely to combine a stationary block. We note that for a higher AR model order, the online BIC requires a bigger block length.

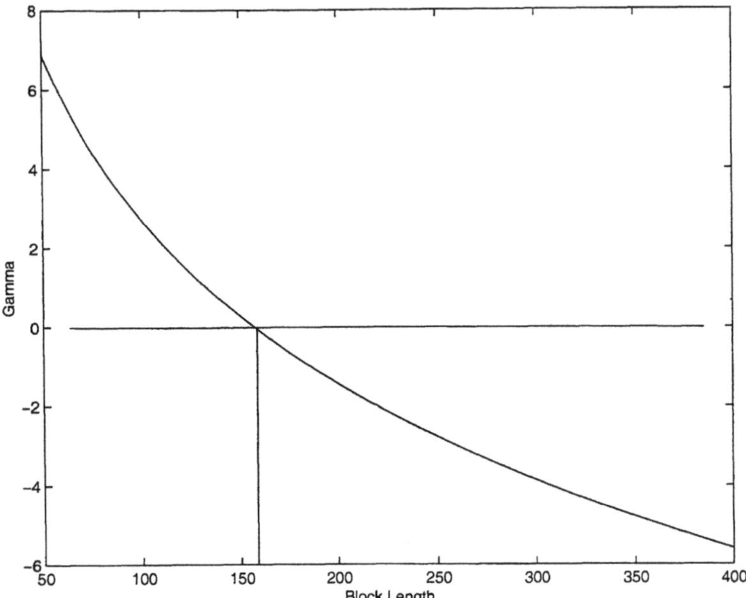

FIG. 3. *Plot of γ (Gamma) for $p = 4$. Negative values indicate that online BIC method is more likely to than Takanami method to make the right decision, i.e., to declare no change between adjacent blocks when in fact the combined block is stationary.*

4. Analysis of seismic waves. Seismic waves or seismograms are vibrations propagated through the interior of the earth during a seismic activity such an explosion or an earthquake. These waves are recorded by seismometers. Our time series dataset consists of seismograms that were recorded during the same time at 2 observation sites. We used one component to obtain the order of the AR model and then used the second component for analysis.

The objective of our analysis was to detect changes in the seismic waves as they occur. We applied the order selection algorithm and obtained the AR order $p = 4$. Similar to Takanami [23], we used block length $N = 100$. The online BIC method detected changes at the time points $t = 800, 1000, 1200, 1400, 1600, 1800, 2000, 2200$ and 2400. The Takanami method also indicated changes at the same time points. This is particularly reassuring because the Takanami method, based on our simulations for $N = 100$, are quite sensitive to actual changes in the time series. We also analyzed this same seismic dataset using $N = 200$ and observed very similar results. We report our findings only for $N = 100$.

Referring to Figure 1, the arrows indicate the change points detected by the online BIC method. It is important to note that the proposed method was able to detect a change at $t = 800$ which is believed to be the approximate arrival of the P wave.

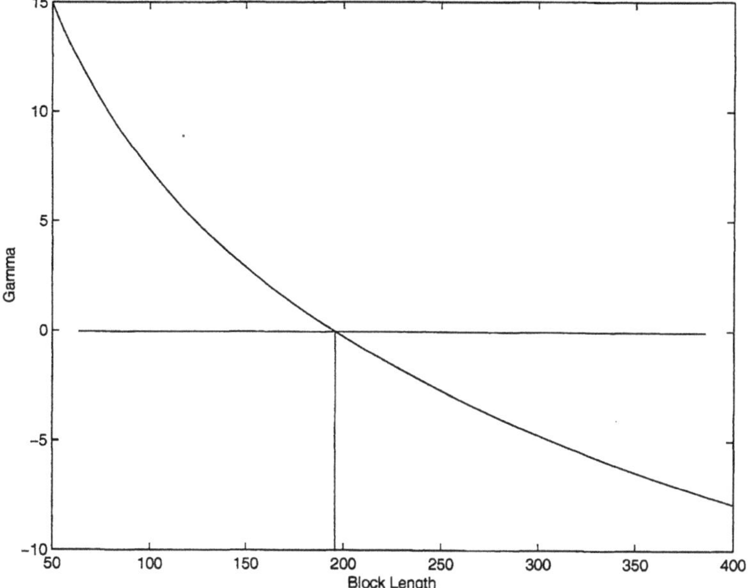

FIG. 4. *Plot of γ (Gamma) for $p = 10$. Negative values indicate that online BIC method is more likely to than Takanami method to make the right decision, i.e., to declare no change between adjacent blocks when in fact the combined block is stationary.*

The estimate of the time-varying coefficients of the AR(4) model in Figure 5 tell us that the estimate of a_1 changed quite dramatically at $t = 800$. The estimates of the other coefficients did not change in the same magnitude as a_1. During the earthquake, we see that estimates of a_1 and a_3 changed over time in a similar magnitude but opposite direction as the estimate of a_2. The estimate of a_4 remained relatively constant and close to 0.

The spectrum in Figure 6 demonstrates that the distribution of power of seismic waves change over time. Prior to the arrival of the P waves, i.e., at the first block ($t = 1 - 800$), the spectrum has a sharp peak that is located at the low frequency band. Shortly after the arrival of the P waves, i.e., at the second block ($t = 801 - 1000$), the power shifts to the middle frequency band. At the third block ($t = 1001 : 1200$), the spectrum displays two peaks, namely at the middle and high frequency bands. From $t = 1401 - 2500$, the peak of the spectrum was gradually restored to the low frequency bands.

Upon the suggestion of a reviewer, we also analyzed the same seismic dataset using the online CUSUM algorithm discussed in Basseville and Nikiforov [3]. Both the online BIC method and the CUSUM algorithms were able to detect the arrival of the P wave. However, signal changes subsequent to the arrival of the P wave were no longer detected by the

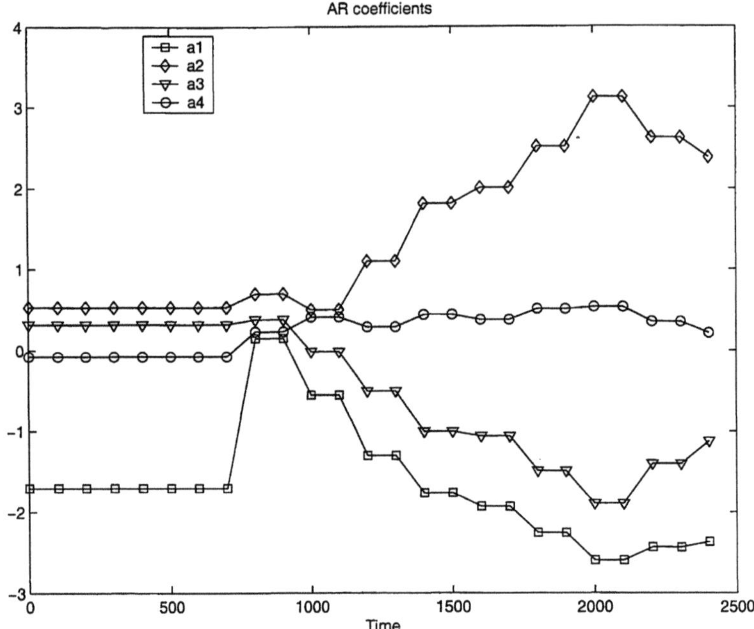

FIG. 5. *Estimates of the time-varying AR parameters of the seismic waves using the online BIC method.*

CUSUM algorithm. The CUSUM algorithm is sensitive when the variance of the error increases from one block to another. When the variance decreases, it is unable to detect that change. Arguably, the main interest is the arrival of the P wave which is also characterized by increased wave amplitude. Thus for this reason the CUSUM algorithm is useful. However, when one is interested in changes following the arrival of the P wave then the CUSUM algorithm is no longer useful.

5. Conclusion and future work. In this paper, we proposed the online BIC method for analyzing online non-stationary time series. Our method fits an AR model to blocks of time series as they are recorded. We then use the BIC to determine whether combine or split adjacent blocks. Our method parallels that of Takanami [23] which uses the AIC as a criterion to determine how to split the time series. We showed in our simulations that when p is large and N is small then the Takanami method is more sensitive to actual changes than the online BIC method. However, it also has a greater tendency to split a block even when it is stationary. On the other hand, when p is small and N is large, then the online BIC is slightly more sensitive to actual changes than the Takanami method. Furthermore, it has a greater tendency to keep a stationary block combined. Finally, the

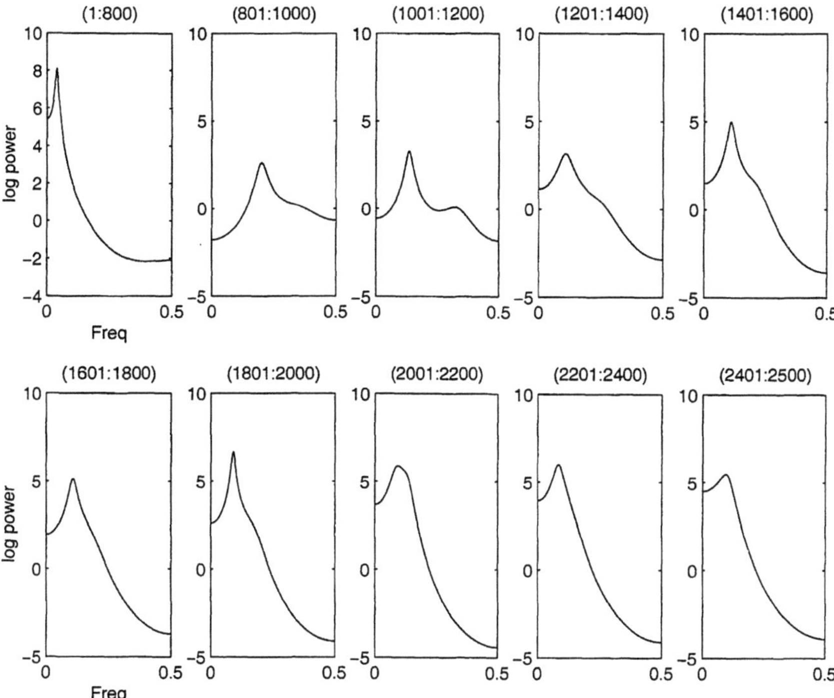

FIG. 6. *Estimate of the time-varying log spectrum of the seismic waves using the online BIC method.*

seismic waves examples demonstrated that the online BIC method is able to capture actual changes and gives reasonable spectral estimates.

Both the Takanami and the BIC methods are parametric methods that are based on the time domain. We are now investigating an alternative method that is nonparametric and based on the frequency domain. In our current investigation, we fit a nonparametric estimate to the spectrum of a block of time series. There are many possible non-parametric estimators of the spectrum. One such estimator is obtained by applying a kernel smoother on the Fourier periodograms. An important step in the smoothing periodograms is the selection of the span. An automatic span selection method for periodogram smoothing is given in Ombao, et al. [13]. Alternatively, one can use wavelets or splines to estimate the time-varying spectrum. The nonparametric spectral estimates at adjacent blocks can then be compared using an objective distance measure. Possible distance measures are the Kolmogorov-Smirnov distance and the Cramér-von Mises distance which are discussed in Priestley [17].

REFERENCES

[1] S. ADAK , *Time dependent spectral analysis of non-stationary time series*, Journal of the American Statistical Association, **93** (1998), pp. 1488–1501.

[2] H. AKAIKE, *Information theory and an extension of the maximum likelihood principle*, 2nd International Symposium on Information Theory (eds. B. Petrov and F. Csaki) (1973), pp. 267–281.

[3] M. BASSEVILLE AND I. NIKIFOROV, *Detection of Abrupt Changes — Theory and Applications*, Prentice-Hall, Englewood, Cliff, New Jersey, 1993.

[4] J. CAVANAUGH AND A. NEATH, *Generalizing the derivation of of the Schwartz information criterion*, Communications in Statistics — Theory and Methods, **28** (1999), pp. 49–66.

[5] R. DAHLHAUS, *Fitting time series models to nonstationary processes*, Annals of Statistics, **25** (1996), pp. 1–37.

[6] R. DAVIS, D. HUANG, AND Y. YAO, *Testing for a change in the parameter values and order of an autoregressive model*, Annals of Statistics, **23** (1995), pp. 282–304.

[7] D. HAUGHTON, *On the choice of a model to fit data from an exponential family*, Annals of Statistics, **6** (1988), pp. 342–355.

[8] T. INOUYE, H. SAKAMOTO, K. SHINOSAKI, S. TOI, AND S. UKAI, *Analysis of rapidly changing EEGs before generalized spike and wave complexes*, Electroencephalography and clinical Neurophysiology, **76** (1990), pp. 205–221.

[9] G. KITAGAWA AND H. AKAIKE, *Procedure for the Modeling of Non-Stationary Time Series*, Annals of the Institute of Statistical Mathematics, **30** (1978), pp. 351–363.

[10] G. KITAGAWA AND W. GERSCH, *Smoothness Priors Analysis of Time Series*, Lecture Notes in Statistics #116, New York: Springer Verlag, 1996.

[11] A. NEATH AND J. CAVANAUGH, *Regression and Time Series Model Selection Using Variants of the Schwarz Information Criterion*, Communications in Statistics — Theory and Methods, **26** (1997), pp. 559–580.

[12] H. OMBAO, J. RAZ, R. VON SACHS, AND B. MALOW, *Automatic Statistical Analysis of Bivariate Non-Stationary Time Series*, Journal of the American Statistical Association, **96** (2001), pp. 543–560.

[13] H. OMBAO, J. RAZ, R. STRAWDERMAN, AND R. VON SACHS, *A simple generalised cross validation method of span selection for periodogram smoothing*, Biometrika, **88** (2001), Vol. 4, pp. 1186–1192.

[14] H. OMBAO, J. RAZ, R. VON SACHS, AND W. GUO, *The SLEX Model of a Non-Stationary Random Process*, Annals of the Institute of Statistical Mathematics (2002), Vol. 1, in press.

[15] T. OZAKI AND H. TONG, *On the fitting of non-stationary autoregressive models in the time series analysis*, Proceedings of the 8th Hawaii International Conference on System Science, Western Periodical Hawaii (1975), pp. 224–226.

[16] M. PAGANI, *Power spectral analysis of beat-to-beat heart and blood pressure variability as a possible marker of sympatho-vagal interaction in man and conscious dog*, XS Circulation Research, **59** (1986), p. 178.

[17] M. PRIESTLEY, *Spectral Analysis and Time Series*, London: Academic Press, 1981.

[18] K. SATO AND K. ONO, *Component activities in the autoregressive activity of physiological systems*, International Order of Neuroscience, **7** (1977), pp. 239–249.

[19] G. SCHWARZ, *Estimating the dimension of a model*, Annals of Statistics, **6** (1978), pp. 461–464.

[20] R. SHUMWAY AND D. STOFFER, *Time Series Analysis and Its Applications*, New York:Springer, 2000.

[21] T. TAKANAMI AND G. KITAGAWA, *A new efficient procedure for the estimation of onset times of seismic waves*, Journal of Physics of the Earth, **36** (1988), pp. 267–290.

[22] T. TAKANAMI AND G. KITAGAWA, *Estimation of the arrival times of seismic waves by multivariate time series model*, Annals of the Institute of Statistical Mathematics, **43** (1991), pp. 403–433.

[23] T. TAKANAMI, *High Precision Estimation of Seismic Waves Arrival Times*, The Practice of Time Series Analysis (eds. Akaike and Kitagawa), New York: Springer-Verlag, 1999.

[24] T. WADA, S. SATO, AND N. MATUO, *Applications of multivariate autoregressive modeling for analyzing chloride-potassium-bicarbonate relationship in the body*, Med. Biol. Eng. Comput., **31** (1993), pp. 99–107.

[25] M. WEST, R. PRADO, AND A. KRYSTAL, *Evaluation and Comparison of EEG Traces: Latent Structure in Non-Stationary Time Series*, Journal of the American Statistical Association, **94** (1999), pp. 1083–1094.

NONSTATIONARY TIME SERIES ANALYSIS OF MONTHLY GLOBAL TEMPERATURE ANOMALIES

T. SUBBA RAO* AND E.P. TSOLAKI[†]

Abstract. In recent years modelling climatic variables has attracted the attention of many researchers. The scientific assessment of the Intergovernmental Panel on Climate Change (IPCC) (Folland et al. (1990)) concluded that despite limitations in the quality and quantity of the available temperature data, there is evidence to a real but irregular warming in the climate. Here, our object is to analyze three important temperature sets using evolutionary spectral methods. We test for stationarity, Gaussianity and linearity and based on the conclusions we fit nonstationary time series models. We also consider forecasting aspects.

Key words. Climatic temperatures, Test for stationarity, Gaussianity and linearity, Structural change points, Nonstationary models, Forecasting.

AMS(MOS) subject classifications. 62M10, 62M15.

1. The monthly global temperature anomalies. The three sets of data analyzed in the paper are global, northern hemisphere and southern hemisphere monthly temperature anomalies from January 1856 to December 1998. By anomalies we mean the difference in temperature values from some reference value. The reference value has been calculated over the period 1961 to 1990. Data were obtained from the University of East Anglia website: `http://www.cru.uea.ac.uk/` (for more details we refer to this website). They are all monthly averages obtained by merging two data sets; namely land air temperature anomalies provided by Jones (1994) and sea surface temperature anomalies provided by Parker et al. (1995) on a 5° x 5° grid-box basis (see Parker et al. (1994) and Jones et al. (2001)). The plots of the data sets are given in Figure 1. We have subtracted the sample mean from each set of the original data and thus the data plotted correspond to these mean deleted observations.

We see from the data plots that there is a linear upward trend indicating an increase in the temperature over the years and also possibly some periodic variation. Over some time periods, we also see wild fluctuations indicating a change in the variance which suggests that the data could be nonstationary. Our object here is to examine these aspects through the evolutionary spectral methods. We fit nonstationary time series models to the data sets and use them for prediction purposes.

2. Evolutionary spectral analysis. Consider a zero mean discrete parameter nonstationary time series $\{X_t\}$ having a representation of the form

---

*University of Manchester Institute of Science and Technology (UMIST), Department of Mathematics, P.O. Box 88, Manchester M60 1QD, UK (tata.subbarao@umist.ac.uk).

†Department of Computing and Mathematics, Manchester Metropolitan University, Manchester M1 5GD, UK (e.tsolaki@mmu.ac.uk).

                    T. SUBBA RAO AND E.P. TSOLAKI

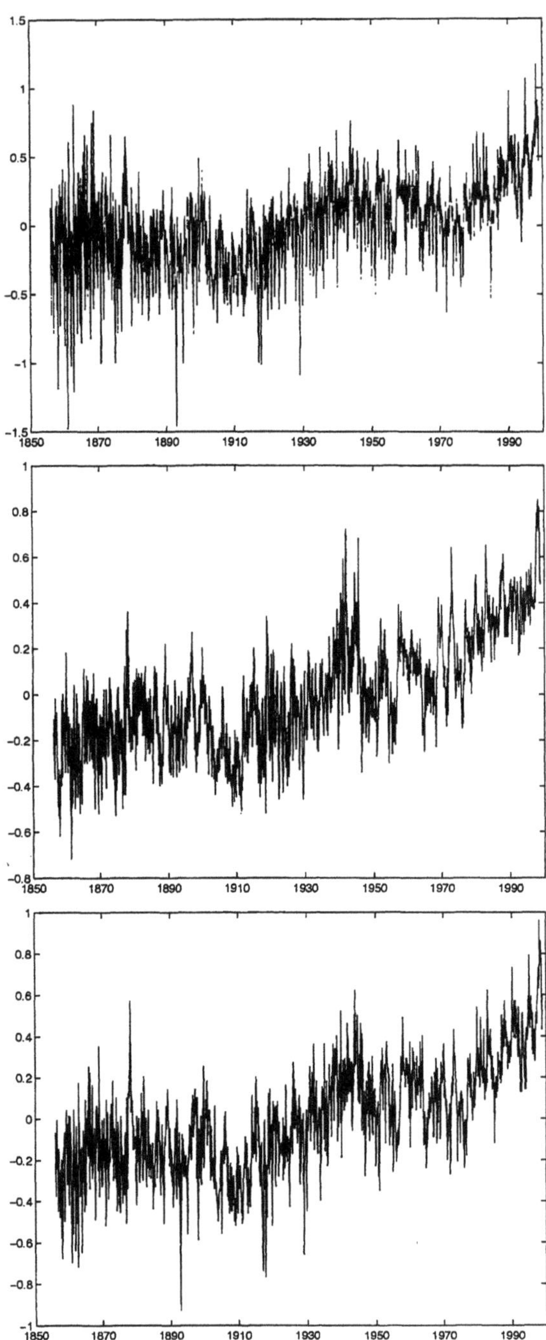

FIG. 1. *Temperature anomalies covering northern hemisphere (top), southern hemisphere (center) and global temperature anomalies (bottom).*

$$(2.1) \qquad X_t = \int_{-\pi}^{\pi} e^{it\omega} A_t(\omega) dZ(\omega)$$

where $Z(\omega)$ is an orthogonal random process, with

$$E[|dZ(\omega)|^2] = d\mu(\omega)$$

and $A_t(\omega)$ has a Generalized Fourier transform (see Lighthill (1964)) whose absolute maximum occurs at origin. A process $\{X_t\}$ that has the representation of the form (2.1) is called an oscillatory process.

The spectrum of a nonstationary process is defined (see Priestley (1965)) as

$$h_t(\omega)d\omega = |A_t(\omega)|^2 d\mu(\omega)$$

where $d\mu(\omega) = E[|dZ(\omega)|^2]$. The function $h_t(\omega)$ is called *the evolutionary spectral density function*.

We now briefly describe a method to estimate the evolutionary spectral density function $h_t(\omega)$. Let $(X_1, X_2, \ldots, X_N)$ be a sample from a zero mean discrete parameter oscillatory process $\{X_t\}$. Let

$$U_t(\omega) = \sum_{u=-\infty}^{\infty} g_u X_{t-u} e^{-i\omega(t-u)}$$

where $\{g_u\}$ is a filter with width B_g satisfying the following conditions:
- It is square summable and normalized so that

$$2\pi \sum_{u=-\infty}^{\infty} |g_u|^2 = \int_{-\pi}^{\pi} |\Gamma(\omega)|^2 d\omega = 1$$

where

$$\Gamma(\omega) = \sum_{u=-\infty}^{\infty} g_u e^{-iu\omega}$$

and
- $\sum_{u=-\infty}^{\infty} |u||g_u| = B_g$.

Also, choose a weight function $w_{T',t}$ with parameter T' such that
- $w_{T',t} \geq 0$ for all t, T'
- $w_{T',t}$ decays to zero as $|t| \to \infty$, for all T'
- $\sum_{t=-\infty}^{\infty} w_{T',t} = 1$, for all T'
- $\sum_{t=-\infty}^{\infty} w_{T',t}^2 < \infty$, for all T'.

Define

$$W_{T'}(\lambda) = \sum_{t=-\infty}^{\infty} w_{T',t} \, e^{-i\lambda t}$$

and assume that there exists a constant C such that

$$(2.2) \qquad \lim_{T' \to \infty} T' \int_{-\pi}^{\pi} |W_{T'}(\lambda)|^2 d\lambda = C.$$

Now $h_t(\omega)$ can be estimated by

$$\hat{h}_t(\omega) = \sum_{v=-\infty}^{\infty} w_{T',v} |U_{t-v}(\omega)|^2.$$

The sampling properties of the estimated evolutionary spectral density function $\hat{h}_t(\omega)$, have been investigated by Priestley (1966). The mean and variance of $\hat{h}_t(\omega)$ are approximately given respectively by

$$(2.3) \qquad E\{\hat{h}_t(\omega)\} \sim \int_{-\pi}^{\pi} \bar{h}_t(\omega + \omega_0)|\Gamma(\omega)|^2 d\omega,$$

and

$$T' Var\{\hat{h}_t(\omega)\} \sim (1 + \delta_{0,\omega} + \delta_{\pi,\omega}) C \tilde{h}_t^2(\omega) \int_{-\pi}^{\pi} |\Gamma(\theta)|^4 d\theta$$

where

$$\bar{h}_t(\omega) = \sum_{v=-\infty}^{\infty} w_{T',v} h_{t-v}(\omega),$$

$$\tilde{h}_t^2(\omega) = \frac{\sum_{v=-\infty}^{\infty} h_{t-v}^2(\omega) w_{T',v}^2}{\sum_{v=-\infty}^{\infty} w_{T',v}^2}$$

and $\delta(.)$ is the Kronecker delta function.

Priestley (1966) has shown that $Cov(\hat{h}_t(\omega), \hat{h}_{t'}(\omega')) \simeq 0$, if either
1. $|\omega_1 \pm \omega_2| \gg$ bandwidth of $|\Gamma(\omega)|^2$ or
2. $|t - t'| \gg$ 'width' of the function $\{W_{T'}(u)\}$.

To establish the sampling distribution of $\hat{h}_t(\omega)$ theoretically is quite difficult. Recently, Tsolaki (2001) has studied the sampling properties of $\hat{h}_t(\omega)$ through Monte Carlo methods. She has shown that $\hat{h}_t(\omega)$ is approximately normal and by taking the logarithm of $\hat{h}_t(\omega)$ one can see that $\ln \hat{h}_t(\omega)$ tends to normality faster which is consistent with the conjecture made by Priestley and Subba Rao (1969).

A special example of an oscillatory process is the *uniformly modulated process* which is given by

$$Y_t = C_t X_t^{(0)},$$

where $\{X_t^{(0)}\}$ is a stationary process with mean 0 and C_t' is a deterministic function which has a generalized Fourier transform whose modulus has an absolute maximum at the origin. Imposition of this condition ensures that the process $\{Y_t\}$ is an oscillatory process as defined by Priestley (1965).

The evolutionary spectral density function of $\{Y_t\}$ is given by

$$h_{t,Y}(\omega) = |C_t|^2 h(\omega)$$

where $h(\omega)$ is the spectral density function of the stationary process $\{X_t^{(0)}\}$.

3. Testing for stationarity.

An assumption usually made in time series analysis is that of stationarity. For the three sets of temperature data we investigate if such an assumption is valid by applying the Priestley and Subba Rao (1969) test. If the test rejects the stationarity hypothesis, we then want to investigate where the structural changes occur. This is done by using the CUSUM test constructed by Subba Rao (1981).

Let $\{X_t,\ t = 1, 2, \ldots, N\}$ be a sample from an oscillatory process and let $\hat{h}_t(\omega)$ be the estimated evolutionary spectral density function. Let

$$(3.1) \qquad \ln \hat{h}_t(\omega) = Y_t(\omega) = \ln\{h_t(\omega)\} + e_t(\omega),$$

where $E[e_t(\omega)] = 0$ for all t and ω,

$$Var[e_t(\omega)] \simeq \begin{cases} \sigma^2 & \text{if} \quad \omega \neq 0,\ \pi \\ 2\sigma^2 & \text{if} \quad \omega = 0,\ \pi \end{cases}$$

and

$$(3.2) \qquad \sigma^2 = Var\{Y_t(\omega)\} = \frac{C}{T'} \int_{-\pi}^{\pi} |\Gamma(\omega)|^4 d\omega.$$

We calculate $Y_{i,j} = \ln \hat{h}_{t_i}(\omega_j)$, $(i = 1, 2, \ldots, I,\ j = 1, 2, \ldots, J)$ where the time $\{t_i\}$ and frequency points $\{\omega_j\}$ are chosen so that $\{\hat{Y}_{t_i}(\omega_j)\}$ are mutually independent.

We write

$$(3.3) \qquad H_1: \quad Y_{i,j} = \mu + \alpha_i + \beta_j + \gamma_{i,j} + e_{i,j},$$

where $i = 1, \ldots, I$ and $j = 1, \ldots, J$, $\sum_{i=1}^{I} \alpha_i = 0$ and $\sum_{j=1}^{J} \beta_j = 0$. The parameters $\{\alpha_i\}$ and $\{\beta_j\}$ may be interpreted as the 'main effects' of the time and frequency respectively and $\{\gamma_{i,j}\}$ as the 'interaction' between these two factors.

When the time series $\{X_t\}$ is second order stationary (i.e $\alpha_i = 0$ for all i and $\gamma_{i,j} = 0$ for all i, j) then the model (3.3) reduces to

$$H_0: \quad Y_{i,j} = \mu + \beta_j + e_{i,j}.$$

Therefore, we may perform a test for stationarity of the process $\{X_t\}$ by testing the model H_0 against H_1. It was noted in Priestley and Subba Rao (1969) that the presence of the interaction terms $\{\gamma_{i,j}\}$ can be tested, using a χ^2 test, even with one observation per 'cell' since the variance of $\{e_{i,j}\}$, i.e. σ^2, is known *a priori* and is given by (3.2).

We can now use the standard analysis of variance methods (see Scheffé (1959)) for testing the hypothesis $H_0^{(1)}$: $\alpha_i = 0$, $\gamma_{i,j} = 0$ for all i and j and $H_0^{(2)}$: $\beta_j = 0$ for all j. Since the procedure is well known we omit the statistical details. We calculate the sources of variations due to various factors summarized in Table 1 where $\bar{Y}_{\cdot,j} = \frac{1}{I}\sum_{i=1}^{I} Y_{i,j}$, $\bar{Y}_{i,\cdot} = \frac{1}{J}\sum_{j=1}^{J} Y_{i,j}$ and $\bar{Y}_{\cdot,\cdot} = \frac{1}{IJ}\sum_{i=1}^{I}\sum_{j=1}^{J} Y_{i,j}$.

TABLE 1

Analysis of Variance table for two factor analysis model.

Source	d.f.	Sum of Squares
Times	$I-1$	$S_T = J\sum_{i=1}^{I}(\bar{Y}_{i,\cdot} - \bar{Y}_{\cdot,\cdot})^2$
Frequencies	$J-1$	$S_F = I\sum_{j=1}^{I}(\bar{Y}_{\cdot,j} - \bar{Y}_{\cdot,\cdot})^2$
Interaction +Residual	$(I-1)(J-1)$	$S_{I+R} = \sum_{i=1}^{I}\sum_{j=1}^{J}(Y_{i,j} - \bar{Y}_{i,\cdot} - \bar{Y}_{\cdot,j} + \bar{Y}_{\cdot,\cdot})^2$
Total	$IJ-1$	$S_0 = \sum_{i=1}^{I}\sum_{j=1}^{J}(Y_{i,j} - \bar{Y}_{\cdot,\cdot})^2$

To test the above hypotheses we proceed in two stages. First we test for the presence of the interaction terms $\{\gamma_{i,j}\}$ in the model (3.3). If $\gamma_{i,j} = 0$, for all i, j, then S_{I+R}/σ^2 is distributed as a χ^2 with $(I-1)(J-1)$ degrees of freedom. We reject the null hypothesis if $S_{I+R}/\sigma^2 > \chi^2_{\alpha,(I-1)(J-1)}$. Otherwise, we proceed to the next stage which is a test for the presence of the time dependent terms $\{\alpha_i\}$ in model H_1 using the result $S_T/\sigma^2 \simeq \chi^2_{I-1}$. If the null hypothesis is rejected then

$$Y_{i,j} = \mu + \alpha_i + \beta_j + e_{i,j}$$

which implies that the series is nonstationary but it is a uniformly modulated process; otherwise, the process $\{X_t\}$ is stationary.

We now apply the above test for stationarity to the three temperature data sets. To estimate the spectral density function of the three data sets we use a filter response function $\{g_u\}$ of the form

$$(3.4) \qquad g_u = \begin{cases} 1/\sqrt{2\pi(2h+1)}, & |u| \leq h, \\ 0, & \text{otherwise} \end{cases}$$

with $h = 7$ and a weight function $\{w_{T',v}\}$ of the form

$$(3.5) \qquad w_{T',v} = \begin{cases} 1/(T'+1), & -T'/2 \le v \le T'/2, \\ 0, & \text{otherwise}, \end{cases}$$

with $T' = 300$. We note here that the theoretical analysis is still valid for smaller values of T' but we are interested in changes in the temperatures that occur over 25 years which implies $T' = 300$. For our choice of $\{g_u\}$ we have $\sigma^2 \simeq 0.1$. The window $|\Gamma(\omega)|^2$ has approximate bandwidth π/h and the window $\{w_{T',u}\}$ has width 300. Thus, we need to choose time and frequency points sufficiently apart (at least 300 for the time points and $\pi/7$ for the frequency points) so that the estimates of the spectral density function are approximately uncorrelated (see Priestley (1965)).

We estimate the spectral density function at time points 170, 530, 890 and 1250 which correspond to the month of February of the years 1870, 1900, 1930 and 1960 respectively. Frequency points are chosen with spacing $3\pi/20$ i.e. $0\left(\frac{3\pi}{20}\right)\pi$ (omitting the frequencies on the boundaries as the variance at these points is double).

2-D plots of the logarithmically transformed spectral density function at different time points with frequency points $0\left(\frac{\pi}{20}\right)\pi$ are shown in Figures 2, 3 and 4 for each set of data. From the graphs we see that the estimates of the spectral density functions seem to be changing with time indicating non-stationarity in the series. To establish this statistically we need to test the series for stationarity. We also note that the logarithm of the spectral density function has large values in the low frequency range. This may indicate 'long memory' behaviour and the long range parameter may be time dependent. There are prominent peaks in the spectral density function which indicate the presence of harmonic components in the series (the significance of these peaks cannot be assessed at this point as we have obtained smoothed spectral estimates and relevant statistical tests are not yet available). Since the series correspond to monthly temperatures this is not surprising. We investigate the presence of periodicities when we fit models to the three series.

We now test the hypothesis of stationarity using the method described earlier. We calculate the sum of squares due to 'time' and 'frequency' and the results are summarized in the form of ANOVA in Tables 2, 3 and 4 for the three sets of data.

For all three sets of data, we compare the above values with the χ^2 5% critical value. There is strong evidence (in all three data sets) to reject the null hypothesis of stationarity. However, there is no evidence to reject the null hypothesis that the time series is uniformly modulated.

If we look at the plots of the spectral density estimates (Figures 2, 3 and 4) we see that the 'pattern' looks similar (i.e. the three series exhibit similar behaviour). In Tsolaki (2001), a statistical test was proposed to test the hypothesis for the equality of evolutionary spectra. An application

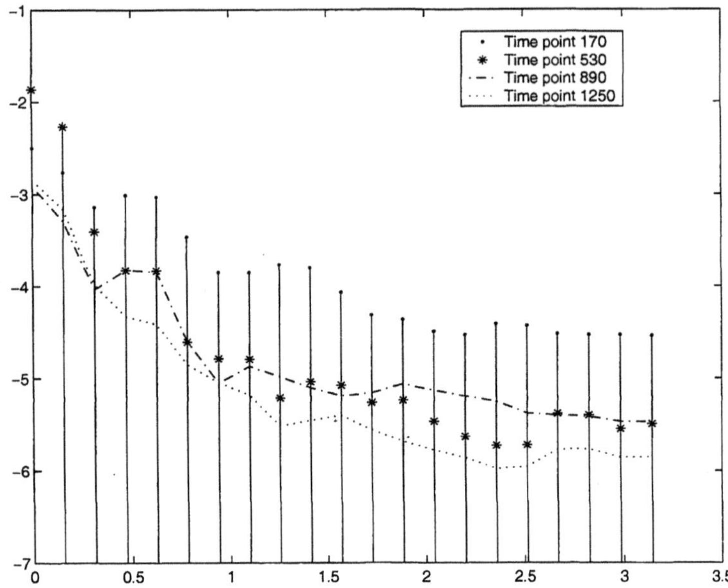

FIG. 2. *Estimated logarithmically transformed spectral density function for northern hemisphere data.*

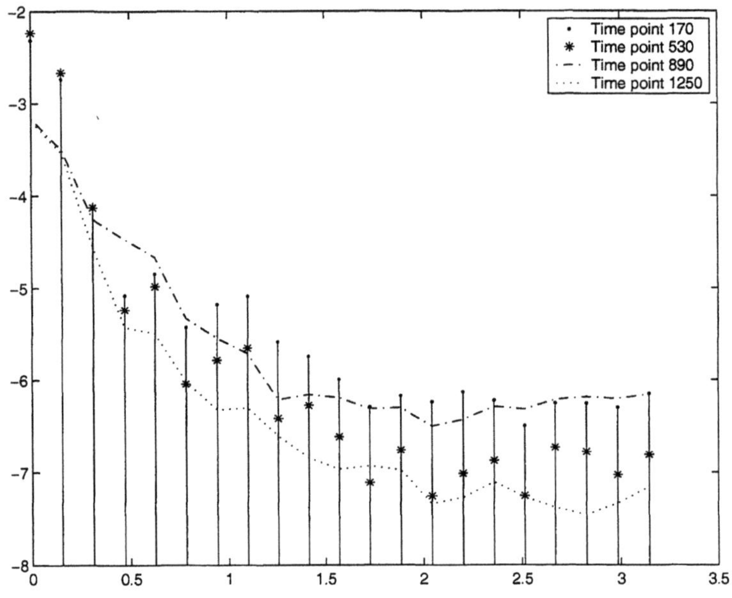

FIG. 3. *Estimated logarithmically transformed spectral density function for southern hemisphere data.*

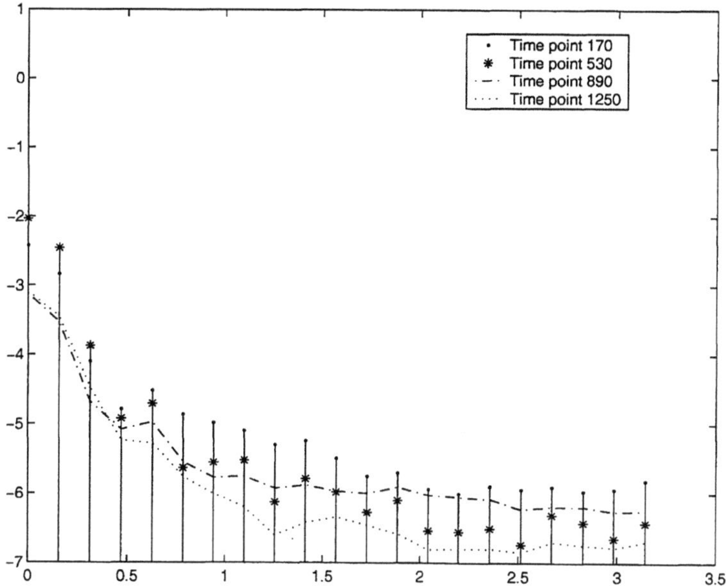

FIG. 4.   *Estimated logarithmically transformed spectral density function for global data.*

TABLE 2
Analysis of Variance table for northern hemisphere data.

Source of var.	d.f.	Sum of Squares	Statistic	$\chi^2(0.05)$
Between times	3	6.280	62.658	7.815
Between frequencies	5	6.900	68.846	11.070
Interaction + Residual	15	0.346	3.452	24.996

of this test to the above data led us to conclude that there are differences between the evolutionary spectra of the three series.

From the above analysis it is clear that all three series are nonstationary. It is, therefore, interesting to see where the structural changes are occurring.

4. Testing for structural change (in the variance- covariance). There are several tests for detecting changes in the mean but not many for detecting changes in the covariance structure. We can only detect such changes in blocks of data and not from a single observation. In view of this, when we calculate the observations $Y_{i,j}$ ($i = 1, \ldots, I$ and $j = 1, \ldots, J$) we are calculating the evolutionary spectrum over various blocks. We have

TABLE 3

Analysis of Variance table for southern hemisphere data.

Source of var.	d.f.	Sum of Squares	Statistic	$\chi^2(0.05)$
Between times	3	3.348	33.4070	7.815
Between frequencies	5	8.232	82.1390	11.070
Interaction + Residual	15	0.536	5.3492	24.996

TABLE 4

Analysis of Variance table for global data.

Source of var.	d.f.	Sum of Squares	Statistic	$\chi^2(0.05)$
Between times	3	2.275	22.695	7.815
Between frequencies	5	5.160	51.487	11.070
Interaction + Residual	15	0.367	3.660	24.996

$$Y_{i,j} = \log\{\hat{h}_{t_i}(\omega_j)\} = \mu + \alpha_i + \beta_j + \gamma_{i,j} + e_{i,j},$$

where $i = 1,\ldots,I$ and $j = 1,\ldots,J$. For a stationary time series we have seen that the time dependent parameters $\{\alpha_i\}$ and $\{\gamma_{i,j}\}$ are zero and hence $Y_{i,j}$ can be written in the form

$$Y_{i,j} = \mu + \beta_j + e_{i,j}$$

where i denotes the time and j the frequency component.

As mentioned in Subba Rao (1981), if a process is stationary up to time $t = t_0$ (i.e. a change in the variance-covariance structure occurs from the time $t_0 + 1$ onwards), then $Y_{i,j}$ can be written in the form

$$Y_{i,j} = \begin{cases} \mu + \beta_j + e_{i,j}, & (i = 1,\ldots,t_0) \\ \mu + \alpha_i + \beta_j + \gamma_{i,j} + e_{i,j}, & (i = t_0 + 1,\ldots) \end{cases}$$

and thus as seen above testing for a change in the variance-covariance of the time series $\{X_t\}$ is the same as testing for a change in the mean of the sequence of independent observations $\{Y_{i,j}\}$ (for each j). Hence, after choosing a reference value k, we obtain the cumulative sum

$$S_m = \sum_{i=1}^{m}(Y_{i,.} - k), \qquad (m = 1, 2, \ldots)$$

where $Y_{i,.}$ is the average of $Y_{i,j}$ over all computed frequencies and k is the reference value (chosen *a priori*) and is taken as the mean of the stationary part of the process $\{Y_{i,j}\}$ which can be estimated from a segment of the data (where there are no changes). Note that in the cumulative sum, the average over all frequency points is used as we are mainly interested in any changes that occur in the time domain.

To detect the change points one uses the fact that $E[S_m]$ is approximately zero when there are no changes in the covariance structure of the process $\{X_t\}$. A way to detect a change in the slope of the CUSUM path is by using the V-mask technique (see Ewan (1963), Johnson and Leone (1964) and Woodward and Goldsmith (1964) for more details).

In this section we apply the CUSUM test to the temperature data to detect any changes in the variance-covariance structure of the series. We evaluate the reference value k using the monthly data from January 1856 to December 1869 which correspond to the first 14 years of our series. Although the data are not so reliable at these years, they are the only available to use in order to assess any changes compared to previous years as many meteorologists believe that rapid changes are occurring in the 20th century. We assume that there are no changes during these years. We then perform the CUSUM test, starting from the time points corresponding to February 1870 and then using data points for every ten years for the same month thereafter. We have chosen data from February as meteorologists are more interested in changes the occur in lower temperatures obtained during winter months.

To find k, we estimate the stationary spectral density function of the monthly data (Jan 1856-Dec 1869) using the method described in Priestley (1981, Section 6.1). We use the Bartlett-Priestley window (see Priestley (1981), Section 6.2.3) with smoothing parameter $m = 10$. The spectral density function is estimated at frequencies $0 \left(\frac{\pi}{168} \right) \pi$ and is shown in Figure 5 for the three sets of data.

The average of the logarithm of the spectral density function over all frequencies (excluding those on the boundaries) is used as the reference value k for the CUSUM test. The reference values k for the three data sets are given below

$$(4.1) \qquad k = \begin{cases} -3.5988, & \text{northern hemisphere} \\ -5.6157, & \text{southern hemisphere} \\ -5.1295, & \text{global data.} \end{cases}$$

We perform the CUSUM test for the three sets of data and the results are shown in Tables 5 and 6.

We have $\sigma^2 = Var\{Y_{i,j}\} = 0.1$ and thus $\sigma = 0.32$. We also plot the values of S_m against $m = 1, 2, \ldots, 12$ with scale 0.6 units on the vertical scale per unit on the horizontal scale. We then use a V-mask with parameters $d = 2$ and $\tan\theta = 0.7$ for the northern hemisphere data which will detect changes of more than 5σ (for the choice of d and θ see Johnson and Leone (1964)). There is a cut on the upper limb of the V-mask by

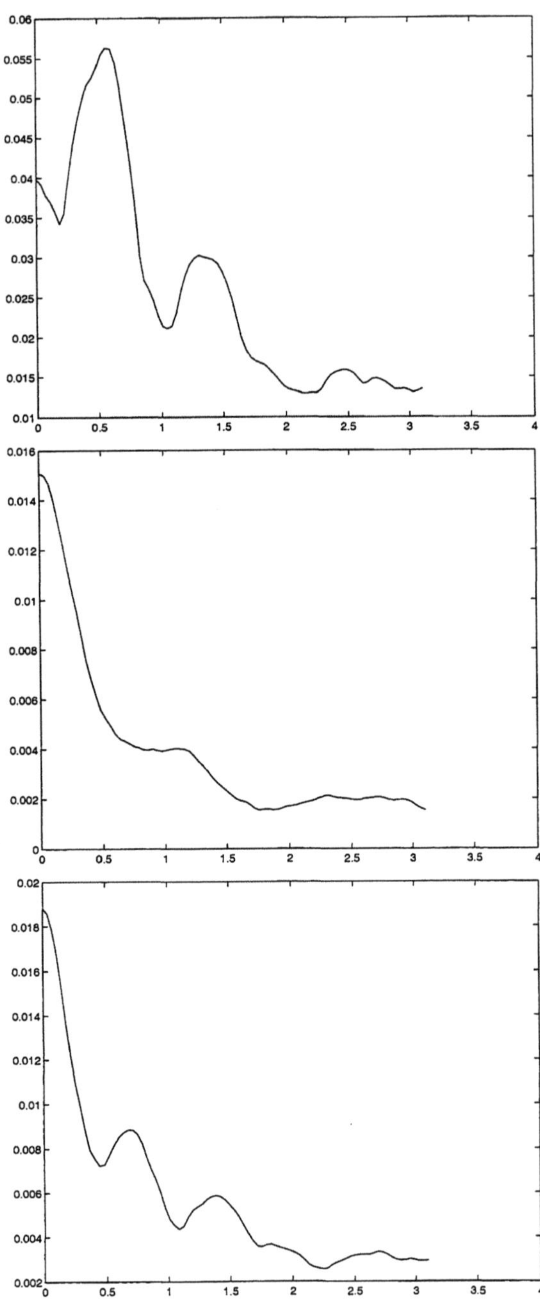

FIG. 5. *Estimated spectral density function of the three data sets for the first 14 years, northern hemisphere (top), southern hemisphere (bottom) and global temperature data (center).*

TABLE 5
CUSUM test for the northern and southern hemisphere data.

Year	$\bar{Y}_{i,.}$	$\bar{Y}_{i,.} - k$	S_m	$\bar{Y}_{i,.}$	$\bar{Y}_{i,.} - k$	S_m
1870	-3.99	-0.39	-0.39	-5.77	-0.15	-0.15
1880	-4.56	-0.96	-1.35	-6.07	-0.45	-0.61
1890	-4.88	-1.28	-2.63	-6.37	-0.76	-1.37
1900	-5.00	-1.40	-4.03	-6.28	-0.67	-2.04
1910	-4.96	-1.36	-5.40	-6.01	-0.39	-2.43
1920	-4.99	-1.39	-6.78	-5.91	-0.29	-2.72
1930	-4.95	-1.35	-8.13	-5.83	-0.21	-2.93
1940	-4.98	-1.38	-9.51	-5.95	-0.34	-3.27
1950	-5.19	-1.59	-11.10	-6.00	-0.39	-3.66
1960	-5.38	-1.78	-12.88	-6.69	-1.08	-4.73
1970	-5.51	-1.91	-14.79	-6.52	-0.91	-5.64
1980	-5.40	-1.81	-16.60	-6.09	-0.47	-6.11
	Northern hemisphere			Southern hemisphere		

TABLE 6
CUSUM test for the global data.

Year	$\bar{Y}_{i,.}$	$\bar{Y}_{i,.} - k$	S_m
1870	-5.43	-0.30	-0.30
1880	-5.82	-0.69	-0.99
1890	-5.95	-0.82	-1.81
1900	-5.88	-0.75	-2.56
1910	-5.74	-0.61	-3.18
1920	-5.79	-0.66	-3.84
1930	-5.82	-0.69	-4.52
1940	-5.81	-0.68	-5.21
1950	-5.92	-0.79	-5.99
1960	-6.30	-1.17	-7.16
1970	-6.40	-1.27	-8.43
1980	-6.11	-0.98	-9.41
	Global data		

the CUSUM path between years 1960–1970 confirming a change in the series (see Figure 6). For the southern hemisphere the V-mask is cut by the CUSUM path (we choose $\tan\theta = 0.3$ which detects changes of 2σ) between 1960 and 1970 (see Figure 7). A cut at the same interval is noticed for the global data if we choose $\tan\theta = 0.4$ which corresponds to changes of 2.5σ (see Figure 8).

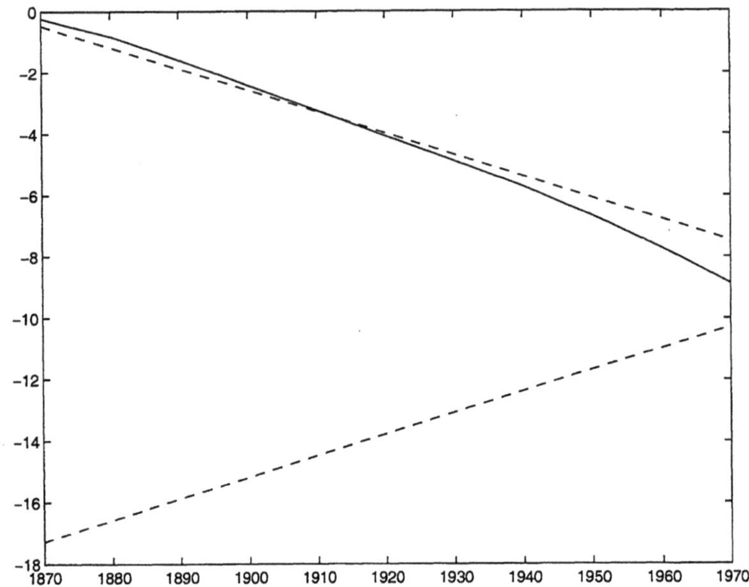

FIG. 6. *CUSUM path and corresponding V-mask for northern hemisphere data.*

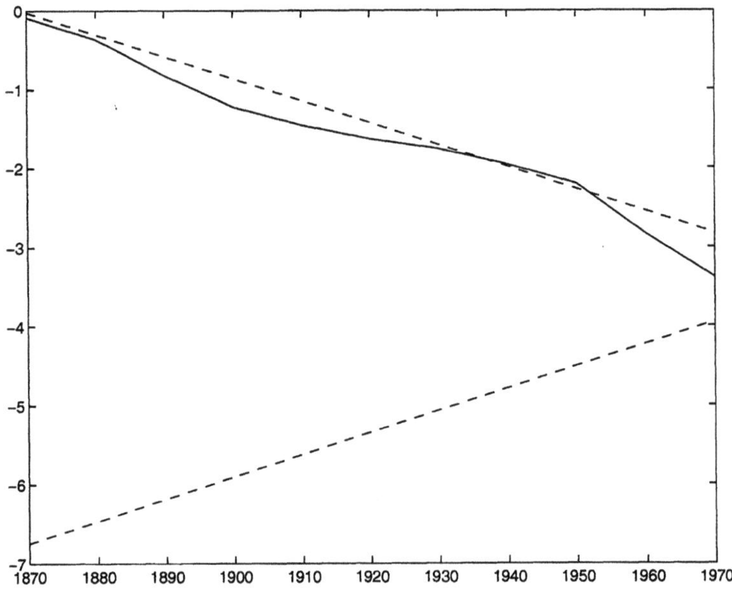

FIG. 7. *CUSUM path and corresponding V-mask for southern hemisphere data.*

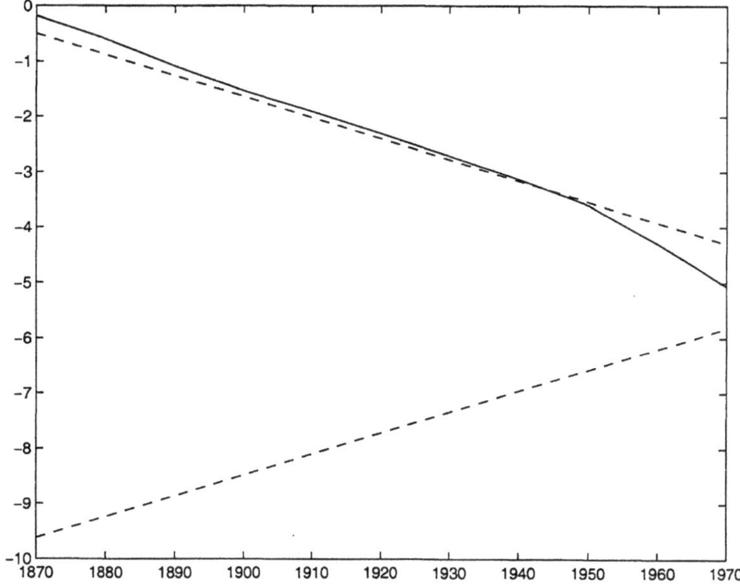

FIG. 8. *CUSUM path and corresponding V-mask for global temperature data.*

On the basis of the above preliminary analysis we can conclude that there is an overall structural change in the northern hemisphere temperature data around $1950 - -1960$.

5. Testing for Gaussianity and linearity. From the above analysis we have concluded that the three series are nonstationary and thus it would be interesting to check whether there is any evidence to suggest that the series is non-Gaussian or nonlinear. Linearity of a nonstationary time series is defined as follows:

DEFINITION 5.1. *We say that a nonstationary process $\{X_t\}$ is linear if it can be written in the form*

$$(5.1) \qquad X_t = \sum_{u=0}^{\infty} g_{t,u} e_{t-u},$$

where $\{e_t\}$ are independent, identically distributed random variables with $E[e_t] = 0$, $E[e_t^2] = \sigma_e^2$ and $E[e_t^3] = \mu_3$ (Abdrabbo and Priestley (1967)).

Subba Rao and Gabr (1980) have proposed a test for Gaussianity and linearity of a stationary time series based on the bispectral density function. An important property of the bispectral density function is that it is zero for a Gaussian process and thus it can be used as a tool for detection of departure from Gaussianity of a series (see Brillinger (1965)). Extensions of the above tests were considered in Tsolaki (2001) for the case of nonstationary processes and were then applied to the temperature data.

Briefly we review the conclusions drawn when the tests were applied to the data.

The bispectrum of a nonstationary time series is defined by Priestley and Gabr (1993) as follows:

DEFINITION 5.2. *The time dependent bispectrum of a discrete parameter oscillatory process $\{X_t\}$ which has a representation of the form (2.1) is given by*

$$h_t(\omega_1, \omega_2)d\omega_1 d\omega_2 = A_t(\omega_1)A_t(\omega_2)A_t(-\omega_1 - \omega_2)d\mu(\omega_1, \omega_2),$$

where $d\mu(\omega_1, \omega_2) = E[dZ(\omega_1)dZ(\omega_2)dZ(-\omega_1 - \omega_2)]$.

The test for Gaussianity is based on the fact that for a Gaussian nonstationary time series $\{X_t\}$, $h_t(\omega_1, \omega_2) = 0$ for all ω_1, ω_2 and t. Also, the test for linearity is based on the fact that when the process is linear as defined in 5.1, we have

$$|f_t(\omega_1, \omega_2)| = \left| \frac{h_t(\omega_1, \omega_2)}{(h_t(\omega_1)h_t(\omega_2)h_t(\omega_1 + \omega_2))^{1/2}} \right| = \frac{\mu_3}{\sqrt{2\pi}\sigma_e^3}$$

which is independent of t and ω. For more details we refer to Tsolaki (2001) and Subba Rao (1997).

To apply the tests for Gaussianity and linearity to the temperature data the spectral and bispectral density functions were estimated at time points 170, 530, 890 and 1250 corresponding to February of the years 1870, 1900, 1930, 1960 and frequency points 0 to π with step $\pi/6$. The bispectrum was estimated following the method given in Priestley and Gabr (1993). They used an approach similar to that for the stationary case, but rather than computing finite Fourier transforms over the complete record, they use 'local' transforms via a moving time domain window.

Figure 9 shows an example of the estimated moduli of the bispectrum of the northern, southern hemisphere and global temperature anomalies data at time point 170.

We notice that for all series the bispectrum has a peak at frequencies close to $(0,0)$. This observation is similar to the one for the estimated evolutionary spectral density function for the same data. We note that temperature data are anomalies and thus numbers are small in their magnitude.

The tests for Gaussianity and linearity when applied to the temperature data showed that there was evidence to suggest that the three sets of data are nonstationary linear but non-Gaussian.

6. Data analysis. On the basis of evolutionary spectral analysis carried out on the temperature data we have concluded that the data are nonstationary. Also the analysis of variance on the estimates of the evolutionary spectral density function (see section 3) led us to conclude that the data are probably of uniformly modulated type. The plot of the

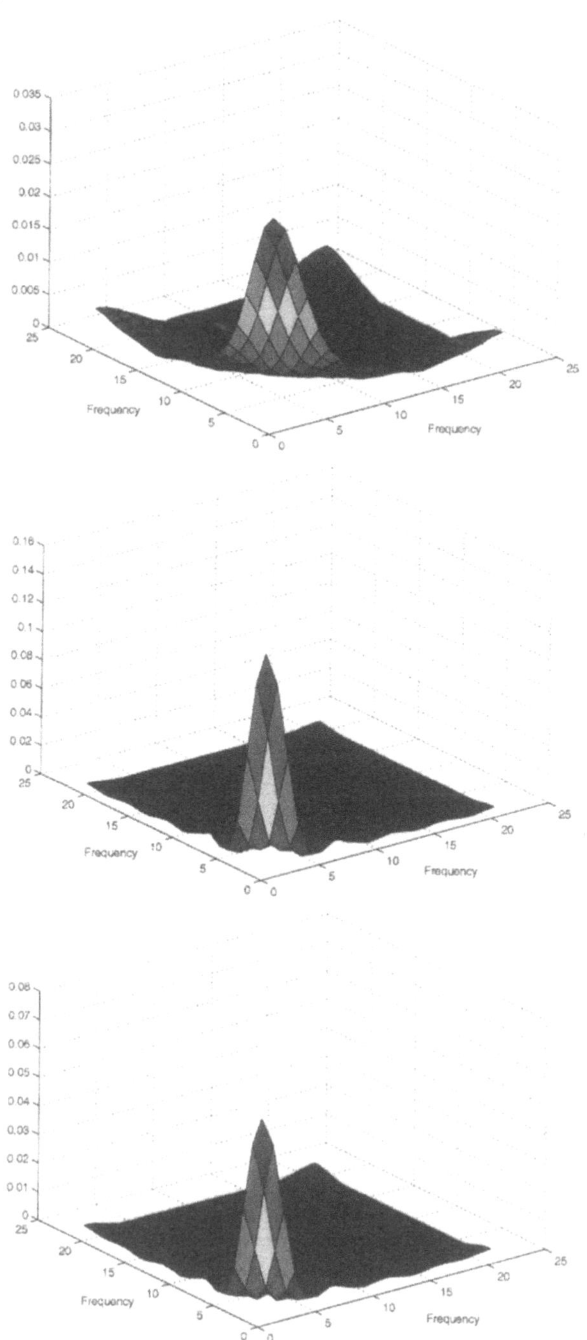

FIG. 9. *Modulus of the bispectrum estimate at $t = 170$ of the northern hemisphere (top) southern hemisphere (center) and global data (bottom).*

data suggests that there is some trend and in view of the fact that they are monthly temperatures there is a possibility of the presence of sinusoid components. These conclusions have led us to propose the additive model given below for the temperature data where $\{Y_t\}$ are the observed time series.

$$(6.1) \quad Y_t = c_0 + c_1 t + \sum_{i=1}^{m} r_i \cos(\omega_i t + \phi_i) + f(t,\underline{\theta}) X_t, \quad t = 1,\ldots,N$$

where c_0, c_1, r_i, ω_i and ϕ_i with $i = 1,\ldots,m$, are parameters to be estimated, m is the number of sinusoids in the model and $f(t,\underline{\theta}) > 0$ for $1 \leq t \leq N$, is a bounded parametric deterministic function of t whose form will be discussed later (parameter $\underline{\theta}$ is also estimated). Also, we assume that the stationary time series $\{X_t\}$ can be represented by an $ARMA(k,l)$ model of the form

$$(6.2) \quad X_t + a_1 X_{t-1} + a_2 X_{t-2} + \ldots + a_k X_{t-k} = e_t + b_1 e_{t-1} + b_2 e_{t-2} + \ldots + b_l e_{t-l}$$

where $\{e_t, \ t = 1,\ldots,N\}$ is a sequence of independent identically distributed random variables with mean zero and variance σ_e^2. We assume that the roots of the polynomials $\phi(z) = 1 + \alpha_1 z + \alpha_2 z^2 + \ldots + \alpha_k z^k$ and $\psi(z) = 1 + b_1 z + b_2 z^2 + \ldots + b_l z^l$ lie outside the unit circle. We now consider the estimation of the parameters of the nonstationary model (6.1), the parameters of the deterministic function $f(t,\underline{\theta})$ as well as the orders (k,l) and the parameters of the stationary $ARMA$ model (6.2).

We first consider the estimation of the parameters c_0 and c_1 by the method of ordinary least square (OLS). Let

$$\zeta_t = \sum_{i=1}^{m} r_i \cos(\omega_i t + \phi_i) + f(t,\underline{\theta}) X_t, \qquad t = 1,\ldots,N.$$

Then

$$(6.3) \qquad Y_t = c_0 + c_1 t + \zeta_t, \qquad t = 1,\ldots,N.$$

In matrix notation,

$$c = [c_0 \quad c_1]', \qquad Y = [Y_1 \quad Y_2 \quad \cdots \quad Y_N]', \qquad Z = [\zeta_1 \quad \zeta_2 \quad \cdots \quad \zeta_N]'$$

and

$$S_1 = \begin{bmatrix} 1 & 1 \\ 1 & 2 \\ \vdots & \vdots \\ 1 & N \end{bmatrix}$$

where $'$ denotes the transpose of a matrix and thus we can write (6.3) as $Y = S_1 c + Z$. We estimate the parameters c_0 and c_1 by the method of ordinary least squares. Toyooka (1977, 1980) showed that under some conditions on the function $f(t,\underline{\theta})$ and on the regression variables, ordinary least squares can give consistent estimates of the parameters even in the case where the errors are nonstationary and uniformly modulated. We have the least squares estimate $\hat{c}$ of c as $\hat{c} = (S_1^T S_1)^{-1} S_1^T Y$.

Having estimated the parameters c_0 and c_1 we can obtain the 'residuals' $\{Z_t\}$ from

$$Z_t = Y_t - \hat{c}_0 - \hat{c}_1 t, \qquad t = 1, \ldots, N.$$

We now estimate the frequencies $\{\omega_i\}$, the phases $\{\phi_i\}$, the amplitudes $\{r_i\}$ and the parameter m using the 'residuals' Z_t, where Z_t, for each t, is given by

$$(6.4) \qquad Z_t = \sum_{i=1}^{m} r_i \cos(\omega_i t + \phi_i) + \xi_t, \qquad t = 1, \ldots, N,$$

where

$$\xi_t \approx f(t,\underline{\theta}) X_t, \qquad t = 1, \ldots, N.$$

Another form of model (6.4) is

$$(6.5) \qquad Z_t = \sum_{i=1}^{m} [a_i \sin(\omega_i t) + b_i \cos(\omega_i t)] + \xi_t, \qquad t = 1, \ldots, N$$

where $r_i = \sqrt{a_i^2 + b_i^2}$, $\cos(\phi_i) = a_i/r_i$ and $\sin(\phi_i) = b_i/r_i$, $i = 1, \ldots, m$.

The estimation of the frequencies $\{\omega_i\}$ is quite complicated. The error is a nonstationary uniformly modulated process and thus classical techniques are not strictly valid. Therefore we employ a two stage approach assuming initially that the errors are i.i.d random variables and thus use the method of ordinary least squares. To estimate $\{\omega_i\}$ we choose the locations of the m greatest maxima of $I(\omega)$ (*the periodogram*) ignoring local maxima close to others which occur due to 'sidelobes' (for details we refer to Priestley (1981) and Quinn and Hannan (2001)). At the true frequencies the periodogram is expected to exhibit large 'peaks'.

In the case when the errors are i.i.d random variables one can test for the significance of the maxima of the periodogram using the Walker's large sample test or Fisher's g-statistic (Walker (1914), Fisher (1929)). Because of the nonstationary structure of the error term these tests are not applicable and thus the number of sinusoids to be included in the model will be determined by the number of 'large peaks' in the periodogram. In this way we obtain initial estimates of the frequencies $\{\omega_i\}$ and the number m of sinusoids.

Now let,

$$Z = [Z_1 \quad Z_2 \quad \ldots \quad Z_N]', \qquad R = [a_1 \quad b_1 \quad \ldots \quad a_{\hat{m}} \quad b_{\hat{m}}]',$$

$$\Xi = [\xi_1 \quad \xi_2 \quad \ldots \quad \xi_N]'$$

and

$$
D = \begin{bmatrix}
\sin(\hat{\omega}_1) & \cos(\hat{\omega}_1) & \ldots & \sin(\hat{\omega}_{\hat{m}}) & \cos(\hat{\omega}_{\hat{m}}) \\
\sin(2\hat{\omega}_1) & \cos(2\hat{\omega}_1) & \ldots & \sin(2\hat{\omega}_{\hat{m}}) & \cos(2\hat{\omega}_{\hat{m}}) \\
\vdots & \vdots & \ldots & \vdots & \vdots \\
\sin(N\hat{\omega}_1) & \cos(N\hat{\omega}_1) & \ldots & \sin(N\hat{\omega}_{\hat{m}}) & \cos(N\hat{\omega}_{\hat{m}})
\end{bmatrix}
$$

where $\hat{\omega}_i$, $i = 1, \ldots, \hat{m}$ and $\hat{m}$ are the estimates of ω_i ($i = 1, \ldots, m$) and m respectively.

Estimates of a_i, b_i can then be obtained by ordinary least squares assuming that $\{\omega_i\}$ and m are known. Thus (6.5) is written as $Z = DR + \Xi$ and the least square estimates $\hat{a}_i$ and $\hat{b}_i$ of the unknown parameters a_i and b_i respectively are given by $\hat{R} = (D^T D)^{-1} D^T Z$. Initial estimates of the amplitudes r_i and phases ϕ_i (in the interval $[-\pi, \pi]$) can now be calculated by $\hat{r}_i = \sqrt{\hat{a}_i^2 + \hat{b}_i^2}$ and $\hat{\phi}_i = \cos^{-1}(\hat{a}_i/\hat{r}_i)$ for $i = 1, \ldots, \hat{m}$.

Given the initial estimates of $\{\omega_i\}$, $\{\phi_i\}$, $\{r_i\}$ and m we can proceed to estimate the parameters of the function $f(t, \underline{\theta})$ (described later). We assume $f(t, \underline{\theta}) > 0$, for all t which is satisfied by the chosen functions. By replacing the parameters of the function $f(t, \underline{\theta})$ with their estimates $\hat{\underline{\theta}}$ we obtain the model

$$
\frac{Z_t}{f(t, \hat{\underline{\theta}})} = \frac{1}{f(t, \hat{\underline{\theta}})} \sum_{i=1}^{\hat{m}} [\hat{a}_i \sin(\hat{\omega}_i t) + \hat{b}_i \cos(\hat{\omega}_i t)] + X_t, \qquad t = 1, \ldots, N.
$$

Although this is a case of mixed spectral estimation, the form of $\{X_t\}$ is unknown. Therefore we proceed assuming that $\{X_t\}$ are i.i.d random variables and re-estimate $\{\omega_i\}$, $\{\phi_i\}$ and $\{r_i\}$ following the above procedure; the only difference being that the new estimates of a_i and b_i will be given by $\hat{R} = (D_{(2)}^T D_{(2)})^{-1} D_{(2)}^T Z_{(2)}$ where $Z_{(2)}$ and $D_{(2)}$ are obtained from Z and D where each element is divided by $f(\cdot, \underline{\theta})$.

Our objective here is to find a suitable functional form of $f(t, \underline{\theta})$ to describe our temperature data. We simulated several data sets for various forms of $f(t, \underline{\theta})$ and then tried to compare them with the temperature data sets. On the basis of the simulation study we arrived at the conclusion that the choice $f(t, \underline{\theta}) = t^{c_2}$ ($t > 0$) with $c_2 < 0$ seems to be most suitable. From the model (6.1) we have

$$
Y_t - c_0 - c_1 t - \sum_{i=1}^{m} r_i \cos(\omega_i t + \phi_i) = f(t, \underline{\theta}) X_t, \qquad t = 1, \ldots, N.
$$

Let

$$
(6.6) \qquad W_t = Y_t - \hat{c}_0 - \hat{c}_1 t - \sum_{i=1}^{\hat{m}} \hat{r}_i \cos(\hat{\omega}_i t + \hat{\phi}_i), \qquad t = 1, \ldots, N,
$$

where $\hat{\omega}_i$, $\hat{\phi}_i$, $\hat{r}_i$ and $\hat{m}$ are the estimates of ω_i, ϕ_i, r_i and m respectively. Hence,

$$(6.7) \qquad \ln|W_t| = c_2 \ln t + \mu + \eta_t, \qquad t = 1, \ldots, N,$$

where $\mu = E[\ln|X_t|]$ is a constant to be estimated and $\eta_t = \ln|X_t| - \mu$ is a random variable with mean zero and variance σ_η^2. In matrix notation, equation (6.7) can be written as

$$Wl = S\theta + E$$

where

$$\theta = [\mu \quad c_2]', \qquad Wl = [\ln|W_1| \quad \ln|W_2| \quad \ldots \quad \ln|W_N|]',$$

$$E = [\eta_1 \quad \eta_2 \quad \ldots \quad \eta_N]'$$

and

$$S = \begin{bmatrix} 1 & \ln 1 \\ 1 & \ln 2 \\ \vdots & \vdots \\ 1 & \ln N \end{bmatrix}$$

Least square estimates of μ and c_2 are given by

$$\hat{\theta} = (S^T S)^{-1} S^T Wl.$$

Since $c_2 < 0$, the chosen function $f(t) = t^{c_2}$ will tend to 0 as $t \to \infty$. This means that as $t \to \infty$ the model for Y_t will reduce to a deterministic function. However, if we assume that $f(t, \underline{\theta}) = t^{c_2} + c_3$, $c_2 < 0$ as $t \to \infty$ then $f(t, \underline{\theta})X_t \to c_3 X_t$ (in probability). Then the process $\{Y_t\}$ is asymptotically stationary.

We consider the estimation of the parameters c_2 and c_3. We have

$$(6.8) \qquad \ln|W_t| = \ln|t^{c_2} + c_3| + \mu' + \eta_t', \qquad t = 1, \ldots, N,$$

where $\{W_t\}$ is given by (6.6). Also μ' and η_t' are defined as in (6.7). The parameters c_2 and c_3 can be estimated using non-linear techniques. Unconstrained non-linear optimization can be achieved using a build-in function of the statistical section of 'MATLAB' software package which is based on a Nelder-Mead simplex algorithm (Lagarias et al. (1998)). In non-linear regression, it is important to choose initial estimates carefully. It is well known that if the minimizing function has many local minima, a 'bad' choice of the initial values may result in missing the global minimum. For our estimation purposes we use the estimates of c_2 and μ from model (6.7) as starting points.

The final stage of the estimation of the parameters of model (6.1) is to fit an $ARMA$ model to $\{W_t/f(t,\hat{\underline{\theta}})\}$ where $f(t,\hat{\underline{\theta}}) = t^{\hat{c}_2}$ or $f(t,\hat{\underline{\theta}}) = t^{\hat{c}_2} + \hat{c}_3$. The function $\{W_t/f(t,\hat{\underline{\theta}})\}$ is approximately an estimate of $\{X_t\}$ and thus we can assume that the resulting series is stationary and satisfies an $ARMA$ model. We use Bayes Information Criterion (BIC) to determine the orders of the $ARMA$ model and then estimate the parameters. Also, in order to estimate the parameters of the $ARMA$ models, we follow the Hannan and Rissanen (1982) procedure.

7. Time series models to the temperature anomalies. Following the method described above we fitted models of the form (6.1) to the three sets of temperature data for

$$f(t,\underline{\theta}) = \begin{cases} t^{c_2} , & c_2 < 0 \\ t^{c_2} + c_3 , & c_2 < 0 . \end{cases}$$

We note here that from the analysis obtained in the previous sections we could fit two different stationary (or nonstationary models) to the data taking into account the changing point identified in the CUSUM test. The whole analysis though is based on the assumption that we are dealing with semi-stationary processes and thus we are interested in fitting a single model that is slowly varying. In addition, the remaining data after the change that was identified around 1950–1960 are not sufficient to carry out a separate analysis in order to fit a model.

We summarize the obtained models for each set of data.

• **Northern hemisphere.**

<u>Model 1</u>

$$\begin{aligned} Y_t = &-0.28329 + 0.00033t + 0.14140 \cos{(0.00812t - 1.86878)} \\ &+ 0.10939 \cos{(0.52323t - 3.06732)} + t^{-0.16178} X_t, \end{aligned} \tag{7.1}$$

for $t = 1,\ldots,1716$ where $\{X_t\}$ is given by

$$X_t - 0.75498X_{t-1} = e_t - 0.35094e_{t-1}, \qquad t = 1,\ldots,1716 \tag{7.2}$$

and $\{e_t,\ t = 1,\ldots,N\}$ are i.i.d with mean zero and variance σ_e^2. An estimate of the variance of σ_e^2 is 0.3613.

<u>Model 2</u>

$$\begin{aligned} Y_t = &-0.28329 + 0.00033t + 0.14131 \cos{(0.00812t - 1.85882)} \\ &+ 0.08163 \cos{(0.52323t - 3.06662)} + (t^{-0.12331} - 0.09824)X_t, \end{aligned} \tag{7.3}$$

for $t = 1,\ldots,1716$ where $\{X_t\}$ is given by

$$X_t - 0.76040X_{t-1} = e_t - 0.36322e_{t-1}, \qquad t = 1,\ldots,1716 \tag{7.4}$$

and $\{e_t,\ t = 1,\ldots,N\}$ are i.i.d with mean zero and variance σ_e^2. An estimate of the variance of σ_e^2 is $\hat{\sigma}_e^2 = 0.3215$.

- **Southern hemisphere.**

<u>Model 1</u>

$$
\begin{aligned}
Y_t = &- 0.31521 + 0.00037t + 0.08687\cos{(0.00945t + 3.07338)} \\
&+ 0.03451\cos{(0.52301t + 0.30123)} \\
&+ 0.01570\cos{(1.04728t - 1.44703)} + t^{-0.09225}X_t,
\end{aligned}
$$
(7.5)

for $t = 1,\ldots,1716$ where $\{X_t\}$ is given by

$$X_t - 0.87731X_{t-1} = e_t - 0.36602e_{t-1}, \qquad t = 1,\ldots,1716$$

and $\{e_t,\ t = 1,\ldots,N\}$ are i.i.d with mean zero and variance σ_e^2. An estimate of the variance of σ_e^2 is $\hat{\sigma}_e^2 = 0.0395$.

<u>Model 2</u>

$$
\begin{aligned}
Y_t = &- 0.31521 + 0.00037t + 0.08709\cos{(0.00941t + 3.11590)} \\
&+ 0.03470\cos{(0.52302t + 0.29577)} \\
&+ 0.01586\cos{(1.04728t - 1.45448)} + (t^{-0.32870} + 0.50592)X_t,
\end{aligned}
$$
(7.6)

for $t = 1,\ldots,1716$ where $\{X_t\}$ is given by

$$X_t - 0.87697X_{t-1} = e_t - 0.36749e_{t-1}, \qquad t = 1,\ldots,1716$$

and $\{e_t,\ t = 1,\ldots,N\}$ are i.i.d with mean zero and variance σ_e^2. An estimate of the variance of σ_e^2 is $\hat{\sigma}_e^2 = 0.0303$.

- **Global data.**

<u>Model 1</u>

$$
\begin{aligned}
Y_t = &- 0.30755 + 0.00036t + 0.11584\cos{(0.00834t - 2.06385)} \\
&+ 0.02452\cos{(0.52304t - 3.00267)} + t^{-0.07032}X_t,
\end{aligned}
$$
(7.7)

for $t = 1,\ldots,1716$ where $\{X_t\}$ is given by

$$(7.8)\quad X_t - 0.86939X_{t-1} = e_t - 0.44099e_{t-1}, \qquad t = 1,\ldots,1716$$

and $\{e_t,\ t = 1,\ldots,N\}$ are i.i.d with mean zero and variance σ_e^2. An estimate of the variance of σ_e^2 is $\hat{\sigma}_e^2 = 0.0394$.

<u>Model 2</u>

$$
\begin{aligned}
Y_t = &- 0.30755 + 0.00036t + 0.11588\cos{(0.00834t - 2.06595)} \\
&+ 0.02449\cos{(0.52304t - 3.00305)} + (t^{-0.07682} + 0.02550)X_t,
\end{aligned}
$$
(7.9)

for $t = 1,\ldots,1716$ where $\{X_t\}$ is given by

$$(7.10)\quad X_t - 0.86954X_{t-1} = e_t - 0.44079e_{t-1}, \qquad t = 1,\ldots,1716$$

and $\{e_t, \; t = 1, \ldots, N\}$ are i.i.d with mean zero and variance σ_e^2. An estimate of the variance of σ_e^2 is $\hat{\sigma}_e^2 = 0.0395$.

For each of the three sets of temperature data we fitted two different models of the form (6.1). All series exhibit the same periodicity of 1 year which is expected as we are dealing with monthly temperature data (the periodicity of 1 year corresponds to the hight frequency estimate for all models). They also exhibit a low frequency cycle corresponding to around 65 years for the northern hemisphere data around 63 years for the global data and around 55 years for the southern hemisphere data. The southern hemisphere data exhibit an extra periodicity of 6 months corresponding to the third estimated cycle component (1.04728).

As estimates of the constant c_1 are small values, we calculate the standard errors. Thus, we have

$$\left\{\begin{array}{ll} \text{Northern data} & s.e(\hat{c}_1) = 3.8568 \times 10^{-6} \\ \text{Southern data} & s.e(\hat{c}_1) = 1.4473 \times 10^{-6} \\ \text{Global data} & s.e(\hat{c}_1) = 1.6880 \times 10^{-6} \; . \end{array}\right.$$

Also, from the slope of the linear trend fitted to each data set we notice that there is an increase in the temperature which is calculated from

$$Increase = Slope \times number \; of \; months$$

where the number of months for each set of data is 143.

Thus we have an increase of

$$\left\{\begin{array}{ll} 0.566^0\text{C} & \text{northern hemisphere} \\ 0.625^0\text{C} & \text{southern hemisphere} \\ 0.618^0\text{C} & \text{global data.} \end{array}\right.$$

These results agree with Jones et al. (1999). They found an increase of 0.57^0C over the period 1861–1997 and 0.62^0C over 1901–1997 for the global data. They also noted that over both periods the warming was slightly greater in the southern hemisphere than in the northern hemisphere.

8. Prediction. Our aim is to forecast the future values of the temperature data, that is, monthly forecasts for years 1999 and 2000. A good forecast must be close to the original values. Thus, given a series $\{X_t\}$ which is zero mean, second order stationary, a sensible criterion would be to minimize the mean square error given by

$$(8.1) \qquad M(m) = E[(X_{N+m} - \tilde{X}_N(m))^2]$$

where m is the step in the future at which we want to predict our series and $\tilde{X}_N(m)$ is the prediction of X_{N+m}. It can be shown (see Wei (1990)) that the optimum predictor is the conditional expectation of the future given the past, i.e.

$$(8.2) \qquad E[X_{N+m}/X_s, \quad s \leq N]$$

which we denote by $\hat{X}_N(m)$. The forecast error $e_N(m)$ is given by $e_N(m) = X_{N+m} - \hat{X}_N(m)$.

To forecast values from a fitted model, one has to assume that its parameters are known. By replacing the parameters c_0, c_1, r_i, ω_i and ϕ_i for $i = 1, \ldots, m$ as well as the parameters of the function $f(t, \underline{\theta})$ in model (6.1) by its estimates we can write

$$\hat{X}_t = [Y_t - \hat{c}_0 - \hat{c}_1 t - \sum_{i=1}^{\hat{m}} \hat{r}_i \cos(\hat{\omega}_i t + \hat{\phi}_i)]/f(t, \hat{\underline{\theta}})$$

where $\{X_t\}$ is an $ARMA$(k,l) stationary process of the form

$$X_t + \hat{a}_1 X_{t-1} + \ldots + \hat{a}_k X_{t-k} = e_t + \hat{b}_1 e_{t-1} + \ldots + \hat{b}_l e_{t-l}$$

and $\{e_t, \quad t = 1, \ldots, N\}$ is a Gaussian white noise process with mean zero and variance σ_e^2. The order of the $ARMA$ model for each set of data have been considered in the previous chapter. For all sets of data we have selected an $ARMA(1,1)$ model for X_t. Therefore, to obtain forecasts we consider only this $ARMA$ model.

Let as assume that $\{X_t, \quad t = 1, \ldots, N\}$ satisfies one of the $ARMA$ models fitted. We also assume that $\{e_t\}$ are zero mean, mutually independent errors. We are interested in finding the optimal forecasts $\hat{X}_{N+m}$ in each case for $m \geq 1$. As shown above the optimum predictor is given by (8.2). Thus we have that for $ARMA(1,1)$, i.e. $X_t + \hat{a}_1 X_{t-1} = e_t + \hat{b}_1 e_{t-1}$, $t = 1, \ldots, N$ the one step ahead predictor is given by

(8.3) $$\hat{X}_N(1) = -\hat{a}_1 X_N + \hat{b}_1 e_N,$$

where

(8.4) $$e_N = X_N - \hat{X}_{N-1}(1)$$

and thus

$$\hat{X}_N(1) = -\hat{a}_1 X_N + \hat{b}_1(X_N - \hat{X}_{N-1}(1)).$$

We obtain monthly forecasts for the three temperature sets of data. Each time, we predict one step ahead i.e. for the following month. We consider the two models (7.1) and (7.3) for the northern hemisphere temperature data, model (7.7) and model (7.9) for the global temperature data, model (7.5) and model (7.6) for the southern hemisphere temperature data.

The one step ahead forecasts we obtained, seem to justify our time series analysis and the models we have fitted. In general the predicted values are quite close to the observed values. The prediction only seems to miss the big changes in the real data. Although we have considered nonstationary models we must remember that we assumed that the series is changing slowly and thus it is expected that sudden changes in the original series cannot be observed by nonstationary linear models.

In Figures 10–15, we plot the real data of years 1990–2000 and the predicted values for years 1999 and 2000 for all sets of data.

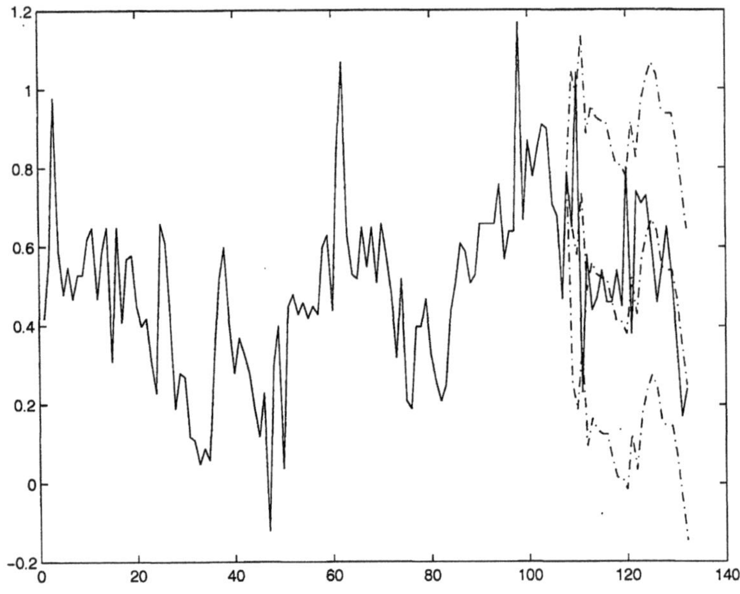

FIG. 10. *Prediction of northern hemisphere temperatures using $f(t, \hat{\underline{\theta}}) = t^{\hat{c}_2}$ (years 1999–2000) with confidence intervals.*

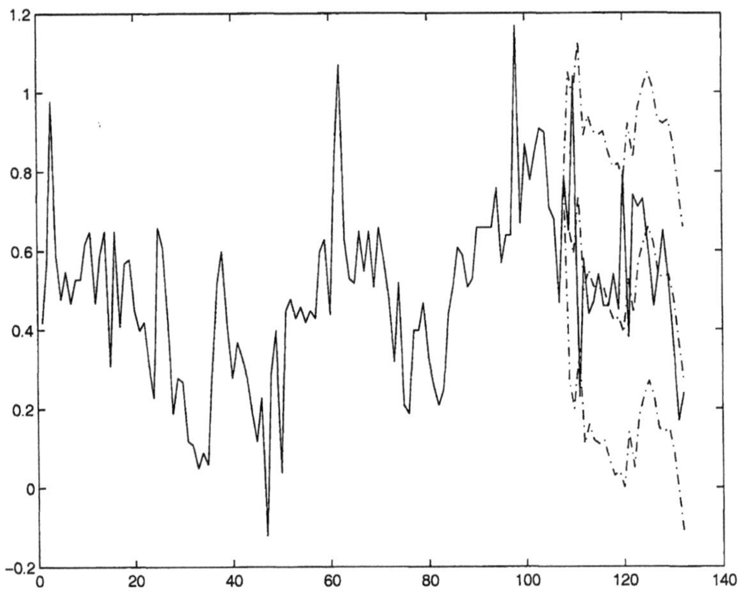

FIG. 11. *Prediction for northern hemisphere temperatures using $f(t, \underline{\theta}) = t^{\hat{c}_2} + \hat{c}_3$ (years 1999–2000) with confidence intervals.*

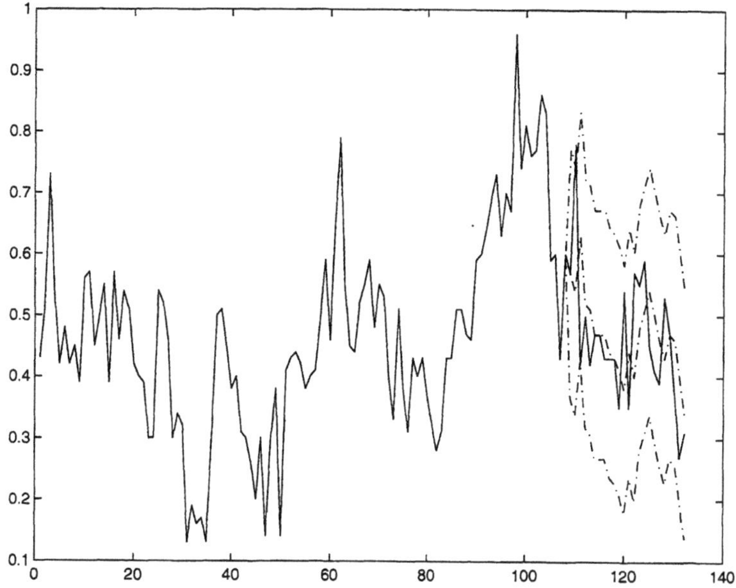

FIG. 12. *Prediction of global temperatures using $f(t, \hat{\underline{\theta}}) = t^{\hat{c}_2}$ (years 1999–2000) with confidence intervals.*

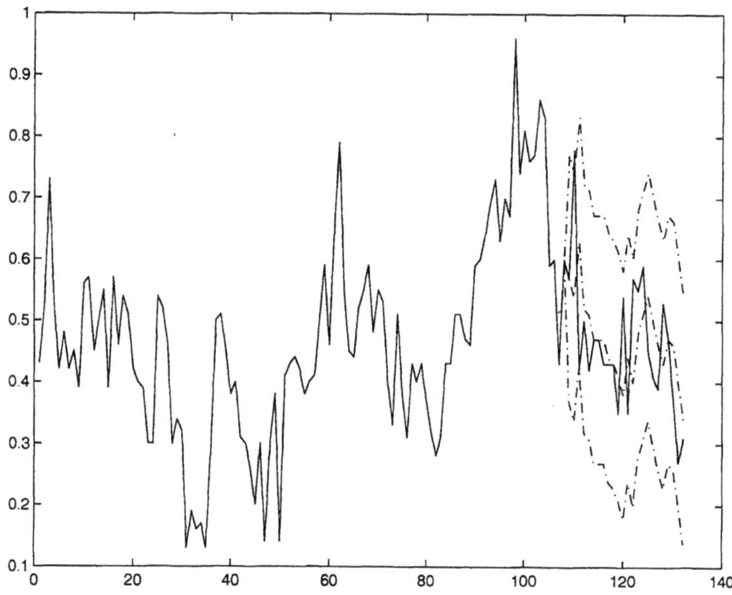

FIG. 13. *Prediction of global temperatures using $f(t, \underline{\theta}) = t^{\hat{c}_2} + \hat{c}_3$ (years 1999–2000) with confidence intervals.*

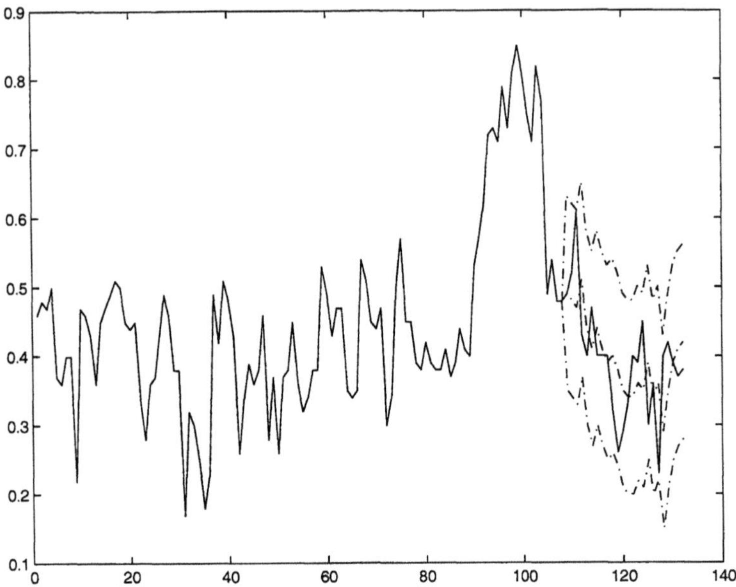

FIG. 14. *Prediction of southern hemisphere temperatures using* $f(t, \hat{\underline{\theta}}) = t^{\hat{c}_2}$ *(years 1999–2000) with confidence intervals.*

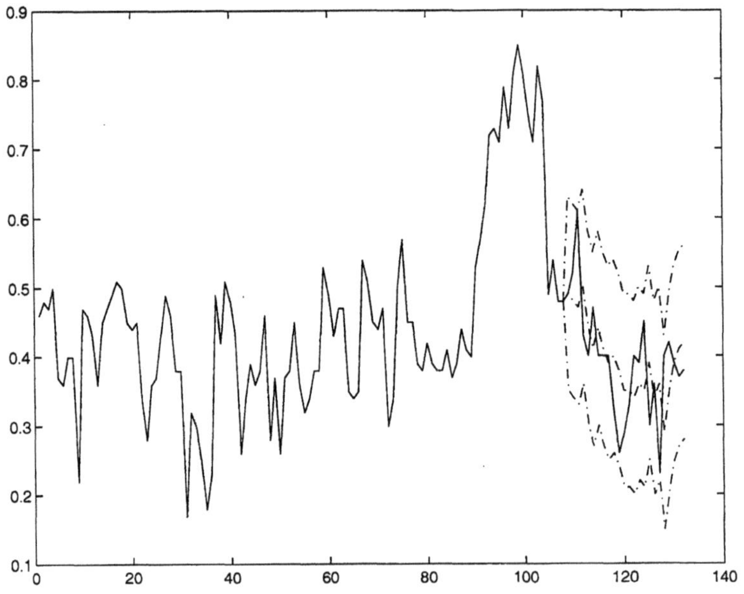

FIG. 15. *Prediction for southern hemisphere temperatures using* $f(t, \underline{\theta}) = t^{\hat{c}_2} + \hat{c}_3$ *(years 1999–2000) with confidence intervals.*

9. Conclusions. Modelling climatic variables has attracted the attention of many researchers from different areas. One of these climatic variables, namely temperature, is known to affect everyday life and thus its prediction is of great importance. We are also interested in knowing whether there is Global Warming.

In this paper we analyzed three sets of temperature data; covering the northern hemisphere, the southern hemisphere and global temperature anomalies and propose models under realistic assumptions. We employed nonstationary techniques for the analysis. By nonstationarity we refer to a more general class of processes namely the semi-stationary processes as defined by Priestley (1965). These processes, although by definition nonstationary, are 'locally stationary;' i.e., the time series characteristics vary slowly.

We applied the test of stationarity (Priestley and Subba Rao (1969)) to our data sets. We concluded that the series are nonstationary and can be adequately represented by a uniformly modulated model. We then looked for change points, using CUSUM test proposed by Subba Rao (1981), in the covariance structure of the series. A highly significant change (more than 5σ) is detected for northern hemisphere data between 1950–1960. A smaller change (about 2.5σ) is detected for the southern hemisphere and global data around 1960–1970. This is consistent with the observations made by meteorologists (see Parker et al. (1994)).

The conclusions of the analysis of these data lead us to propose nonstationary, uniformly modulated linear models for the temperature data. We proposed an additive model consisting of a linear trend, sinusoid terms and finally a uniformly modulated process (of two different forms). The parameters were estimated for the three sets of data and used later for prediction for years 1999–2000.

The findings from the fitted models reveal an expected periodicity of 12 months for all sets of data. Another interesting feature is the presence of a low frequency component corresponding to around 65 years for the northern hemisphere data around 63 years for the global data and around 55 years for the southern hemisphere data. This periodicity can be seen from the plots of the data. The global data exhibits an extra periodicity of 6 months.

From the forecasts we obtained we see that the time series models are adequate for representing the series.

Acknowledgement. We are very grateful to the referee for reading through the paper very carefully and providing us with a list of corrections and making many constructive and helpful comments.

REFERENCES

[1] N.A. ABDRABBO AND M.B. PRIESTLEY, On the prediction of non-stationary processes. *J. Roy. Statist. Soc. Ser.* B, **29**, 1967, 570–585.

[2] D.R. BRILLINGER, An Introduction to Polyspectra. *Annals of Mathematical Statistics*, **36**, 1965, 1351–1374.

[3] W.D. EWAN, When and how to use cusum charts. *Technometrics*, **5**, 1963, 1–22.

[4] R.A. FISHER, Tests of significance in harmonic analysis. *Proc. Roy. Soc. Ser.* A, **125**, 1929, 54–59.

[5] C.K. FOLLAND, T.R. KARL, AND K.Y. VINNIKOV, Observed climate variations and change. Section 7 in *Climate Change, the IPCC Scientific Assesment*, edited by J.T. Houghton, G.J. Jenkins and J.J. Ephraums, Cambridge University Press, New York, 1990, 195–238.

[6] E.J. HANNAN AND J. RISSANEN, Recursive estimation of mixed autoregressive-moving average order. *Biometrika*, **69**, 1982, 89–94. Correction (1983), **70**, p. 303.

[7] N.L. JOHNSON AND F.C. LEONE, *Statistics and experimental design in engineering and the physical sciences*, 2 vols. Wiley, New York, London, 1964.

[8] P.D. JONES, Hemispheric surface air temperature variations: a reanalysis and an update to 1993. *Journal of Climate*, **7**, 1994, 1794–1802.

[9] P.D. JONES, M. NEW, D.E. PARKER, S. MARTIN, AND I.G. RIGOR, Surface air temperature and its changes over the past 150 years. *Reviews of Geophysics*, **37**, 1999, 173–199.

[10] P.D. JONES, T.J. OSBORN, K.R. BRIFFA, C.K. FOLLAND, E.B. HORTON, L.V. ALEXANDER, D.E. PARKER, AND N.A. RAYNER, Adjusting for sampling density in grid box land and ocean surface temperature time series. *Journal of Geophysical Research*, **106**, 2001, 3371–3380.

[11] J.C. LAGARIAS, J.A. REEDS, M.H. WRIGHT, AND P.E. WRIGHT, Convergence properties of the Nelder-Mead simplex method in low dimensions. *SIAM Journal of Atmospheric Science*, **9**(1), 1998, 112–147.

[12] M.J. LIGHTHILL, *Introduction to Fourier Analysis and Generalized functions.* Cambridge University Press, Great Britain, 1964.

[13] D.E. PARKER, C.K. FOLLAND, AND M. JACKSON, Marine surface temperature: observed variations and data requirements. *Climatic Change*, **31**, 1995, 559–600.

[14] D.E. PARKER, P.D. JONES, A. BEVAN, AND C.K. FOLLAND, Interdecadal changes of surface temperature since the late 19th century. *Journal of Geophysical Research*, **99**, 1994, 14373–14399.

[15] M.B. PRIESTLEY, Evolutionary spectra and non-stationary processes. *J.R. Statistical Society, Series B*, **27**, 1965, 204–237.

[16] M.B. PRIESTLEY, Design relations for non-stationary processes. *J. Roy. Statist. Soc. Ser.* B, **28**, 1966, 228–240.

[17] M.B. PRIESTLEY, *Spectral Analysis and Time Series*, 2 vols. Academic Press, London and New York, 1981.

[18] M.B. PRIESTLEY AND M.M. GABR, Bispectral analysis of non-stationary processes. In *Multivariate Analysis: Future Directions.* edited by C.R. Rao, Elsevier Science Publishers, Chapter 16, 1993, 295–317.

[19] M.B. PRIESTLEY AND T. SUBBA RAO, A test for stationarity of time series. *J. Roy. Statist. Soc. Ser.* B, **31**, 1969, 140–149.

[20] B.G. QUINN AND E.J. HANNAN, *The estimation and tracking of frequency.* Cambridge University Press, Great Britain, 2001.

[21] H. SCHEFFÉ, *The analysis of variance.* Wiley, New York, 1959.

[22] T. SUBBA RAO, A cumulative sum test for detecting change in time series. *Int. J. Control*, **34**, 1981, 285–293.

[23] T. SUBBA RAO, Statistical analysis of nonlinear and nonGaussian time series. In *Stochastic differential and difference equations*, edited by Csiszar Gy Michaletzky, Birkhauser, Boston, 1997, 285–298.

[24] T. SUBBA RAO AND M.M. GABR, A test for linearity of stationary time series. *J. Time Series Anal.*, 1, 1980, 145–158.

[25] Y. TOYOOKA, A note on weak consistency of simple least squares estimators in polynomial regression with a nonstationary error process. *Keio Eng. Rep.*, **30**, 1977, 35–44.

[26] Y. TOYOOKA, An asymptotically efficient estimation procedure in time series regression model with uniformly modulated error process. *Math. Japonica*, **25**, 1980, 525–532.

[27] E.P. TSOLAKI, *Non-stationary time series analysis of monthly global temperature data*, unpublished PhD thesis, *University of Manchester Institute of Science and Technology (UMIST)*, 2001.

[28] G. WALKER, On the criteria for the reality of relationships or periodicities. *Calcutta Ind. Met. Memo.*, **21**, 1914, p. 9.

[29] W. WEI, *Time series analysis. Univariate and multivariate methods*. Addison Wesley, 1990.

[30] R.H. WOODWARD AND P.L. GOLDSMITH, *Cumulative sum techniques*. Oliver and Boyd, Edinburgh, 1964.

A TEST FOR DETECTING CHANGES IN MEAN*

WEI BIAO WU[†]

Abstract. In the classical time series analysis, a process is often modeled as three additive components: long-time trend, seasonal effect and background noise. Then the trend superimposed with the seasonal effect constitute the mean part of the process. The issue of mean stationarity, which is generically called *change-point* problem, is usually the first step for further statistical inference. In this paper we develop testing theory for the existence of a long-time trend. Applications to the global temperature data and the Darwin sea level pressure data are discussed. Our results extend and generalize previous ones by allowing dependence and general patterns of trends.

1. Introduction. Many time series models assume the form

$$(1) \qquad X_k = \psi_k + \theta_k + Z_k, \qquad k = 1, \ldots, n,$$

where ψ_k is the long-time trend, θ_k is the seasonal component and Z_k is the random noise. For example, in the study of global warming, the temperature data consists of the long-time warming trend, seasonal variations within years and random fluctuations (cf. Figure 5). A fundamental issue associated with (1) is to test the existence of a long-time trend ψ based on the observations $X_1, \ldots, X_n$. We refer this problem as *the change problem* instead of the widely accepted term *change-point problem* since an abrupt change-point may not be well-defined in examples like global warming where gradual changes are expected. The change-point analysis has been one of the central issues of statistical inference for several decades.

As a special case of (1), the model $X_k = \psi_k + Z_k$ in which $\{Z_k\}$ are independent and $\{\psi_k\}$ only take two possible values has been widely studied; see Bhattacharya (1994) and Siegmund (1986) for reviews. Here we assume $\psi_k = f(k/n)$, where f is a piecewise continuous function with finitely many segments. Let $\{Z_k\}$ be a stationary process with mean zero and absolutely summable covariances; namely $\{Z_k\}$ is short-range dependent. We consider *a posteriori* testing in that $\mathbf{X} = (X_1, \ldots, X_n)$ is already available before the analysis. The hypothesis testing problem can now be formulated as

$$(2) \qquad H_0 : f = \text{constant} \qquad \text{against} \qquad H_A : f \text{ is not constant.}$$

This setup generalizes the one adopted in Wu, Woodroofe and Mentz (2001) where $\theta \equiv 0$ and f is assumed to be non-decreasing. An isotonic regression based test is proposed in the latter paper and this test is shown to be more powerful than some existing ones (Brillinger 1989). Our extensions allow more complicated trend patterns without imposing any parametric form such as linearity and quadratic.

---

*Partially supported by the U.S. Army Research Office.
†Department of Statistics, University of Chicago, Chicago, IL 60637.

"""

The remainder of the paper is organized as follows. Some notation and model assumptions are introduced in Section 2. In Section 3, isotonic and Kolmogorov-Smirnov tests are proposed and a power study is performed. In Section 5 our approach is applied to the global warming data and the Darwin sea level pressure data. Proofs are given in Section 6.

2. Preliminaries. For a function H on $[0,1]$, define its *greatest convex minorant* (GCM)

$$\underline{H} = \sup\{G : G \text{ is convex and } G(x) \le H(x), \ x \in [0,1]\}.$$

The *least concave majorant* (LCM) $\overline{H}$ can be similarly defined. Clearly, $-\underline{H} = \overline{G}$, where $G = -H$. Denote by $\underline{h}$ and $\overline{h}$ the left-hand derivatives of $\underline{H}$ and $\overline{H}$ respectively. Let

$$
\begin{aligned}
&\Omega(H;n) \\
(3)\quad &= \sup\left\{ \sum_{k=1}^{n} |H(t_k) - H(t_{k-1})| : \ 0 = t_0 < t_1 < \ldots < t_{n-1} < t_n = 1 \right\}.
\end{aligned}
$$

Then $\Omega(H;\infty)$ is the total variation of H on $[0,1]$. Let $\gamma(k) = \mathbb{E}(Z_0 Z_k)$ and write

$$(4)\qquad \sigma^2 = \sum_{k=-\infty}^{\infty} \gamma(k).$$

For $\sigma > 0$, we define the piecewise linear function $\mathcal{T}_{nt}$, $t \in [0,1]$ such that $\mathcal{T}_{nt} = T_k/(\sigma\sqrt{n})$ at $t = k/n$, where $T_k = Z_1 + \ldots + Z_k$. Suppose that the invariance principle

$$(5)\qquad \{\mathcal{T}_{nt}, 0 \le t \le 1\} \Longrightarrow \{W(t), 0 \le t \le 1\}$$

holds, where $W(\cdot)$ is a standard Brownian motion. Assumption (5) is crucial in the analysis of the asymptotic behavior of our test statistics.

Consider at the outset the special model $X_k = \psi_k + Z_k$, where $\psi \in \Xi = \{\psi = (\psi_1, \ldots, \psi_n) \in \mathbb{R}^n : \ \psi_1 \le \psi_2 \le \ldots \le \psi_n\}$. Let $X_{1,r} = X_1 + r\sqrt{n}$, $X_{n,r} = X_n - r\sqrt{n}$ and $X_{i,r} = X_i$ for $2 \le i \le n-1$. Wu, Woodroofe and Mentz (2001) proposed the test statistic

$$(6)\qquad \underline{\Lambda}(n,r) = \frac{1}{\sigma_n^2} \sum_{k=1}^{n} (\underline{\psi}_{k,r} - \bar{X}_n)^2,$$

where σ_n^2 is a consistent estimator for σ^2, $\bar{X}_n = \sum_{k=1}^{n} X_k/n$ and

$$(7)\qquad \underline{\psi}_{k,r} = \max_{i \le k} \min_{j \ge k} \frac{X_{i,r} + \ldots + X_{j,r}}{j - i + 1}.$$

If Z_k were iid $N(0, \sigma^2)$ random variables, then $\underline{\psi}_{k,r}$ maximize the penalized log-likelihood function $\ell_{n,r}(\psi|\mathbf{X}) = \ell_n(\psi|\mathbf{X}) - r\sqrt{n}(\psi_n - \psi_1)/\sigma^2$, where

$\ell(\psi|\mathbf{X}) = -(2\sigma^2)^{-1} \sum_{i=1}^{n} (X_i - \psi_i)^2 + C$. Penalization is introduced to overcome the spiking problem where ψ_1 and ψ_n are estimated with bias. Unfortunately, for the general case where Z_k are dependent and may have distributions other than normal, the likelihood function is complicated and the maximum likelihood estimator may not have a close form. More seriously, a numeric solution can also be very difficult to be obtained since it involves the computationally intensive high dimensional maximization problem. In such cases, under the invariance principle (5), the asymptotic distribution of the easily computable test statistic (6) has a close form (cf Theorem 1), and has a very good power. In Section 3.4, the powers of isotonic and UMP tests are compared. Thus (6) is preferable in our setting.

2.1. A Geometric interpretation. Let $F(x) = \int_0^x f(t)dt$ and $G(x) = F(x) - xF(1)$. Then G is convex under the condition $\psi \in \Xi$. The null hypothesis H_0 corresponds to $G \equiv 0$; and the departure of alternative hypothesis to the null can be geometrically interpreted as the distance between G and 0. This distance can be naturally measured by $d(G,0) = [\int_0^1 g^2(t)dt]^{1/2}$, where g is the left-handed derivative of G.

Set $H_{n,r}(t) = \sqrt{n}\{G_{n,r}(t) - \bar{X}t\}/\sigma_n$, where $G_{n,r}(t)$ is a continuous, piecewise linear function such that $G_{n,r}(k/n) = \sum_{i=1}^{k} X_{i,r}/n$ for $k = 1, \ldots, n$. Then $d^2(\underline{H}_{n,r}, 0) = \int_0^1 \underline{h}_{n,r}^2(t)dt$, where $\underline{h}_{n,r}$ is the left-hand derivative of $\underline{H}_{n,r}$, the LCM of $H_{n,r}$. Interestingly enough, $d^2(\underline{H}_{n,r}, 0)$ is exactly the log-likelihood ratio $\ell_{n,r}(\psi|\mathbf{X}) - \ell_{n,r}(\bar{X}_n|\mathbf{X}) = \underline{\Lambda}(n,r)$ (cf. Equation (16) in Wu, Woodroofe and Mentz, 2001).

3. Testing for a long-time trend. In this section we assume throughout that $X_k = \psi_k + Z_k$, namely $\theta_k \equiv 0$. The constancy of a function f can be characterized by $\overline{F} = \underline{F}$, or equivalently $\overline{G} = \underline{G} = 0$, where $G(x) = F(x) - xF(1)$. The geometric consideration in Section 2.1 motivates that the characterization of constancy can be utilized to test for a change of general pattern. To this end, it is natural to consider the distance

$$(8) \qquad d(G,0) = \left\{ \int_0^1 [\overline{g}^2(t) + \underline{g}^2(t)]dt \right\}^{1/2},$$

where $\overline{g}$ and $\underline{g}$ are the left-handed derivatives of $\overline{G}$ and $\underline{G}$ respectively. As shown in the proof of Proposition 3, the distance d has the property that $d(G,0) \le d(H,0)$ if $H(0) = H(1) = 0$ and $0 \le G(x) \le H(x)$ holds for all $0 \le x \le 1$. Namely d preserves the order. Another characterization of constancy is given by $d'(G,0) = \sup_{t\le 1} |G(t)|$, which entails the Kolmogorov-Smirnov test (see Section 3.3).

Based on the distance (8), we propose the test statistic $d^2(\underline{H}_{n,r}, 0) + d^2(\overline{H}_{n,-r}, 0)$, or equivalently

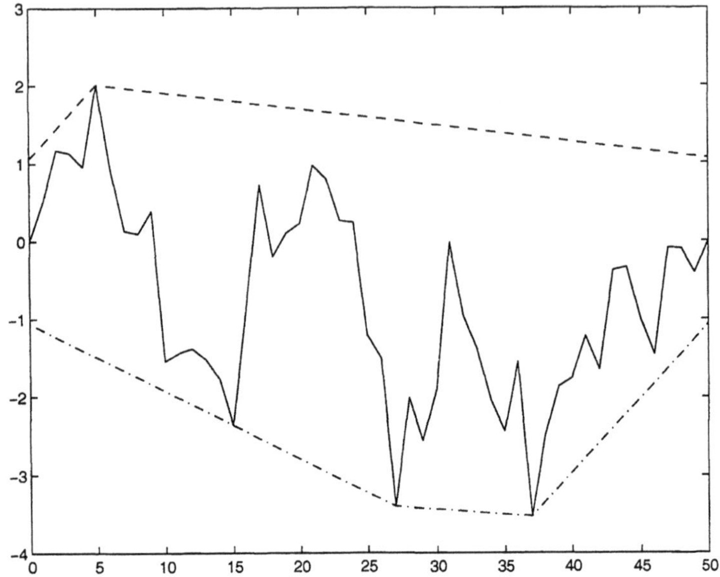

FIG. 1. *The LCM $\overline{Q}_n$ ($--$) and the GCM $\underline{Q}_n$ ($-\cdot$) form a convex hull that envelopes Q_n.*

$$
(9) \quad
\begin{aligned}
\Lambda(n,r) &= \frac{1}{\sigma_n^2}\sum_{k=1}^{n}(\underline{\psi}_{k,r} - \bar{X}_n)^2 + \frac{1}{\sigma_n^2}\sum_{k=1}^{n}(\overline{\psi}_{k,r} - \bar{X}_n)^2 \\
&=: \ \underline{\Lambda}(n,r) + \overline{\Lambda}(n,r)
\end{aligned}
$$

where

$$
(10) \quad \overline{\psi}_{k,r} = \min_{i \leq k}\max_{j \geq k}\frac{X_{i,-r}+\ldots+X_{j,-r}}{j-i+1}.
$$

Let $Q_n(t)$ be piecewise linear function such that $Q_n(i/n) = (T_i - i\bar{X}_n)/(\sigma\sqrt{n})$ at $i = 0,\ldots,n$ (Figure 1). Let $c = r/\sigma > 0$ and $\overline{Q}_n$ be the LCM of the $n+1$ points $(0,c), (i/n, Q_n(i/n)), 1 \leq i \leq n-1$ and $(1,c)$; let $\underline{Q}_n$ be the GCM of $n+1$ points $(0,-c), (i/n, Q_n(i/n)), 1 \leq i \leq n-1$ and $(1,-c)$. Then $\overline{Q}_n$ and $\underline{Q}_n$ form a convex hull that envelopes Q_n. This convex hull is expected to be thin if the trend is constant.

Consider the local alternative

$$
(11) \quad f(t) = \mu + \sigma\phi(t)/\sqrt{n},
$$

where ϕ is a right-continuous function on $[0,1]$ for which $\int_0^1 \phi(t)dt = 0$. Recall that W is a standard Brownian motion. For $c > 0$ let $\mathbb{B}_c^\phi(t) = W(t) - tW(1) + \int_0^t \phi(s)ds + c\mathbf{1}_{(0,1)}(t)$, $\underline{b}_c^\phi(t)$ and $\overline{b}_{-c}^\phi(t)$ be the left-hand derivatives of $\underline{\mathbb{B}}_c^\phi(t)$ and $\overline{\mathbb{B}}_{-c}^\phi(t)$ respectively.

TABLE 1
Critical values for $\lambda_{.15}(\alpha)$.

n	$\lambda_{.15}(.05)$	$\lambda_{.15}(.01)$
10	5.95	8.94
20	7.02	10.28
40	7.73	11.05
80	8.51	11.95
160	8.90	12.17
320	9.32	12.79
640	9.68	13.19
1000	9.76	13.46
2000	9.95	13.60
4000	10.08	13.69

THEOREM 1. *Assume (11) and let $\widehat{\sigma}_n^2$ be a consistent estimator of σ^2. Then (5) implies*

$$(12) \qquad \Lambda(n,r) \Longrightarrow \int_0^1 [\underline{b}_c^\phi(t)]^2 + [\overline{b}_{-c}^\phi(t)]^2 dt.$$

Proof. See Theorem 1 in Wu, Woodroofe and Mentz (2001) for a proof of the convergence of $\underline{\Lambda}(n,r)$ in (9) to $\int_0^1 [\underline{b}_c^\phi(t)]^2$ in (12). It is easily seen that the arguments therein lead to (12). To see this heuristically, observe that $\underline{h}_{n,r}$ and $\overline{h}_{n,-r}$ are left-hand side derivatives of $\underline{Q}_n$ and $\overline{Q}_n$ respectively. Therefore by Robertson, Wright and Dykstra (p. 7, 1988),

$$\Lambda(n,r) = \int_0^1 \underline{h}_{n,r}^2(t)dt + \int_0^1 \overline{h}_{n,-r}^2(t)dt,$$

which intuitively converges to the right side of (12) in view of (5) and (11). $\blacksquare$

3.1. Critical values. The limiting distribution in (12) does not seem to have a close form even under the null hypothesis $f \equiv$ constant. Table 1 gives the simulated critical values $\lambda_c(\alpha)$ for $c = .15$ and two levels $\alpha = .05$ and .01. The simulation is based on $50,000$ Brownian motions for each $n = 10, 20, 40, 80, 160, 320, 640, 1000, 2000$ and 4000.

3.2. Monotonic trends. Proposition 1 implies the coherence property of the test statistics Λ_n in that if the alternative is a strictly increasing function, then $\int_0^1 [\overline{b}_{-c}^\phi(t)]^2 dt$, which corresponds to the LCM part of $\Lambda(n,r)$, vanishes asymptotically.

PROPOSITION 1. *Let $\psi_k = \psi + \delta_n \phi(k/n)\sigma/\sqrt{n}$, where ϕ is a nondecreasing function such that $\phi(0) < 0, \phi(1) > 0$ and $\int_0^1 \phi(t)dt = 0$. Then as $\delta_n \to \infty$,*

$$\lim_{n\to\infty} \mathbb{P}\left\{ \int_0^1 [\bar{b}_c^{\phi\delta_n}(t)]^2 dt = 0 \right\} = 1.$$

Proof of Proposition 1. It suffices to show that

$$\lim_{n\to\infty} \mathbb{P}\left\{ \sup_{0\le t\le 1} \left[W(t) - tW(1) + \delta_n \int_0^t \phi(u)du \right] \le c \right\} = 1$$

since $\overline{B}_c^\phi(t) \equiv c$ under the event inside the probability measure. Observing that $\phi(0) < 0$ and $\phi(1) > 0$, there exists $C > 0$ such that for any $x \in [1/\sqrt{\delta_n}, 1 - 1/\sqrt{\delta_n}]$, $\delta_n \int_0^t \phi(u) \le -C\sqrt{\delta_n}$. Hence

$$\sup_{0\le t\le 1} \left[W(t) - tW(1) + \delta_n \int_0^t \phi(u)du \right]$$

$$\le \max\left\{ \sup_{0\le t\le 1/\sqrt{\delta_n}} [W(t) - tW(1)], \right.$$

$$\sup_{1/\sqrt{\delta_n}\le t\le 1-1/\sqrt{\delta_n}} [W(t) - tW(1) - C\sqrt{\delta_n}],$$

$$\left. \sup_{1-1/\sqrt{\delta_n}\le t\le 1} [W(t) - tW(1)] \right\}.$$

Then almost surely the second term diverges to $-\infty$ and the first and the third terms converges to 0 as $\delta_n \to \infty$. Thus the proof is completed. ∎

3.3. Test of the Kolmogorov-Smirnov type. Another characterization of the constancy of f is $\max_{0\le t\le 1} |G(t)| = 0$, which leads to the cumulative sum (CUSUM) test of the Kolmogorov-Smirnov type

$$(13) \qquad K_n = \frac{1}{\sigma_n\sqrt{n}} \max_{k\le n} \left| \sum_{i=1}^k X_i - \frac{k}{n} \sum_{i=1}^n X_i \right|.$$

If σ_n is a consistent estimator of σ, then K_n converges to $\sup_{0\le t\le 1} |W(t) - tW(1)|$. The well-known blocking technique can be employed to estimated σ^2 (cf (17) and (18) in Theorem 2). The limiting distribution is well-known (Shorack and Wellner, 1986):

$$\mathbb{P}\left[\sup_{0\le t\le 1} |W(t) - tW(1)| > x \right] = 2\sum_{j=1}^\infty (-1)^{j+1} e^{-2j^2 x^2}.$$

In Section 3.4, the powers of the isotonic-based test (9) and the Kolmogorov-Smirnov test (13) are compared.

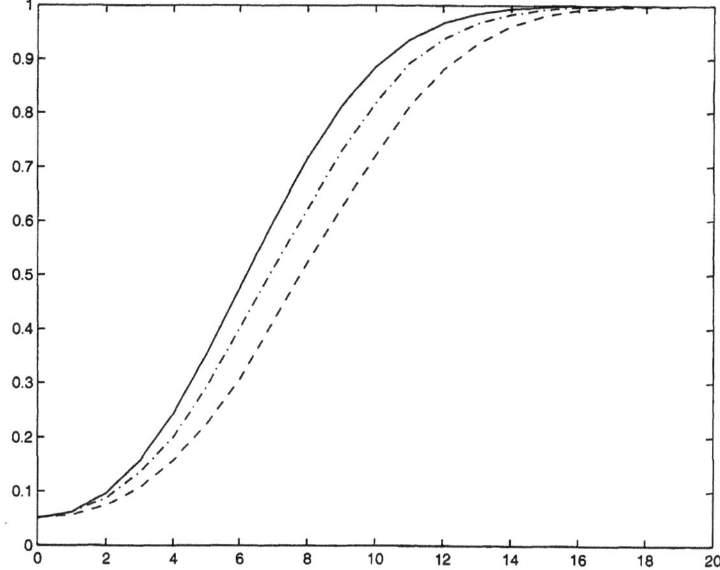

FIG. 2. *Power curves for the uniformly most powerful test (solid line), the isotonic test (dashed line) and the Kolmogorov-Smirnov test (dashdot line) for the step function* $f_1(u) = \mathbf{1}(u \le 1/2)$.

3.4. Power study. To study the powers of the isotonic and the Kolmogorov-Smirnov tests, we consider the model

$$X_k = \frac{\delta}{\sqrt{n}}\phi\left(\frac{k}{n}\right) + Z_k, \quad 1 \le k \le n,$$

where Z_k are iid standard normal random variables, ϕ is a nonzero function defined on $[0, 1]$ and $\delta \ge 0$ measures the distance to the null hypothesis. The simulated power curves for three different functions (i) $\phi_1(u) = \mathbf{1}(u \le 1/2)$ (ii) $\phi_2(u) = \mathbf{1}(u \le 1/16) + \mathbf{1}(u \ge 15/16)$ and (iii) $\phi_3(u) = \sin(\pi u)$ are displayed in Figures 2, 3 and 4 respectively. In all figures $n = 160$ and $\delta = (l - 1)/20$, $l = 1, \ldots, 20$. Since Z_i are iid normal and the parameter of interest δ is one-dimensional, the uniformly most powerful (UMP) test exists. The simulation study shows that the power of the isotonic test is not far away from that of the UMP test. The powers for all tests approach 1 if and only if $\delta = \delta_n \to \infty$. Figure 2 suggests that the Kolmogorov-Smirnov test has a slightly higher power than the isotonic one, while in Figure 3 the former has a severely less power than the latter.

To explain the above-mentioned phenomena, observe that for large δ, the major term in the Kolmogorov-Smirnov test is $\delta \sup_{0 \le u \le 1} |G(u)|$ while in the isotonic test the term $\delta^2 \int_0^1 [\bar{g}^2(u) + \underline{g}^2(u)]du$ has a dominated contribution, where $G(x) = \Phi(x) - x\Phi(1)$, $\Phi(x) = \int_0^x \phi(t)dt$. Proposition 3

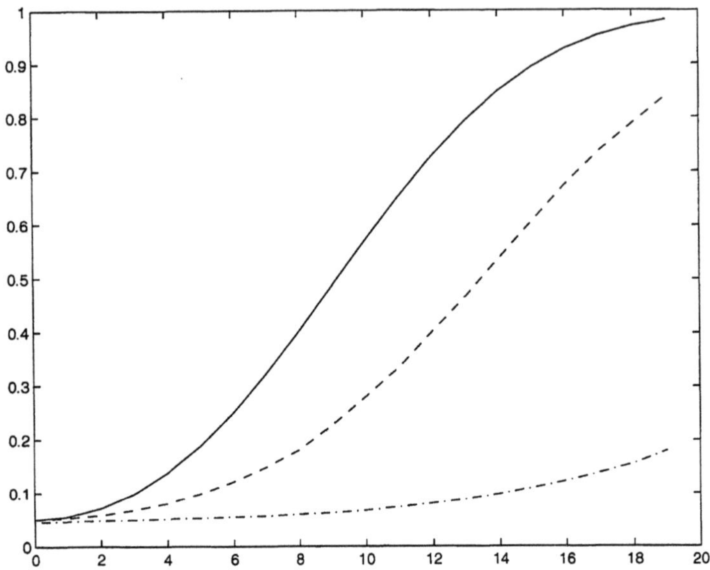

FIG. 3. *Power curves for the uniformly most powerful test (solid line), the isotonic test (dashed line) and the Kolmogorov-Smirnov test (dashdot line) for the step function* $f_2(u) = \mathbf{1}(u \leq 1/16) + \mathbf{1}(u \geq 15/16)$.

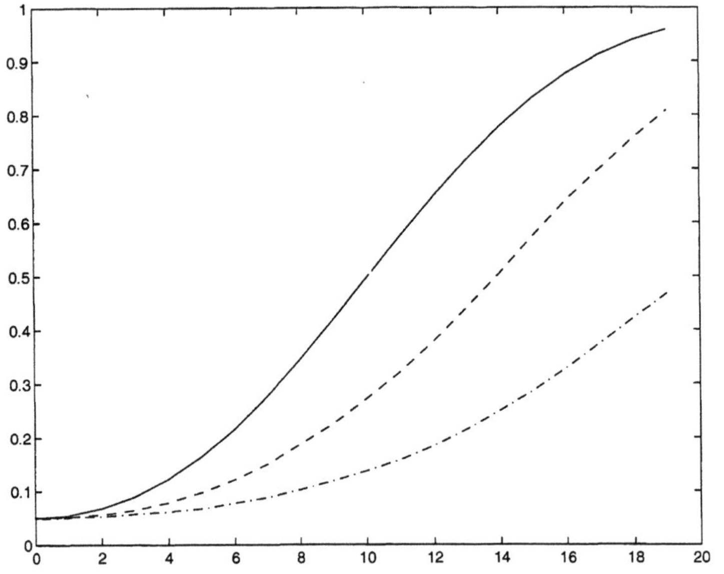

FIG. 4. *Power curves for the uniformly most powerful test (solid line), the isotonic test (dashed line) and the Kolmogorov-Smirnov test (dashdot line) for the sine function* $f_3(u) = \sin(\pi u)$.

asserts that in any case the power of the isotonic test will not be significantly inferior to that of the Kolmogorov-Smirnov test. On the other hand, since inequality (14) cannot be reversed even if a constant multiple is allowed, the Kolmogorov-Smirnov test may have very low power for certain alternatives. For example, for $\lambda \in (0,1)$ let $G(x) = \int_0^x g(u)du$, where

$$g(u) = C(\lambda)\left[-\frac{1}{\lambda}\mathbf{1}(u \leq \lambda) + \frac{1}{1-\lambda}\mathbf{1}(u > \lambda)\right] \text{ and } C(\lambda) = [\lambda(1-\lambda)]^{\frac{1}{3}}.$$

Then it is easily seen that $\max_{0 \leq u \leq 1}|G(u)| = C(\lambda) \downarrow 0$ and $\int_0^1 [\overline{g}^2(u) + \underline{g}^2(u)]du = 1/C(\lambda) \uparrow \infty$ as $\lambda \downarrow 0$. To summarize, the isotonic test has a uniformly reasonable power in all circumstances.

PROPOSITION 2. *If $\overline{H}$ or $\underline{H}$ is not identically 0, then as $\delta_n \to \infty$,*

$$\lim_{\delta_n \to \infty} \mathbb{P}\left\{ \int_0^1 [\overline{b}_c^{\rho_n\phi}(t)]^2 dt + \int_0^1 [\underline{b}_c^{\rho_n\phi}(t)]^2 dt > \lambda_c(\alpha) \right\} = 1,$$

where $\rho_n = \delta_n/\sigma$.

Proof. See Wu, Woodroofe and Mentz (2001). ∎

PROPOSITION 3. *Let G be a function defined on $[0,1]$ such that $G(0) = G(1) = 0$. Then*

$$(14) \qquad 4 \sup_{0 \leq u \leq 1} G^2(u) \leq \int_0^1 [\overline{g}^2(u) + \underline{g}^2(u)]du,$$

where the equality holds if and only if $|G(1/2)| = \sup_{0 \leq u \leq 1}|G(u)|$ and the graph of G is contained in the triangle by points $(0,0), (1,0)$ and $(1/2, G(1/2))$.

4. Estimating σ^2. In Wu, Woodroofe and Mentz (2001), the model $X_k = \psi_k + Z_k$ is considered in which the trend is assumed to be nondecreasing. Then lag-windows type of estimators are constructed based on the the estimated residuals $\widehat{Z}_k = X_k - \widehat{\psi}_k$, where $\widehat{\psi}$ is the isotonic regression estimator. Here monotonicity assumption is not imposed and we shall estimate σ^2 in the presence of θ_k. Recall that $T_k = \sum_{i=1}^k Z_i$ and $f(k/n) = \psi_k$. Let $S_k = \sum_{i=1}^k X_i$, $\Theta_k = \sum_{i=1}^k \theta_i$ and $\Psi_k = \sum_{i=1}^k \psi_i$.

THEOREM 2. *Let $m \to \infty$, $m = \mathcal{O}(n^{1/3})$, $b = \lfloor n/m \rfloor$. Assume that*

$$(15) \qquad \sup_{k \geq 0} |\Theta_{k+m} - \Theta_k| = o(\sqrt{m})$$

and

$$(16) \qquad \Omega(f;b) = o(\sqrt{b}).$$

Then

$$(17) \qquad \sigma^2_{\text{Block}}(Z) = \frac{1}{2n} \sum_{k=2}^b \left[T_{km} - T_{(k-1)m}\right]^2 \to_{\mathbb{P}} \sigma^2,$$

implies

$$(18) \qquad \sigma^2_{\mathrm{Block}}(X) = \frac{1}{2n} \sum_{k=2}^{b} \left[S_{km} - S_{(k-1)m} \right]^2 \to_{\mathbb{P}} \sigma^2.$$

We say that a function f is Hölder continuous with index $h > 0$ if there exist $L > 0$ such that for all $0 \le x, y \le 1$, $|f(x) - f(y)| \le L|x - y|^h$. Clearly (16) holds for piecewise Hölder continuous functions with index $h > 1/2$. In the case that Z_k are iid, Hall, Kay and Titterington (1990) considered the difference-based estimation of $\sigma^2 = \mathbb{E}(Z_1^2)$ from the model $Y_j = f(j/n) + Z_j$, $j = 1, \ldots, n$ by assuming f is Hölder continuous with $h > 1/2$. Our concise estimator $\sigma^2_{\mathrm{Block}}(X)$ uses first order differences when Z_k are allowed to be dependent. To reduce bias, estimators based on higher order differences can be similarly constructed as in Hall *et al.*

REMARK 1. For the commonly used seasonal model, $\theta_k = \sum_{i=1}^{I} A_i \cos(k\omega_i + \alpha_i)$, where $0 < \omega_i < 2\pi$ are frequencies and A_i are amplitudes, it is easily seen that $\sup_{k \ge 0} |\Theta_{k+m} - \Theta_k| = \mathcal{O}(1)$ and hence (15) holds.

5. A separation principle. In this section we shall consider the testing problem proposed in the Introduction, namely we test for "$f = $ constant" in the model $X_k = \psi_k + \theta_k + Z_k$. For the seasonal component, let $\theta_k = \sum_{i=1}^{I} A_i \cos(k\omega_i + \alpha_i)$, where $0 < \omega_i < 2\pi$ are frequencies and $A_i > 0$ are amplitudes. Let $Y_k = \psi_k + Z_k$ be the process without seasonal components and analogously $V_k = \theta_k + Z_k$ be the process without long-time trend. Let $\kappa_n(\omega) = \sum_{k=1}^{n} \exp(\omega k \imath)$; $S_{n,X}(\omega) = \sum_{k=1}^{n} X_k \exp(\omega k \imath)$ and $S_{n,V}(\omega) = \sum_{k=1}^{n} V_k \exp(\omega k \imath)$, where $\imath$ is the imaginary unit. Then for a fixed $\omega \in (0, 2\pi)$, $\sup_{n \ge 0} |\kappa_n(\omega)| \le 2/|1 - \exp(\omega \imath)| = \mathcal{O}(1)$. So if $\Omega(f, n) = o(\sqrt{n})$, then

$$
\begin{aligned}
\frac{|S_{n,X}(\omega) - S_{n,V}(\omega)|}{\sqrt{n}} &= \frac{1}{\sqrt{n}} \left| \sum_{k=1}^{n} f(k/n) \exp[\omega k \imath] \right| \\
&= \frac{1}{\sqrt{n}} \left| \sum_{k=2}^{n} \{ f(k/n) - f((k-1)/n) \} \kappa_k(\omega) \right| + \mathcal{O}\left(\frac{1}{\sqrt{n}} \right) \\
&= \frac{\mathcal{O}[\Omega(f; n)]}{\sqrt{n}} \\
&= o(1),
\end{aligned}
$$

which suggests an interesting feature of the spectral analysis: the periodograms of X_k and V_k have asymptotically negligible differences. Clearly, $S_{n,V}(\omega)$ has a magnitude of order n if ω is one of the frequencies ω_i. The identification of ω_i will require the asymptotic distribution of periodograms (see, for example, Chapter 10 in Brockwell and Davis, 1991). Wu (2002) obtain central limit theorems for the Fourier transform $S_{n,Z}(\omega)$ under mild conditions on Z_k.

On the other hand, since $\sup_{k\geq0}|\Theta_k| = \mathcal{O}(1)$, isotonic regressions based on X_k and Y_k produce asymptotically equivalent estimators for ψ_k. This equivalence in view of the formula (7) is implied by the fact that the invariance principle (5) still holds if we regard $Z_k' = \theta_k + Z_k$ as the new background noises. Recall $Y_k = \psi_k + Z_k$. Similarly as $X_{k,r}$, let $Y_{1,r} = Y_1 + r\sqrt{n}, Y_{n,r} = X_n - r\sqrt{n}$ and $Y_{i,r} = Y_i$ for $2 \leq i \leq n - 1$, and define

$$(19) \quad \underline{\nu}_{k,r} = \max_{i\leq k} \min_{j\geq k} \frac{Y_{i,r}+\ldots+Y_{j,r}}{j-i+1}, \quad \overline{\nu}_{k,r} = \min_{i\leq k} \max_{j\geq k} \frac{Y_{i,-r}+\ldots+Y_{j,-r}}{j-i+1}.$$

THEOREM 3. *Under the condition (15), we have*

$$\sum_{k=1}^{n}(\underline{\psi}_{k,r} - \underline{\nu}_{k,r})^2 + \sum_{k=1}^{n}(\overline{\psi}_{k,r} - \overline{\nu}_{k,r})^2 = o_{\mathbb{P}}(1).$$

To summarize, the spectral analysis and the isotonic regression filter ψ and θ respectively. Programs are available at `http://www.stat.uchicago.edu/faculty/wu.html`.

5.1. Global warming data. The global temperature data consists of monthly temperature anomalies from 1856 to 2000 (cf. Figure 5, `http://cdiac.esd.ornl.gov/trends/temp/jonescru/jones.html`). Now we shall apply the separation principle to the global temperature data. Wu, Woodroofe and Mentz (2001) analyzed the yearly averaged data by using the penalized isotonic regression with $c = 0.15$ (cf. Figure 6) and showed that there exists a substantial increasing trend. The estimated variance is $\widehat{\sigma}^2 = 0.0158$. As shown in Figure 6, the estimated trends based on the monthly data and the yearly data are sufficiently close. Noticing that by taking yearly average is tantamount to eliminating seasonal effects, this comparison suggests the robustness of isotonic regression against seasonal components.

On the other hand, the periodogram plot for this monthly temperature data in which the long-term trend is present indicates a cyclic component with frequency $\omega_1 = 2\pi/12$ (cf. Figure 7). This observation reflects the common sense that the period is 12 months.

It is generally believed that the average global surface temperature has increased $0.4 \sim 0.8\,°C$ since the late 19th century (cf the report by IPCC, 2001). The IPCC report also mentioned that there are two major periods of increment: 1910-1945 and 1976-present. Based on our isotonic regression, the estimated increment is $\underline{\psi}_{145,r} - \underline{\psi}_{1,r} = 0.72\,°C$, where the penalty $c = 0.15$ is used, $\underline{\psi}_{145,r}$ and $\underline{\psi}_{1,r}$ are the estimated mean temperatures of the year 2000 and 1856. Interestingly enough, from Figure 6, the isotonic regression procedure indicates that two major periods of increment are

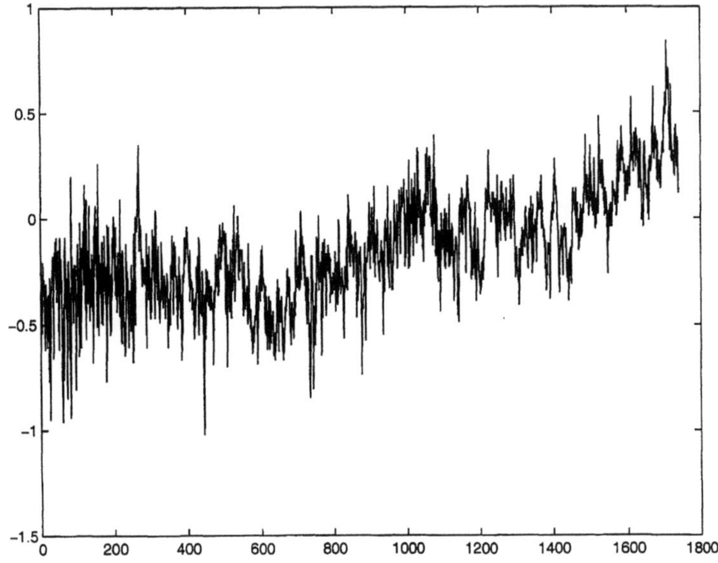

FIG. 5. *Global warming data: monthly temperature anomalies from 1856 to 2000.*

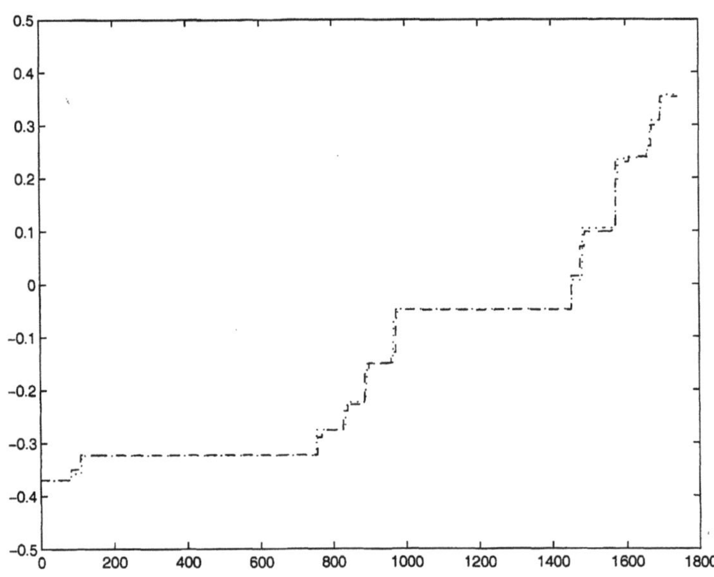

FIG. 6. *Isotonic regression estimators for monthly (dotted line) and yearly (dashed line) temperature data.*

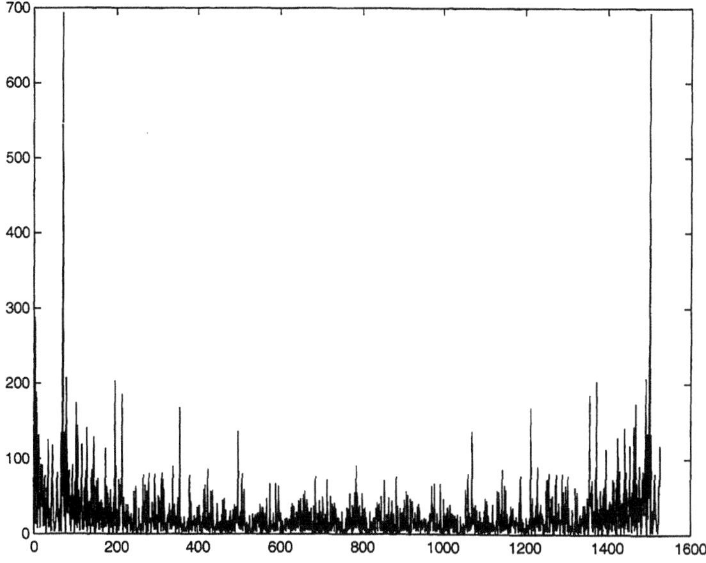

FIG. 7. *Periodogram plot for the global monthly temperature data.*

1920–1935 and 1976–present. Thus our procedure performs well and it appears more versatile than the usual method where the trend is modeled linearly.

5.2. The Darwin sea level pressure data. The sea level pressure data were collected at Darwin, Australia (13S, 131E) from year 1882 to 2001; see the website http://www.cpc.ncep.noaa.gov/data/indices/ (by Climate Prediction Center, National Centers for Environmental Prediction, National Oceanic and Atmospheric Administration) for more detailed information. Yearly and monthly plots are displayed in Figures 8 and 9 respectively. The unit is millibar (MB) with 1000 MB subtracted from the original observations.

For the monthly data, the estimated $\sigma_{\mathrm{Month}} = 2.4176$ and the isotonic test statistic is 7.1789 by choosing the penalty $c = .15$. For the yearly data, $\sigma_{\mathrm{Year}} = 0.6372$ and the test statistic is 6.9013. Both test statistics are very close to each other, and they indicate that the sea level pressure has not undergone a significant change at least in the last century.

6. Proofs.
Proof of Theorem 2. Note that $S_k = T_k + \Theta_k + \Psi_k$. By condition (17) it suffices to establish

$$\mathbb{E}\left|\sum_{k=2}^{b}\left\{\left[S_{km} - S_{(k-1)m}\right]^2 - \left[T_{km} - T_{(k-1)m}\right]^2\right\}\right| = o(n).$$

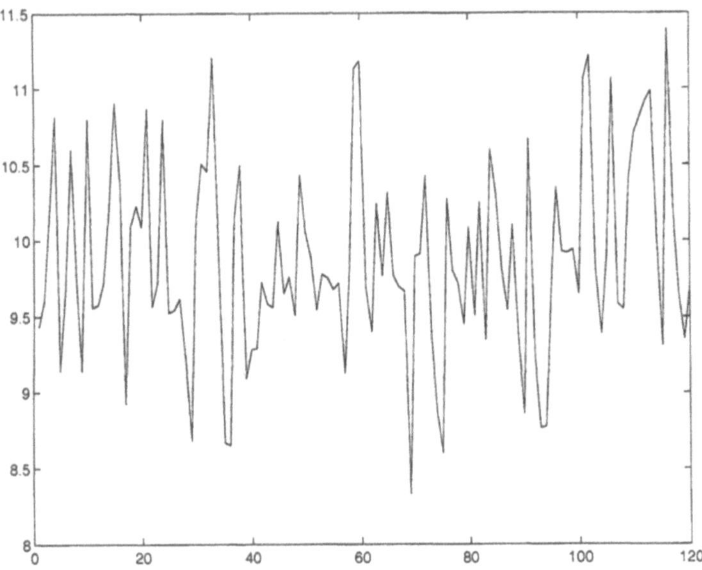

FIG. 8. *Yearly sea level pressure data collected at Darwin, Australia (13S, 131E) from 1882 to 2001.*

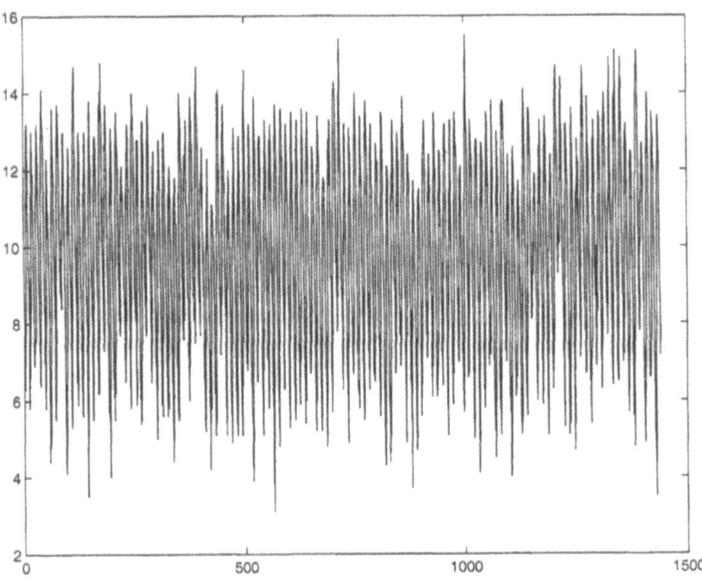

FIG. 9. *Monthly sea level pressure data collected at Darwin, Australia (13S, 131E) from 1882 to 2001.*

This relation clearly follows from

$$\sum_{k=2}^{b} \mathbb{E}\left\{ \left[\Theta_{km}-\Theta_{(k-1)m}\right]^2 + |\Theta_{km}-\Theta_{(k-1)m}||T_{km}-T_{(k-1)m}| \right\} = o(n)$$

and

$$\sum_{k=2}^{b} \mathbb{E}\left\{ \left[\Psi_{km}-\Psi_{(k-1)m}\right]^2 + |\Psi_{km}-\Psi_{(k-1)m}||T_{km}-T_{(k-1)m}| \right\} = o(n).$$

The former results easily from (15) and $\mathbb{E}|T_m| = \mathcal{O}(\sqrt{m})$. For the latter, let $C = \sup_{x\in[0,1]} |f(x)| < \infty$. Then by Cauchy's inequality,

$$
\begin{aligned}
\sum_{k=2}^{b}\left[\Psi_{km}-\Psi_{(k-1)m}\right]^2 &= \sum_{k=2}^{b}\left\{ \sum_{j=1-m}^{0} \left[f\big((km+j)/n\big)-f\big((km-m+j)/n\big)\right]\right\}^2 \\
&\leq \sum_{k=2}^{b} m \sum_{j=1-m}^{0} \left[f\big((km+j)/n\big)-f\big((km-m+j)/n\big)\right]^2 \\
&\leq Cm \sum_{j=1-m}^{0} \sum_{k=2}^{b} \left|f\big((km+j)/n\big)-f\big((km-m+j)/n\big)\right| \\
&\leq Cm^2 \Omega(b) \\
&= o(n).
\end{aligned}
$$

Observe that $\mathbb{E}|T_{km} - T_{(k-1)m}| = \mathbb{E}|T_m| = \mathcal{O}(\sqrt{m})$ and $\sum_{k=2}^{b} |\Psi_{km} - \Psi_{(k-1)m}| = o(\sqrt{n})\sqrt{b}$, we have

$$\sum_{k=2}^{b} |\Psi_{km} - \Psi_{(k-1)m}|\mathbb{E}|T_{km} - T_{(k-1)m}| = \mathcal{O}(\sqrt{m})o(\sqrt{n})\sqrt{b} = o(n)$$

completes the proof. ∎

Proof of Theorem 3. Recall $G_{n,r}(k/n) = \sum_{i=1}^{n} X_{i,r}/n$ and $H_{n,r}(t) = \sqrt{n}[G_{n,r}(t)-\bar{X}_n t]/\sigma$. Analogously, for Y_k let $P_{n,r}(t) = \sqrt{n}[R_{n,r}(t)-\bar{Y}_n t]/\sigma$, where $R_{n,r}(k/n) = \sum_{i=1}^{n} Y_{i,r}/n$. Let $\|F\| = \sup_{0\leq t\leq 1}|F(t)|$. Then by Woodroofe and Sun (1999),

$$\int_0^1 [\underline{\psi}_{k,r}(t) - \underline{\nu}_{k,r}(t)]^2 dt$$
$$\leq \|\underline{P}_{n,r} - \underline{H}_{n,r}\|[\underline{\psi}_{k,r}(1-) - \underline{\psi}_{k,r}(0+) + \underline{\nu}_{k,r}(1-) - \underline{\nu}_{k,r}(0+)].$$

By Marshall's lemma,

$$\|\underline{P}_{n,r} - \underline{H}_{n,r}\| \leq \|P_{n,r} - H_{n,r}\| = \mathcal{O}\left(\sup_{1\leq k\leq n} |\Theta_k|/\sqrt{n} \right) = o(1),$$

which entails the theorem by noticing that $\underline{\psi}_{k,r}(1-)$, $\underline{\psi}_{k,r}(0+)$, $\underline{\nu}_{k,r}(1-)$ and $\underline{\nu}_{k,r}(0+)$ are all stochastically bounded (cf. Lemma A4 in Wu, Woodroofe and Mentz (2001)).   ∎

Proof of Proposition 3. For two functions G_1 and G_2 defined on $[0,1]$, denote by $G_1 \leq G_2$ if $G_1(x) \leq G_2(x)$ holds for all $x \in [0,1]$. Let H be a concave function defined on $[0,1]$ for which $H(0) = H(1) = 0$. Then for any triple $0 \leq x_1 < x_2 < x_3 \leq 1$, by concavity it is easily verified that

$$\left[\frac{H(x_2) - H(x_1)}{x_2 - x_1}\right]^2 (x_2 - x_1) + \left[\frac{H(x_3) - H(x_2)}{x_3 - x_2}\right]^2 (x_3 - x_2)$$
$$\geq \left[\frac{H(x_3) - H(x_1)}{x_3 - x_1}\right]^2 (x_3 - x_1).$$

Let G_1 and G_2 be concave, continuous and piecewise linear (CCPL) functions with values being 0 at the endpoints $x = 0, 1$. Then $G_1 \leq G_2$ entails $d(G_1, 0) \leq d(G_2, 0)$. To see this, a chain of CCPL functions $G_1 = H_1 \leq H_2 \ldots \leq H_q = G_2$ can be constructed such that H_i and H_{i+1} differ only on the interval $I = [x_1, x_3]$ (say), where $H_i(x)$ is a linear function, $H_i(x_1) = H_{i+1}(x_1)$, $H_i(x_3) = H_{i+1}(x_3)$, and H_{i+1} is linear on intervals $[x_1, x_2]$ and $[x_2, x_3]$ respectively. The above inequality on H implies that $d(H_i, 0) \leq d(H_{i+1}, 0)$. Hence $d(G_1, 0) \leq d(G_2, 0)$. Since concave and continuous functions can be approximated by CCPL functions, the general case that $G_1 \leq G_2$ implies $d(G_1, 0) \leq d(G_2, 0)$ follows by taking limits.

Let $\lambda = \mathrm{argmax}_u |G(u)|$. Without loss of generality assume $G(\lambda) > 0$. Then $\overline{G} \geq L$, where $L(0) = L(1) = 0$, $L(\lambda) = G(\lambda)$ and L is linear on intervals $[0, \lambda]$ and $[\lambda, 1]$. Using the monotonicity of $d(\cdot, 0)$, we have $d(\overline{G}, 0) \geq d(L, 0)$, or

$$\int_0^1 \overline{g}^2(u)\,du \geq \left[\frac{L(\lambda)}{\lambda}\right]^2 \lambda + \left[\frac{L(\lambda)}{1-\lambda}\right]^2 (1-\lambda) = \frac{L^2(\lambda)}{\lambda(1-\lambda)} \geq 4G^2(\lambda). \quad ∎$$

Acknowledgment. The author thanks H. Pollack (Department of Geological Sciences, University of Michigan) for the IPCC (2001) report. The suggestions from the reviewer greatly improve the paper.

REFERENCES

BHATTACHARYA P.K., Some aspects of change-point analysis. In *Change-Point Problems*, Ed. E. Carlstein, H. Müller and D. Siegmund, pp. 28–56. Hayward, CA: Inst. Math. Statist., 1994.

BRILLINGER D.R. (1989). Consistent detection of a monotonic trend superposed on a stationary time series. *Biometrika*. **76**, 23–30.

BROCKWELL P.J. AND DAVIS R.A. *Time series: theory and methods.* New York: Springer-Verlag, 1991.

HALL P., KAY J.W., AND TITTERINGTON D.M.. Asymptotically Optimal Difference-Based Estimation of Variance in Nonparametric Regression. *Biometrika,* **77**, No. 3. (Sep., 1990), pp. 521–528.

IPCC. *Climate change 2001: the scientific basis. Contribution of Working Group I to the third assessment report of the Intergovernmental Panel on Climate Change. Edited by* J.T. Houghton *et al.,* Cambridge University Press, 2001.

RICE J, Bandwidth Choice for Nonparametric Regression. *Annals of Statistics,* **12**, 1215–1230, 1984.

SHORACK G.R. AND WELLNER J.A. *Empirical processes with applications to statistics.* New York: Wiley, 1986.

SIEGMUND D. (1986). Boundary crossing probabilities and statistical applications. *Ann. Statist.* **14**, 361–404.

WOODROOFE M. AND SUN J. (1999). Testing uniformity versus a monotone density. *Ann. Statist.* **27**, 338–360.

WU W.B. (2002). Fourier transforms of stationary processes. *Technical Report #514, Department of Statistics, University of Chicago.*

WU W.B., WOODROOFE M., AND MENTZ G., Isotonic regression: another look at the change point problem. *Biometrika,* **88**, 793–804, 2001.

SPATIO-TEMPORAL MODELLING OF TEMPERATURE TIME SERIES: A COMPARATIVE STUDY

T. SUBBA RAO* AND ANA MONICA COSTA ANTUNES*†

Abstract. A special class of linear stationary spatial time series models: Space-Time ARMA(STARMA) models, has been proven useful in modelling observations measured in space and time. A review of STARMA models and modelling procedure is presented. An order determination method and approach for initial estimation of the model parameters are proposed. The STARMA modelling procedure and extensions are implemented and tested using simulated data. Then the performance in forecasting of the STARMA model is compared with that of separate univariate ARMA models. This comparison is performed using real data of monthly mean temperatures from nine meteorological stations around the United Kingdom.

Key words. Forecasting, spatio-temporal models, STARMA models, temperature data.

AMS(MOS) subject classifications. 62M10.

1. Introduction. Spatial time series data can be viewed as time series collected simultaneously at a number of locations assuming that the locations are fixed and so are the distances between them. Space-time models attempt to explain the dependencies across space in systems that exhibit systematic dependencies between the observations at various locations. Applications of spatial statistics cover many areas and much of the original impetus for the area was driven by geostatistics but in recent years the applications have extended to sociology, economics, environmental and ecological sciences. Smith [14] gave a broad overview of current themes in spatial statistics concentrating on the environmental applications which has been the fastest growing field in recent years.

Pfeifer and Deutsch [8] proposed a comprehensive class of space-time models: The STARMA model class. Processes that can be modelled by the space-time autoregressive moving - average (STARMA) models are characterized by a single random variable observed at N fixed sites in space. The dependencies between the N time series is incorporated in the model through hierarchical $N \times N$ weighting matrices, specified by the model builder prior to analyzing the data. These weighting matrices should incorporate the relevant physical characteristics of the system into the model. Each of the N time series is simultaneously modelled as a linear combination of past observations and disturbances as well as weighted past observations and disturbances at neighboring sites.

---

*University of Manchester Institute of Science and Technology (UMIST), Department of Mathematics, P.O. Box 88, Manchester M60 1QD, UK (tata.subbarao@umist.ac.uk)

†a.antunes@student.umist.ac.uk. The work of the author was supported by FCT Grant SFRH/BD/1473/2000 (Fundacao para a Ciencia e a Tecnologia, Portugal) and the author is grateful for the grant.

The objectives of this paper are to review the STARMA model and modelling procedure; propose a new approach for determining initial estimates of the parameters; and apply the procedure. We also see wether the STARMA model gives a better fit and provides better forecasts than separate univariate ARMA models. The approach to be proposed for determining initial estimates of the parameters is an extension of the Hannan and Rissanen procedure for obtaining initial estimates for univariate ARMA models.

The modelling procedure is implemented in MATLAB and tested using simulated data, and the results are very satisfactory. The modelling procedure is applied to real data. The data consists of time series of monthly mean temperatures measured at nine different sites around the United Kingdom spanning 18 years and 8 months from January 1951 to August 1968. The first 192 observations are used for modelling purposes and the last 31 observations are used for forecasting purposes. The forecast test consists of comparing the performance of the STARMA model with the performance of separate univariate ARMA models using the observed data on this period.

2. The STARMA model. Let $z_i(t)$ be an observation at site i $(i = 1, 2, ..., N)$ and time $t(t \in \mathbf{Z})$.

The STARMA model is represented by the difference equation:

$$(2.1) \quad z(t) = -\sum_{k=1}^{p}\sum_{l=0}^{\lambda_k} \phi_{kl} W^{(l)} z(t-k) + \varepsilon(t) + \sum_{k=1}^{q}\sum_{l=0}^{m_k} \theta_{kl} W^{(l)} \varepsilon(t-k)$$

where $z(t)$ is a $N \times 1$ vector of observations at time t, $t = 1, ..., T$ and

p is the autoregressive order,

q is the MA order,

λ_k is the spatial order of the kth AR term,

m_k is the spatial order of the kth MA term,

ϕ_{kl} is the AR parameter at temporal lag k and spatial lag l (scalar),

θ_{kl} is the MA parameter at temporal lag k and spatial lag l (scalar),

$W^{(l)}$ is the N×N matrix of weights for spatial order l,

$\varepsilon(t)$ normally distributed random error at time t with

$$E\left[\varepsilon(t)\right] = 0$$

and

$$E\left[\varepsilon(t)\,\varepsilon(t+s)'\right] = \begin{cases} G & s = 0 \\ 0 & s \neq 0 \,. \end{cases}$$

The model given by eq(2.1) is called $STARMA\left(p_{\lambda_1,\lambda_2,...,\lambda_p}, q_{m_1,m_2,...,m_q}\right)$ model with two special subclasses of the model being STAR $\left(p_{\lambda_1,\lambda_2,...,\lambda_p}\right)$ when $q = 0$ and STMA $\left(q_{m_1,m_2,...,m_q}\right)$ when $p = 0$.

It is assumed that

$$det\left(x^p I + \sum_{k=1}^{p}\sum_{l=0}^{\lambda_k}\phi_{kl}W^{(l)}x^{p-k}\right) = 0$$

for $|x| < 1$, which determines the values of ϕ_{kl} for which the process is stationary. Similarly, the invertibility condition is

$$det\left(x^q I + \sum_{k=1}^{q}\sum_{l=0}^{m_k}\theta_{kl}W^{(l)}x^{q-k}\right) = 0$$

for $|x| < 1$. It is further assumed that the polynomials above have no common roots.

The definition of the STARMA model implies that the space-time process is second order stationary both in space and in time. In this case the mean $\mu(i,t) = \mu$ for all i and all t and

$$Cov(z_i(t), z_j(t+s)) = C(h,s)$$

where C is a spatial-temporal covariance function, $h = x_i - x_j$, x_i and x_j being the coordinates of sites i and j respectively. This means that the space-time process is second order stationary if the spatial-temporal covariance does not depend on the specific locations of the sites, only on their separation vector h and on the time lag s. In most cases it is assumed only that $C(h,s) = C(\|h\|,s)$, where $\|h\|$ is the norm of h. In this case the covariance function is assumed to depend only on the distance between the sites and time lag in which case the process is said to be isotropic.

2.1. Modelling procedure: case $G = \sigma^2 I_N$. The modelling procedure consists of the following stages ([8]): The sample space time autocorrelation function(STACF) and sample space time partial autocorrelation function(STPACF) are analyzed and a number of potentially suitable models are selected. The parameters for each of these models are estimated and the model which minimizes a criterion is selected as the model that best fits the data. Finally the selected model is tested for adequacy to the data in the diagnostic checking stage. If the model is found inadequate the required modifications to the model are undertaken and a more adequate model should result. The parameters should be re-estimated and the diagnostic checking stage repeated. The modelling procedure terminates when an appropriate model is found.

2.1.1. Identification. The primary tools in identification are the space-time autocorrelation and space-time partial autocorrelation functions ([8]).

Space-time autocorrelation function. Assuming $E\left[z\left(t\right)\right] = 0$ the space-time autocorrelation function is defined as [9]:

$$\gamma_{lk}\left(s\right) = E\left\{\frac{\left[W^{(l)}z\left(t\right)\right]'\left[W^{(k)}z\left(t+s\right)\right]}{N}\right\}$$

$$\Longleftrightarrow \quad \gamma_{lk}\left(s\right) = tr\left\{\frac{W^{(k)'}W^{(l)}\Gamma(s)}{N}\right\}$$

where $\Gamma(s) = E[z(t)z(t+s)']$ is the covariance matrix of lag s and $tr[A]$ is the trace of A (sum of the diagonal elements of A).

The sample estimate $\hat{\gamma}_{lk}\left(s\right)$ of $\gamma_{lk}\left(s\right)$ is determined by replacing $\Gamma(s)$ in the above expression by its estimate:

$$\hat{\Gamma}(s) = \sum_{t=1}^{T-s}\frac{(z(t) - \bar{z})(z(t+s) - \bar{z})'}{T-s} \; .$$

An important property of the space-time autocorrelation function is $\gamma_{lk}\left(s\right) = \gamma_{kl}\left(-s\right)$.

The space-time autocorrelation between l^{th} and k^{th} order neighbors s time-lags apart is:

$$\rho_{lk}(s) = \frac{\gamma_{lk}\left(s\right)}{\left[\gamma_{ll}\left(0\right)\gamma_{kk}\left(0\right)\right]^{1/2}}$$

and its sample estimates:

$$\hat{\rho}_{lk}(s) = \frac{\hat{\gamma}_{lk}\left(s\right)}{\left[\hat{\gamma}_{ll}\left(0\right)\hat{\gamma}_{kk}\left(0\right)\right]^{1/2}} \; .$$

For identification purposes $\rho_{l0}(s)$ or simply $\rho_l(s)$ is usually sufficient since we are interested in the space-time autocorrelation function between l^{th} order neighbors and zeroth order neighbors at lag s. Therefore, the function of interest is

$$\hat{\rho}_l(s) = \frac{\hat{\gamma}_{l0}\left(s\right)}{\left[\hat{\gamma}_{ll}\left(0\right)\hat{\gamma}_{00}\left(0\right)\right]^{1/2}} \; .$$

Space-time partial autocorrelation function. The space-time analogue of the Yule-Walker equations for the STAR model have been derived by Pfeifer and Deutsch [8]. By pre-multiplying both sides of the general $STAR(k_{\lambda,\lambda,...,\lambda})$ model

$$z(t) = -\sum_{j=1}^{k}\sum_{l=0}^{\lambda} \phi_{jl} W^{(l)} z(t-j) + \varepsilon(t)$$

by $[W^{(h)} z(t-s)]'$, $s > 0$, taking expectations of both sides and dividing by N, the following equations are obtained:

$$\gamma_{h0}(s) = -\sum_{j=1}^{k}\sum_{l=1}^{\lambda} \phi_{jl}\, \gamma_{hl}(s-j) \qquad s = 1,2,..,k; \quad h = 0,1,...,\lambda$$

which are the analogues of the Yule-Walker equations. The last coefficient $\phi_{k\lambda}$, obtained from solving the system of equations for $\lambda = 0, 1, ...$ *and* $k = 1, 2, ...$ is called the space-time partial autocorrelation function of spatial order λ.

As in the case of univariate time series, STARMA processes are characterized by distinct space-time partial and autocorrelation functions. The relationship between the theoretical space-time partial and autocorrelation functions and the three subclasses of the STARMA model family is summarized in Table 1.

TABLE 1

Characteristics of the theoretical space-time autocorrelation and partial autocorrelation functions for STAR, STMA and STARMA models ([8]).

Process	Space-time Autocorrelation function	Space-time Partial Autocorrelation function
STAR	tails off with both space and time	cuts off after p lags in time and λ_p lags in space
STMA	cuts off after q lags in time and m_q lags in space	tails off with both space and time
STARMA	tails off	tails off

2.1.2. Estimation. The maximum likelihood estimates of

$$\Phi = [\phi_{10}, \phi_{11}, ..., \phi_{1\lambda_1}, ..., \phi_{p0}, \phi_{p1}, ..., \phi_{p\lambda_p}]'$$

and

$$(2.2) \qquad \Theta = [\theta_{10}, \theta_{11}, ..., \theta_{1\lambda_1}, ..., \theta_{q0}, \theta_{q1}, ..., \theta_{p\lambda_q}]'$$

rely on the assumption that the errors ε are normally distributed with mean zero and variance-covariance matrix equal to $\sigma^2 I_N$.

The likelihood function is:

$$f(\varepsilon \mid \Phi, \Theta, \sigma^2) = (2\pi)^{-\frac{TN}{2}} \left|\sigma^2 I_{NT}\right|^{-\frac{1}{2}} \exp\left\{-\frac{1}{2\sigma^2}\varepsilon' I \varepsilon\right\}$$

$$= (2\pi)^{-\frac{TN}{2}} (\sigma^2)^{-\frac{TN}{2}} \exp\left\{-\frac{S(\Phi, \Theta)}{2\sigma^2}\right\}$$

where

$$S(\Phi, \Theta) = \varepsilon' I \varepsilon = \sum_{i=1}^{N}\sum_{t=1}^{T} \varepsilon_i^2(t)$$

is the sum of squares of the errors and

$$\varepsilon' = [\varepsilon_1(1), ..., \varepsilon_1(T), ..., \varepsilon_N(1), ..., \varepsilon_N(T)] \ .$$

Finding the values of the parameters that maximize the likelihood function is equivalent to finding the values of Φ and Θ that minimize the sum of squares $S(\Phi, \Theta)$. Therefore, the problem is reduced to finding the least squares estimates of Φ and Θ.

The errors $\varepsilon(t)$ need to be recursively calculated using the equation:

$$\varepsilon(t) = z(t) + \sum_{k=1}^{p}\sum_{l=0}^{\lambda_k} \phi_{kl} W^{(l)} z(t-k) - \sum_{k=1}^{q}\sum_{l=0}^{m_k} \theta_{kl} W^{(l)} \varepsilon(t-k)$$

for $t = 1, ..., T$ and for given values of the parameters (Φ, Θ).

Because the values of the observations z and of the errors ε are unknown for times previous to time 1, these initial values need to be calculated.

Thus, for any given choice of the parameters (Φ, Θ) and starting values (z_*, ε_*) the set of values $\varepsilon(\Phi, \Theta \mid z_*, \varepsilon_*, W)$ could be calculated successively given a particular data set z. The log likelihood associated with the parameter values (Φ, Θ, σ^2) conditional on the choice of (z_*, ε_*) would be:

$$l_*(\Phi, \Theta, \sigma^2) = -\frac{TN}{2}\ln(2\pi) - \frac{TN}{2}\sigma^2 - \frac{S_*(\Phi, \Theta)}{2\sigma^2} \ .$$

So for fixed σ^2, the conditional maximum likelihood estimates of Φ, Θ are the conditional least squares estimates obtained by finding the values of Φ, Θ that minimize the conditional sum of squares function

$$(2.3) \qquad S_*(\Phi, \Theta) = \sum_{i=1}^{N}\sum_{t=1}^{T} \varepsilon_i(t)^2 \ .$$

A sufficient approximation to the unconditional likelihood is obtained by using the conditional likelihood with suitable values substituted for the elements of z_* and ε_*. One procedure is to set the elements of z_* and

ε_* equal to their unconditional expectations for all values of $z(t)$ and $\varepsilon(t)$ with $t < 1$, as suggested in [8]. The unconditional expectations of the elements of ε_* are zero and if the model contains no deterministic part, and in particular if $\mu = 0$ the unconditional expectations of the elements of z_* will also be zero.

Another reliable approximation is to calculate the ε's from ε_{p+1} onwards, setting previous ε's equal to zero. Thus, actually occurring values are used for the z's throughout. This method implies a slight loss of information but for a long series it is negligible.

The conditional maximum likelihood estimators of (Φ, Θ, σ^2) are the values $(\hat{\Phi}, \hat{\Theta})$ that minimize $S_*(\hat{\Phi}, \hat{\Theta})$ and the estimate of σ^2 is $\hat{\sigma}^2 = \frac{S_*(\hat{\Phi}, \hat{\Theta})}{TN}$.

Conditional maximum likelihood estimation of parameters is considered assuming that the orders of the model, $p, q, \lambda_p, \lambda_q$ are known, but the parameters (Φ, Θ, σ^2) are unknown. Also it is assumed that the STARMA model is stationary and invertible.

Because of the nonlinear nature of the procedure when STMA terms are included in the model explicit expressions for the maximum likelihood cannot be derived and numerical techniques have to be used to minimize $S_*(\Phi, \Theta)$.

2.1.3. Initial estimation of the parameters. Before attempting to effect an iterative non-linear optimization of the likelihood function it is important to obtain good initial estimates. The estimation procedure will be divided in two stages: the first stage being the estimation procedure for finding good initial estimates and the second stage being the efficient estimation of the parameters. Order determination is also included in the procedure.

The following approach is an extension of the Hannan and Rissanen [5] procedure for initial estimation of the parameters of an univariate ARMA model.

First a high order $STAR(k_{\lambda_1, \lambda_2, ..., \lambda_k})$ model is fitted to the data estimating the space-time autoregressive coefficients ϕ_{jl} through the Yule-Walker equations. A restrictive assumption that is now used is that $\lambda_1 = \lambda_2 = ... = \lambda_p = 1$ which means that the spatial orders are fixed at 1. The coefficients $\Phi = [\phi_{10}, \phi_{11}, \phi_{20}, \phi_{21}, ..., \phi_{k0}, \phi_{k1}]'$ of the $STAR(k_{1,1,...,1})$ are obtained solving the Y-W equations for $\lambda = 1$:

$$\gamma_{h0}(s) = -\sum_{j=1}^{k}\sum_{l=1}^{\lambda} \phi_{jl}\, \gamma_{hl}(s-j) \qquad s = 1, 2, .., k; \quad h = 0, 1, ..., \lambda .$$

The BIC ([1]) criterion for order determination for multivariate models takes the form $\ln(|\widetilde{G}|) + m\frac{\ln T}{T}$ (where m is the number of parameters in a multivariate AR model and $\widetilde{G}$ is the estimate of the residual covariance matrix G). This criterion can be used for the determination of the order

k of the STAR model if the spatial order is supposed to be fixed at 1. The modified criterion would be $NT \ln(|\tilde{\sigma}^2|) + 2k \ln T$, where $\tilde{\sigma}^2$ is the corresponding estimate of the residual variance and $2k$ is the number of AR parameters in a $STAR(k_{1,...,1})$.

Once the vector parameter Φ of the $STAR(k_{1,1,...,1})$ model is determined for k large, the residuals are calculated from:

$$\varepsilon(t) = z(t) + \sum_{j=1}^{k} \sum_{l=0}^{\lambda_j} \phi_{jl} W^{(l)} z(t-j) \quad t \geq k+1 \ .$$

So

$$\varepsilon(t) = z(t) + [\phi_{10} z(t-1) + \phi_{11} W z(t-1) + \phi_{20} z(t-2)$$
$$+ \phi_{21} W z(t-2) + \ldots + \phi_{k0} z(t-k) + \phi_{k1} W z(t-k)]$$
$$= z(t) - [z(t-1) \ W z(t-1) \ z(t-2) \ W z(t-2) \ldots z(t-k) \ W z(t-k)]$$
$$= \begin{bmatrix} \phi_{10} \\ \phi_{11} \\ \phi_{20} \\ \phi_{21} \\ \vdots \\ \phi_{k0} \\ \phi_{k1} \end{bmatrix}$$

where each of the z's has dimension $N \times 1$ and $t \geq k+1$.

Now the model can be written in a general linear form as: $Y = X\beta + \varepsilon$ or

$$z(t) = - \sum_{k=1}^{p} \sum_{l=0}^{\lambda_k} \phi_{kl} W^{(l)} z(t-k) + \sum_{k=1}^{q} \sum_{l=0}^{m_k} \theta_{kl} W^{(l)} \varepsilon(t-k) + \varepsilon(t)$$

where $m \leq t \leq T$ and $m = \max(k+p+1, k+q+1)$.

Our vector of parameters to be estimated is now:

$$\beta = [\phi_{10}, \phi_{11}, \phi_{20}, \phi_{21}, ..., \phi_{p0}, \phi_{p1}, \theta_{10}, \theta_{11}, ..., \theta_{q0}, \theta_{q1}]' \ .$$

The present problem is treated henceforth as a problem of general linear regression where the regressor variables are

$$x = \big[- z(t-m) \ - W z(t-m)... - z(t-p)$$
$$- W z(t-p)\varepsilon(t-m) \ W \varepsilon(t-m)...\varepsilon(t-q) \ W \varepsilon(t-q) \big] \ .$$

The least squares normal equations are

$$(X'X)\beta = X'Y$$

and, assuming that $X'X$ is nonsingular, $\hat{\beta} = (X'X)^{-1}X'Y$.

Then, the initial estimates of the parameters for the initial model are taken as the parameter values $(\widetilde{\Phi}, \widetilde{\Theta})$ for which the pair $(\tilde{p}, \tilde{q})$ minimizes

$$NT \ln(\tilde{\sigma}^2) + 2(p+q)\ln T$$

where $\tilde{\sigma}^2$ is the corresponding estimate of the residual variance σ^2 and p and q are the AR and MA orders (respectively) of the model.

Once the order of the model $(\tilde{p}, \tilde{q})$ has been determined, the strongly consistent estimates $(\widetilde{\Phi}, \widetilde{\Theta})$ can be used to initiate any efficient estimation procedure. The procedure chosen for efficient estimation ([8]) is the Marquardt algorithm ([7]). This algorithm is implemented in a routine of Matlab and can be used through the function 'fminsearch' with one of the options set for the routine to choose the Marquardt procedure.

2.1.4. Confidence intervals for the parameters. The sum of squares can be expanded in Taylor Series

$$S(\Phi, \Theta) = S(\delta) \approx S(\hat{\delta}) + (\delta - \hat{\delta})'Q(\delta - \hat{\delta})$$

where

$$\delta' = (\Phi', \Theta')$$

and

$$Q = \frac{1}{2}\left[\frac{\partial^2 S(\delta)}{\partial \delta_i \partial \delta_j}\right]_{\hat{\delta}}$$

for $i = 1, 2, ..., K, j = 1, 2, ..., K; K$ is the dimension of δ, or the total number of parameters.

Since

$$S(\delta) = \sum_{t=1}^{T} \varepsilon(t)'\varepsilon(t)$$

$$(2.4) \qquad \frac{\partial S(\delta)}{\partial \delta_i} = \sum_{t=1}^{TN} 2\varepsilon(t)'\frac{\partial \varepsilon(t)}{\partial \delta_i}|_{\hat{\delta}} = 0$$

$$\frac{1}{2}\left[\frac{\partial^2 S(\delta)}{\partial \delta_i \partial \delta_j}\right]_{\hat{\delta}} = \sum_{t=1}^{T} \varepsilon(t)'\frac{\partial^2 \varepsilon(t)}{\partial \delta_i \partial \delta_j}|_{\hat{\delta}} + \sum_{t=1}^{T} \frac{\partial \varepsilon(t)'}{\partial \delta_i}\frac{\partial \varepsilon(t)}{\partial \delta_j}|_{\hat{\delta}}.$$

Because $\frac{\partial^2 \varepsilon(t)}{\partial \delta_i \partial \delta_j}|_{\hat{\delta}}$ is a function of $\varepsilon(t)$ occurring before time t and since it is expected that if the model fits, $E\left[\varepsilon(t)\varepsilon(t-k)'\right] = 0$ for $k \geq 1$, the first term in 2.4 can be neglected. The matrix Q can be written as

$$Q = X'X$$

where

$$X = \begin{bmatrix} \frac{\partial \varepsilon(1)}{\partial \delta_1}\big|_{\hat{\delta}} & \frac{\partial \varepsilon(1)}{\partial \delta_2}\big|_{\hat{\delta}} & \cdots & \frac{\partial \varepsilon(1)}{\partial \delta_k}\big|_{\hat{\delta}} \\ \frac{\partial \varepsilon(2)}{\partial \delta_1}\big|_{\hat{\delta}} & \frac{\partial \varepsilon(2)}{\partial \delta_2}\big|_{\hat{\delta}} & \cdots & \\ & & \vdots & \\ \frac{\partial \varepsilon(T)}{\partial \delta_1}\big|_{\hat{\delta}} & \frac{\partial \varepsilon(T)}{\partial \delta_2}\big|_{\hat{\delta}} & \cdots & \frac{\partial \varepsilon(T)}{\partial \delta_k}\big|_{\hat{\delta}} \end{bmatrix}$$

Thus the sum of squares is approximated by

$$S(\delta) = S(\hat{\delta}) + (\delta - \hat{\delta})' Q (\delta - \hat{\delta})$$

and an approximate $100(1 - \alpha)\%$ confidence region for $[\Phi, \Theta]' = \delta$ is obtained via ([8]):

$$S(\delta) = S(\hat{\delta}) + \frac{K}{TN - K} S(\hat{\delta}) F_{K, \, TN-K, \, \alpha}$$

where $F_{K, \, TN-K, \, \alpha}$ is the percentage point of the F-distribution with K and $TN - K$ degrees of freedom. The matrix Q must be numerically estimated and $S(\delta)$ should be replaced by the conditional sum of squares $S_*(\delta)$ when the conditional maximum likelihood is used.

Confidence intervals for σ^2 are calculated using:

$$\frac{S_*(\delta)}{(\sigma^2 | z(1), z(2), ..., z(T))} \sim \chi^2_{TN-K} \cdot$$

2.1.5. Diagnostic-Checking. At this stage the objective is to determine if the model does adequately represent the data. If the fitted model adequately represents the data, the residuals should be gaussian white noise, i.e., should be distributed normally with mean zero and variance-covariance matrix equal to $\sigma^2 I_N$. One way of testing for correlation is to calculate the sample space-time autocorrelations of the residuals and check for additional significant structure.

If the model is adequate then ([12])

$$var(\hat{\rho}_{l0}(s)) \approx \frac{1}{N\,(T - s)}$$

where $\hat{\rho}_{l0}(s)$ is the space-time autocorrelation function of the residuals of the fitted model. Thus, the residual space-time autocorrelations, since they are approximately normal, can be standardized and checked for significance. If the residuals are not independent the pattern is identified and the tentative model updated.

Another assumption is that

$$E\left[\varepsilon(t)\,\varepsilon(t+s)'\right] = \begin{cases} \sigma^2 I_N & s = 0 \\ 0 & s \neq 0 \end{cases}.$$

This should be tested ([10]) and if there is evidence indicating that the assumption is not met, a different model should be used and consider the more general form of the variance-covariance matrix of the innovations ([4]).

The estimated parameters can be tested for statistical significance in two ways: Use the confidence regions for the parameters to test the hypothesis that $\Phi = \Theta = 0$ or test the hypothesis that a particular ϕ_{kl} or θ_{kl} is zero with the remaining parameters unrestricted.

Let $\hat{\delta}$ be the least squares estimate of the full parameter vector and $\hat{\delta}^*$ the least squares estimate of the parameter vector with δ_K constrained to zero. The procedure consists of testing the hypothesis $\delta_K = 0$ using the statistic

$$\frac{(TN - K)\left[S_*(\hat{\delta}^*) - S_*(\hat{\delta})\right]}{S_*(\hat{\delta})}$$

which is approximately distributed as an $F_{1, TN-K}$ under the null hypothesis.

Any estimated parameter that proves to be statistically insignificant should be removed from the model and the simpler model should be considered as the candidate model and the estimation stage should be repeated.

2.1.6. Modelling procedure: case $G \neq \sigma^2 I_N$. For the case when the variance-covariance matrix G is not equal to $\sigma^2 I_N$ (the assumption of sphericity is not valid), Deutsch and Pfeifer [4] describe the procedure for building STARMA models as well as the tests developed to test the hypotheses about the form of G. Pfeifer and Deutsch [10] give an extensive explanation of such tests as well as a table with critical values.

3. Weighting matrices. The specification of the hierarchical ordering of neighbors of each site and the selection of an appropriate sequence of weighting matrices is a matter left to the model builder. In many cases the space structure is assumed to form a regularly spaced system in which the sampling points lie on a regular lattice. In the majority of the applications this is only a simplifying assumption since typically the sites are irregularly spaced. The Diagram pictured below shows some spatial order neighbors of a particular site for a two-dimensional grid system. This definition of spatial order represents an ordering in terms of Euclidean distance of all sites surrounding a location of interest.

The weighting matrices adopted as most appropriate for this case are

$$w_{ij}^{(l)} = \begin{cases} 1/n_i^{(l)} & i \text{ and } j \ l^{th} \text{ order neighbors} \\ 0 & otherwise \end{cases}$$

where $n_i^{(l)}$ is the number of the lth order neighbors. In other words, each site is assigned first order neighbors, second order neighbors and so on.

Each of the lth order neighbors is assigned the same weight in relation to that specific site.

$$
\begin{array}{ccccc}
\cdot & 4 & 3 & 4 & \cdot \\
4 & 2 & 1 & 2 & 4 \\
3 & 1 & 0 & 1 & 3 \\
4 & 2 & 1 & 2 & 4 \\
\cdot & 4 & 3 & 4 & \cdot
\end{array}
$$

Diagram. Spatial order in two-dimensional systems.

An example of the use of this approach for defining weighting matrices can be found in ([8]) in their illustration of the space-time procedure in modelling Boston assault arrests.

Rather than define a hierarchical system of neighbors, Pfeifer and Bodily [13] decided to use the driving distances between each location to define a single weighting matrix. Their STARMA model class considers every site a first order neighbor of every other site and uses weights that are inversely proportional to the driving distances. They have chosen this approach for reasons of simplicity and because of an *a priori* belief that most of the benefit of incorporating spatial factors can be captured with a single first-order term.

Another approach is suggested by the analysis of wind speeds by Haslett and Raftery [6]. The authors found the correlation between wind speeds at different places to be strongly related to the distance between them, and suggested that the covariance structure could be reasonably well approximated by the relations:

$$
cov(X_{it}, X_{jt}) = \sigma_X^2 r_{ij}
$$

where

$$
r_{ij} = \left\{
\begin{array}{ll}
1 & if \;\; i = j \\
\alpha \exp(-\beta d_{ij}) & if \;\; i \neq j
\end{array}
\right.
$$

with $0 \leq \alpha \leq 1, \beta \geq 0$ and d_{ij} is the distance between places i and j.

After estimating the parameters α and β this covariance structure could be used to define a weighting matrix for the space-time model. This structure has the advantage that the data can be used to define the weighting matrix for the STARMA model instead of considering predefined weighting matrices.

4. Simulations. The estimation procedures were implemented in MATLAB and tested on simulated data. In this section the results of some simulations are presented. All the simulations reported were performed using pseudo-normal random numbers. The simulations are designed to show how the procedures for order selection, initial estimation and efficient maximum likelihood estimation perform. The data was simulated from a system of nine sites distributed spatially on a regular grid with weighting matrix (for maximum spatial order 1):

$$W = \begin{bmatrix}
0 & \frac{1}{2} & 0 & \frac{1}{2} & 0 & 0 & 0 & 0 & 0 \\
\frac{1}{3} & 0 & \frac{1}{3} & 0 & \frac{1}{3} & 0 & 0 & 0 & 0 \\
0 & \frac{1}{2} & 0 & 0 & 0 & \frac{1}{2} & 0 & 0 & 0 \\
\frac{1}{3} & 0 & 0 & 0 & \frac{1}{3} & 0 & \frac{1}{3} & 0 & 0 \\
0 & \frac{1}{4} & 0 & \frac{1}{4} & 0 & \frac{1}{4} & 0 & \frac{1}{4} & 0 \\
0 & 0 & \frac{1}{3} & 0 & \frac{1}{3} & 0 & 0 & \frac{1}{3} & 0 \\
0 & 0 & 0 & \frac{1}{2} & 0 & 0 & 0 & \frac{1}{2} & 0 \\
0 & 0 & 0 & 0 & \frac{1}{3} & 0 & \frac{1}{3} & 0 & \frac{1}{3} \\
0 & 0 & 0 & 0 & 0 & \frac{1}{2} & 0 & \frac{1}{2} & 0
\end{bmatrix}$$

Tables 2, 3, 4 present the results of 100 replications of a number of simulations for varying time points T ($T = 100$ and $T = 200$) and different parameters. Next to each estimated parameter and in brackets are the standard errors of the estimates.

The tables show that the procedure for initial estimation of the parameters give initial estimates which are close to the true values and the efficient final estimation does not improve the results significantly. The order selection and estimation procedures provide better results the greater the true values of the parameters (within the invertibility and causality regions) and the larger the value of T.

TABLE 2

Frequency of correct order selection(F), mean of initial estimates $\tilde{\tilde{\Theta}} = \begin{bmatrix} \tilde{\tilde{\theta}}_{10} & \tilde{\tilde{\theta}}_{11} \end{bmatrix}'$ *and mean of the efficient estimates* $\bar{\tilde{\Theta}} = \begin{bmatrix} \bar{\tilde{\theta}}_{10} & \bar{\tilde{\theta}}_{11} \end{bmatrix}'$ *for simulated data from a* $STMA(1_1)$ *model.*

	$\Theta = [0.1\ 0.1]'$		$\Theta = [0.4\ 0.4]'$	
T	100	200	100	200
F	31	39	97	98
$\tilde{\tilde{\Theta}}$	0.1076(0.0258)	0.1056(0.0203)	0.3889(0.0328)	0.3901(0.0239)
	0.0997(0.0562)	0.0957(0.0449)	0.3717(0.0542)	0.3814(0.0363)
$\bar{\tilde{\Theta}}$	0.1115(0.0270)	0.1084(0.0200)	0.3978(0.0297)	0.3979(0.0206)
	0.1040(0.0598)	0.0999(0.0457)	0.3898(0.0451)	0.3990(0.0311)

5. Temperature data. The data consists of monthly mean temperatures, recorded in Celsius scale, at nine meteorological stations around the United Kingdom. The data source is the NOAA NCDC GCPS which was accessed through the website of the LDEO/IRI Data Library found in `http://rainbow.ldeo.columbia.edu/`.

There are 223 observations available for the nine sites from January 1951 through to August 1969. The stations are located in the central region of Great Britain from $0.3999939^\circ W$ to $3.100006^\circ W$ and from $53.03^\circ N$ to

TABLE 3

Frequency of correct order selection(F), mean of initial estimates $\tilde{\tilde{\Phi}} = \begin{bmatrix} \tilde{\bar{\phi}}_{10} & \tilde{\bar{\phi}}_{11} \end{bmatrix}'$ and mean of the efficient estimates $\bar{\tilde{\Phi}} = \begin{bmatrix} \bar{\tilde{\phi}}_{10} & \bar{\tilde{\phi}}_{11} \end{bmatrix}'$ for simulated data from a $STAR(1_1)$ model.

	$\Phi = [0.1\ 0.1]'$		$\Phi = [0.4\ 0.4]'$	
T	100	200	100	200
F	41	38	98	100
$\tilde{\tilde{\Phi}}$	0.1243(0.0235)	0.1041(0.0169)	0.3944(0.0281)	0.4014(0.0201)
	0.1241(0.0445)	0.1065(0.0315)	0.4036(0.0426)	0.3950(0.0282)
$\bar{\tilde{\Phi}}$	0.1242(0.0236)	0.1039(0.0166)	0.3947(0.0283)	0.4014(0.0201)
	0.1243(0.0435)	0.1070(0.0307)	0.4027(0.0423)	0.3947(0.0283)

TABLE 4

Frequency of correct order selection(F), mean of initial estimates $\left(\tilde{\tilde{\Theta}}, \tilde{\tilde{\Phi}}\right) = \left[\begin{bmatrix} \tilde{\bar{\theta}}_{10} & \tilde{\bar{\phi}}_{10} \end{bmatrix}' \begin{bmatrix} \tilde{\bar{\theta}}_{11} & \tilde{\bar{\phi}}_{11} \end{bmatrix}'\right]$ and mean of the efficient estimates $\left(\bar{\tilde{\Theta}}, \bar{\tilde{\Phi}}\right) = \left[\begin{bmatrix} \bar{\tilde{\theta}}_{10} & \bar{\tilde{\phi}}_{10} \end{bmatrix}' \begin{bmatrix} \bar{\tilde{\theta}}_{11} & \bar{\tilde{\phi}}_{11} \end{bmatrix}'\right]$ for simulated data from a $STARMA(1_1, 1_1)$ model.

	$(\Phi, \Theta) = \begin{bmatrix} 0.2 & -0.2 \\ 0.2 & -0.2 \end{bmatrix}$		$(\Phi, \Theta) = \begin{bmatrix} 0.4 & -0.4 \\ 0.4 & -0.4 \end{bmatrix}$	
T	100	200	100	200
F	46	73	98	96
$\tilde{\tilde{\Phi}}$	0.2289(0.1160)	0.2069(0.0824)	0.3898(0.0548)	0.3847(0.0411)
	0.1694(0.1867)	0.2015(0.1221)	0.4301(0.0680)	0.4341(0.0552)
$-\tilde{\tilde{\Theta}}$	0.1737(0.1250)	0.1892(0.0804)	0.3905(0.0615)	0.4014(0.0371)
	0.2369(0.1857)	0.1984(0.1303)	0.3380(0.0793)	0.3354(0.0595)
$\bar{\tilde{\Phi}}$	0.2325(0.1159)	0.2086(0.0904)	0.3944(0.0639)	0.3921(0.0468)
	0.1458(0.1659)	0.1851(0.1302)	0.4091(0.0755)	0.4081(0.0624)
$-\bar{\tilde{\Theta}}$	0.1723(0.1253)	0.1888(0.0886)	0.3955(0.0635)	0.4055(0.0436)
	0.2736(0.1713)	0.2189(0.1383)	0.3779(0.0853)	0.3807(0.0621)

$55.1°N$. Table 5 lists the sites and their geographical locations. Figure 1 gives plots of the nine time series ordered as in Table 5.

The data will be used as an example to illustrate the modelling procedure for the Space-Time ARMA model. We will also compare the performance of this approach with the performance of separate univariate ARMA models. The comparison will be in terms of fitting and forecasting. Our belief is that the STARMA model would give a better fit to the data as well as providing better forecasts than the separate univariate models since the former incorporates information on all sites simultaneously. Forecasts for each station obtained using past information from a particular site as well as from the neighboring sites should be better than forecasts based on past information on that single site.

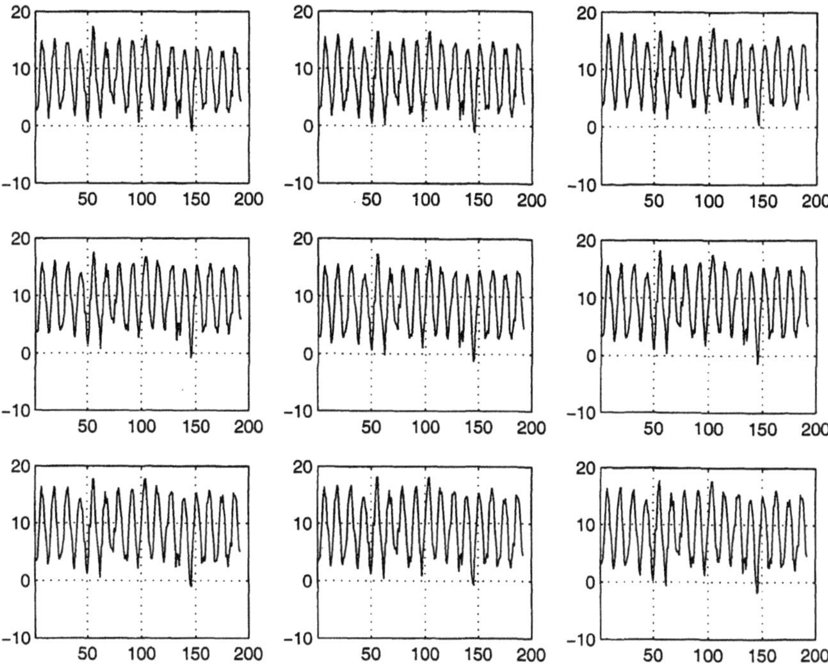

FIG. 1. *Original series of monthly mean temperatures in nine stations in central England.*

TABLE 5
List of sites and their locations.

Station	Longitude	Latitude
Dumfries	3.100006W	55.1N
Durham	1.600006W	54.8N
Scarborough	0.3999939W	54.2N
Stonyhurst	2.5W	53.8N
Bidston	2.899994W	53.4N
Manchester Airport	2.269989W	53.35N
Sheffield	1.500000W	53.40N
York	1.100006W	53.9N
Waddington	0.519989W	53.17N

The first 192 observations are used for model-building purposes using the above two approaches, namely the STARMA modelling approach and the Univariate approach to build an univariate ARMA model to each of the nine stations. The last 31 observations are used for the comparison of forecasts.

5.1. Fitting a STARMA Model. Prior to modelling the data with a STARMA model we need to define a matrix W of weights. We have decided to consider a weighting matrix formed by weights that are inversely proportional to the euclidean distance between sites. That is, the weighting matrix W is first determined such that

$$w_{ij} = \begin{cases} 1/d_{ij} & i \neq j \\ 0 & i = j \end{cases}.$$

The elements are scaled such that

$$\sum_{j=1}^{N} w_{ij} = 1$$

for each i. The euclidean distance(in miles) between the sites is determined using the expression:

$$d_{ij} = 3963\, acos(\sin(lat_i/57.2958)\,\sin(lat_j/57.2958)$$
$$+ \cos(lat_i/57.2958)\cos(lat_j/57.2958)\cos(lon_j/57.2958 - lon_i/57.2958))$$

for $i, j = 1, ..., 9$, where lat_i and lon_i are the latitude and longitude of site i, respectively (www.meridianworlddata.com/Distance-Calculation.asp).

The resulting W matrix is:

$$W = \begin{bmatrix}
0 & .2192 & .1109 & .1173 & .1485 & .1101 & .0928 & .1197 & .0816 \\
.1674 & 0 & .1661 & .0958 & .1352 & .1016 & .0852 & .1614 & .0873 \\
.0759 & .1488 & 0 & .0815 & .1054 & .0981 & .0890 & .2688 & .1325 \\
.0601 & .0642 & .0610 & 0 & .2201 & .2701 & .1665 & .0869 & .0710 \\
.0753 & .0898 & .0782 & .2180 & 0 & .2156 & .1252 & .1219 & .0759 \\
.0470 & .0568 & .0612 & .2252 & .1816 & 0 & .2514 & .0964 & .0804 \\
.0489 & .0588 & .0686 & .1714 & .1302 & .3103 & 0 & .1006 & .1113 \\
.0666 & .1175 & .2185 & .0944 & .1336 & .1256 & .1061 & 0 & .1377 \\
.0616 & .0863 & .1462 & .1047 & .1130 & .1421 & .1593 & .1869 & 0
\end{bmatrix}$$

Having a single weighting matrix means that each site is a neighbor of first order of every other site so that a maximum spatial order of one is defined for the spatial system. Therefore the STARMA model will have maximum spatial order one for each of the autoregressive and moving average terms. By defining the weighting matrix in this way the model is considerably simplified.

From the plots of the series in Figure 1 and from the space-time autocorrelation function in Figure 2 there is evidence of seasonal variability with period 12 which is similar across the sites. The slow decaying of the space-time autocorrelation function at lags which are multiples of 12 suggest that seasonal differencing is necessary in order to remove seasonal nonstationarity. By seasonally differencing the data the differenced series

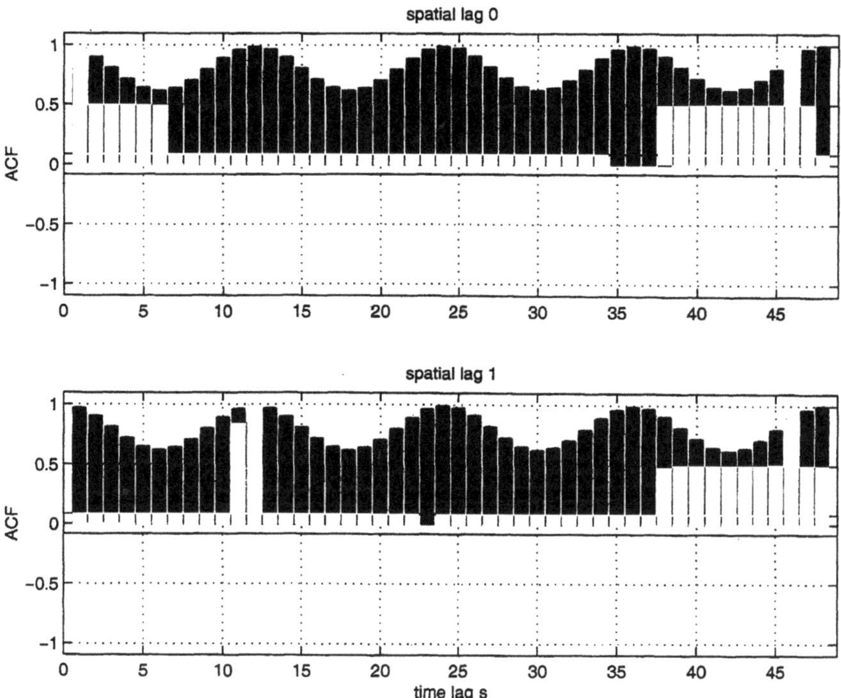

FIG. 2. *Space-time autocorrelation function of the original series in Figure 1.*

TABLE 6
Sample moments of the original series.

Sites	Mean	Variance	Skewness	Kurtosis
site1	8.4479	19.2300	-0.0335	1.7785
site2	8.4396	20.5756	-0.0403	1.7649
site3	9.4266	19.1154	-0.0001	1.7019
site4	9.5427	19.1632	-0.0755	1.8045
site5	8.7497	20.6379	-0.0683	1.7772
site6	9.3687	21.5054	-0.0643	1.7993
site7	9.3875	21.9117	-0.0347	1.7748
site8	9.3937	23.1728	-0.0400	1.7544
site9	9.0786	23.5080	-0.0645	1.8105

plotted in Figure 3 are obtained. Denoting by $x(t)$ the original data at time t then the series that will be modelled is $z(t) = x(t) - x(t - 12)$, a resulting series vector of 180 observations. The assumption of normality of the series is confirmed by tests performed on the skewness and kurtosis measures displayed in Table 6. Details of such tests can be found in [3].

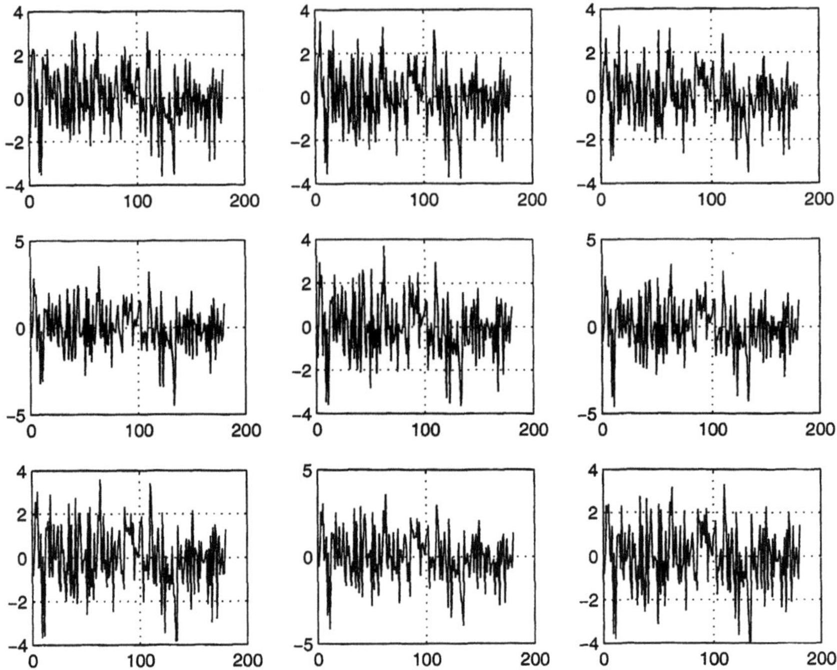

FIG. 3. *Differenced series.*

Inspection of the space-time autocorrelation function(STACF) and space-time partial autocorrelation function(STPACF) (Figure 4 and Figure 5, respectively) suggests that a seasonal STARMA model would be appropriate. The STACF has a significantly large value at time lag 12 cutting off after that and the STPACF values are decreasing in absolute magnitude at lags 12, 24, 36. This indicates the need for a seasonal MA at both spatial lags 0 and 1. The STACF appears to die off across the first three or four lags and there is a significantly large value at lag 1 in the STPACF so a non seasonal AR parameter could be required.

The model suggested by visual inspection of the space-time autocorrelation and partial autocorrelation functions is a Seasonal STARMA $(1_1, 0, 0) \times (0, 1, 1_1)_{12}$ ([11]) of the form:

$$z(t) = -\phi_{10} z(t-1) - \phi_{11} W^{(1)} z(t-1)$$
$$+ \Theta_{10}\varepsilon(t-12) + \Theta_{11} W^{(1)}\varepsilon(t-12) + \varepsilon(t)$$

where $z(t) = x(t) - x(t-12), t = 13, ..., 192$.

For estimating the parameters of the model the estimation procedure described above is applied assuming the errors ε are normally distributed

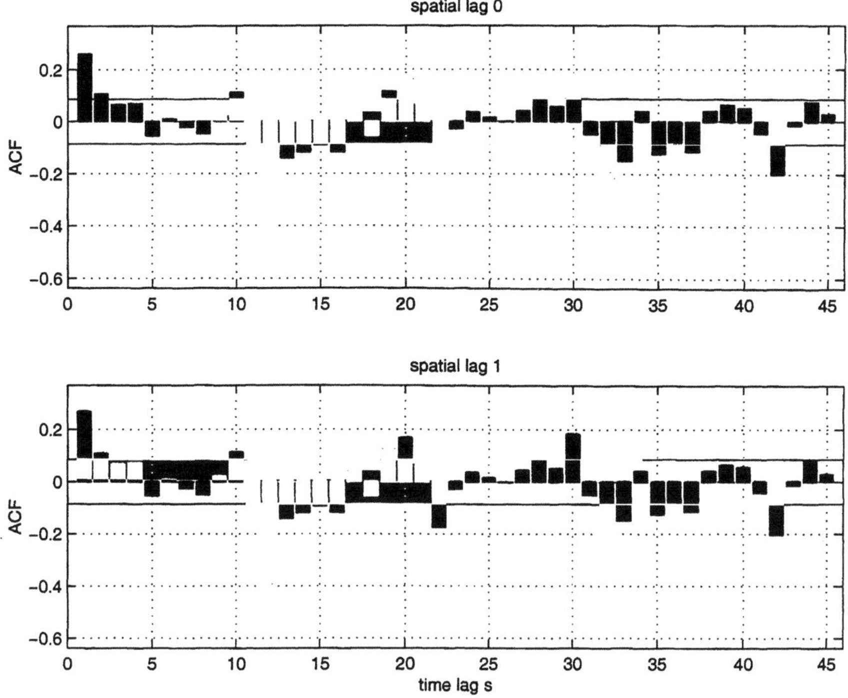

FIG. 4. *Space-time autocorrelation function of differenced series.*

with mean zero and variance-covariance matrix equal to $\sigma^2 I_N$. Using the initial estimation procedure presented above, the initial estimates of the parameters are found to be:

$$\phi_{10} = -0.0022$$
$$\phi_{11} = -0.2633$$
$$\Theta_{10} = -0.5749$$
$$\Theta_{11} = -0.1924 \ .$$

These values are then used to initiate the iterative estimation procedure based on the minimization of the conditional least squares which leads to the following final estimates:

$$\phi_{10} = -0.0024$$
$$\phi_{11} = -0.3113$$
$$\Theta_{10} = -0.7582$$
$$\Theta_{11} = -0.0678 \ .$$

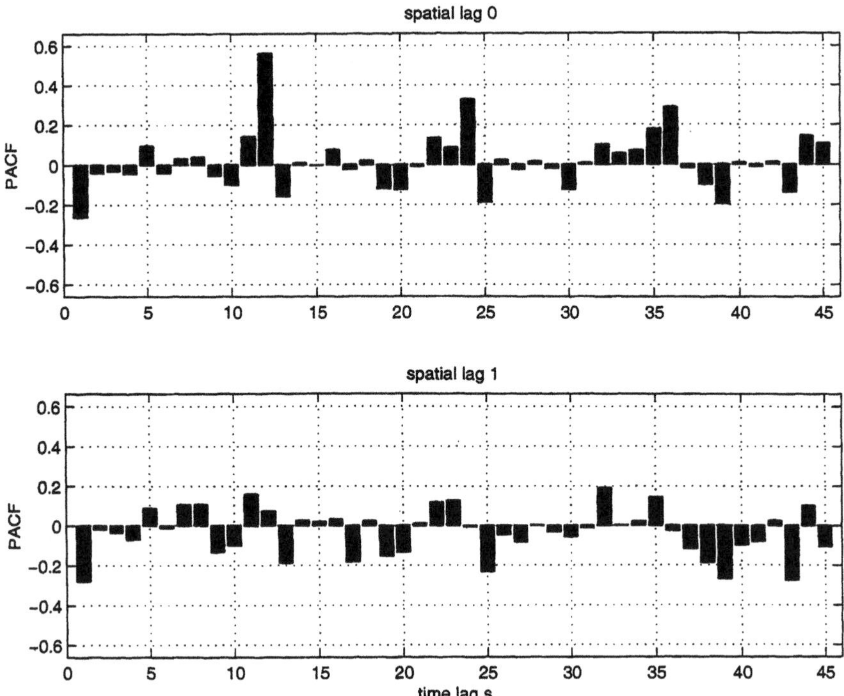

FIG. 5. *Space-time partial autocorrelation function of differenced series.*

A test performed on the form of G, the variance-covariance matrix of the residuals indicates that G is significantly different from a diagonal matrix. The value of $H = 4.1393*10^3$ for the statistic of the test given by the errors of the fitted model, when compared with the critical point of 61.67 from table I in [10], results in the rejection of the hypotheses that $G = D$ (D being a diagonal matrix). Therefore the estimation procedure for a general form of G must be employed.

The transformed space-time autocorrelation and space-time partial autocorrelation functions are calculated based on the knowledge that G is non-diagonal and based on an estimate of G calculated from the data. These functions are displayed in Figures 6 and 7. The inspection of the transformed STACF and STPACF results in an identification of a model identical to the one previously considered for the case $G = \sigma^2 I_N$ in terms of orders. So the model is still a Seasonal STARMA $(1_1, 0, 0) \times (0, 1, 1_1)_{12}$ but the parameters need to be re-estimated. The initial estimates of these parameters are:

$$\phi_{10} = -0.0027$$
$$\phi_{11} = -0.2532$$

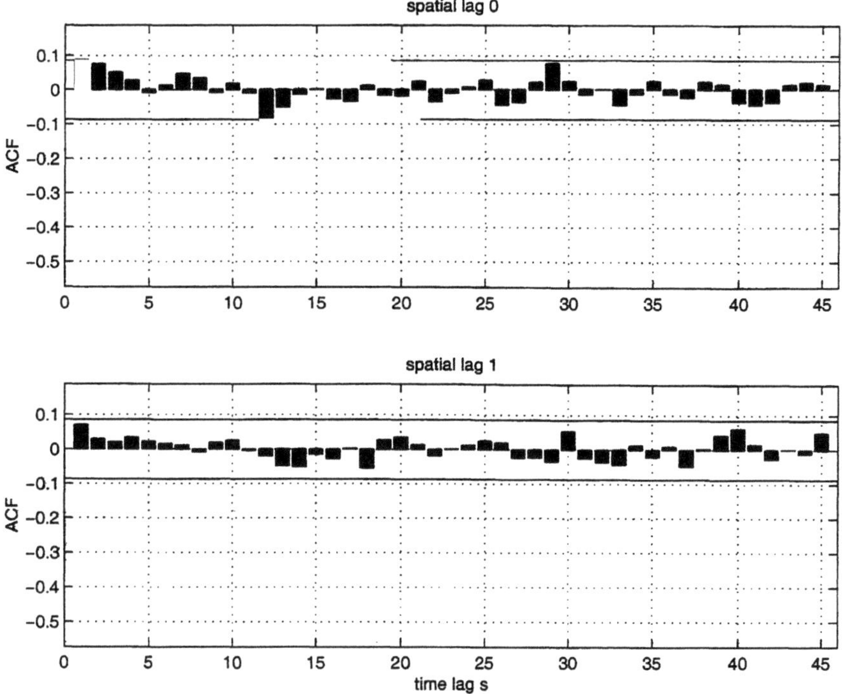

FIG. 6. *Transformed space-time autocorrelation function of differenced series when* $G \neq \sigma^2 I_N$.

$$\Theta_{10} = -0.6325$$
$$\Theta_{11} = -0.1131 \,.$$

These values then used to initiate the iterative estimation procedure lead to the following final estimates:

$$\phi_{10} = -0.1610$$
$$\phi_{11} = -0.1176$$
$$\Theta_{10} = -0.7425$$
$$\Theta_{11} = -0.0656 \,.$$

These final estimates do not differ substantially from the estimates obtained when $G = \sigma^2 I_N$ except for the two space-time AR terms. The transformed space-time autocorrelation function of the residuals, displayed in Figure 8 indicates that the assumption of uncorrelated errors is satisfied by the estimated model.

5.2. Univariate modelling. Now consider univariate ARMA models for each site separately. The nine time series are analyzed and the respective

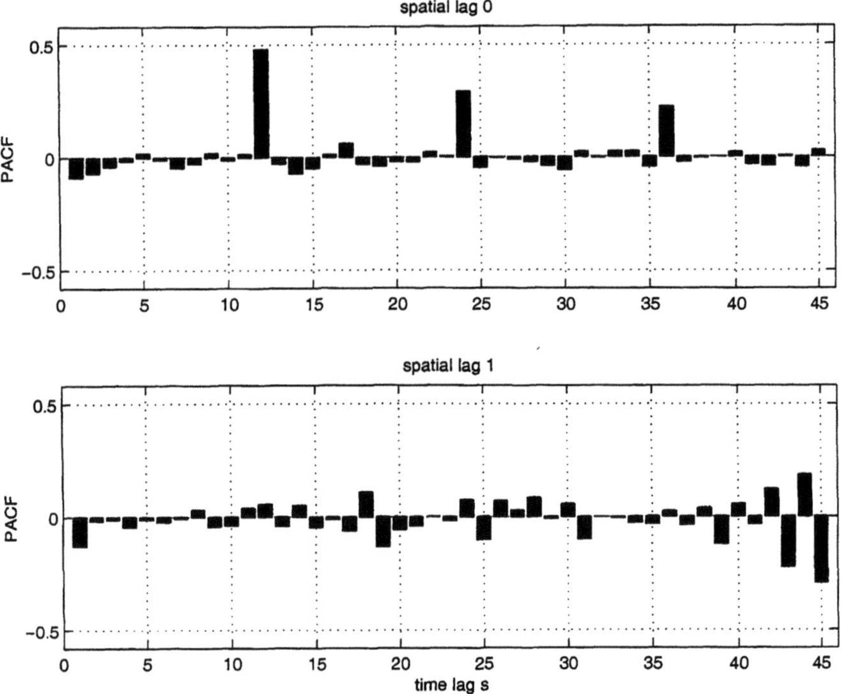

FIG. 7. *Transformed space-time partial autocorrelation function of differenced series when* $G \neq \sigma^2 I_N$.

autocorrelation and partial autocorrelation functions displayed in Figures 9 and 10. The identification stage indicates that a Seasonal ARMA model SARMA$(1,0,0) \times (0,1,1)_{12}$ of the form

$$z_i(t) = -\phi_1 z_i(t-1) + \Theta_1 \varepsilon_i(t-12) + \varepsilon_i(t)$$

$i = 1,..,9$, is the appropriate model to fit to all nine time series. The estimates of the parameters are given in Table 7. The autocorrelation functions of the residuals from these models, shown in Figure 11 indicate that the chosen models adequately represent the respective time series. Table 9 shows the covariance matrix of the residuals from the nine ARMA models and comparison with the covariance matrix of the residuals from the STARMA model (Table 8), the extra parameters in the nine separate models lead to a lower sum of squared residuals for five of the nine sites.

5.3. Comparing the forecasts. The MMSE (minimum mean square error) one step ahead forecasts $\hat{z}_t(1) = E(z_{t+1}|z_s, s \leq t)$ were determined using both the STARMA model and the univariate models. The models were not updated when forecasting so only 192 observations for each site

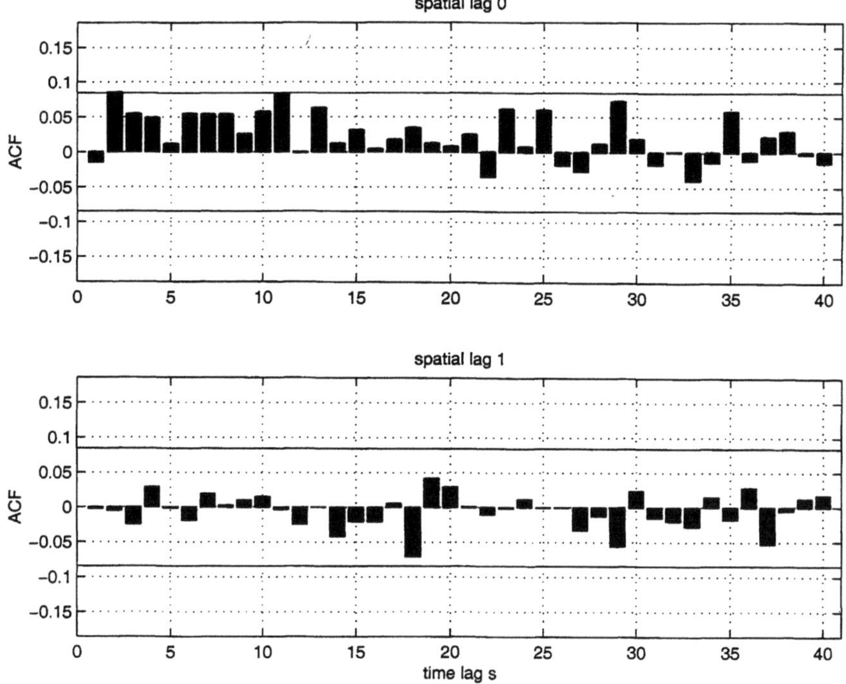

FIG. 8. *Space-time autocorrelation function of the residuals form the fitted seasonal STARMA with $G \neq \sigma^2 I_N$.*

where used for modelling and then the resulting model used to determine 31 one-step ahead forecasts. In order to compare the forecasts from the two models, the covariance matrices of the one-step-ahead forecast errors (obtained using the hold-out sample) are displayed in Tables 10 and 11. The diagonal elements of these matrices are the MSE of the forecast errors. The prediction mean square error is the average squared forecast error computed from [2]:

$$MSE = \sum_{t=T+1}^{T+m} \varepsilon(t)^2 / m$$

where m is the total number of one step ahead forecasts. It is used to compare forecasts from different models. Comparing the diagonals from the two matrices it can be seen that for four of the nine sites the prediction MSE is smaller for the STARMA model forecasts than for the separate ARMA model forecasts. The trace of the matrix in Table 10 ($=16.2554$) is inferior to the trace of the matrix in Table 11 ($=16.3057$), indicating an overall better performance of the STARMA model.

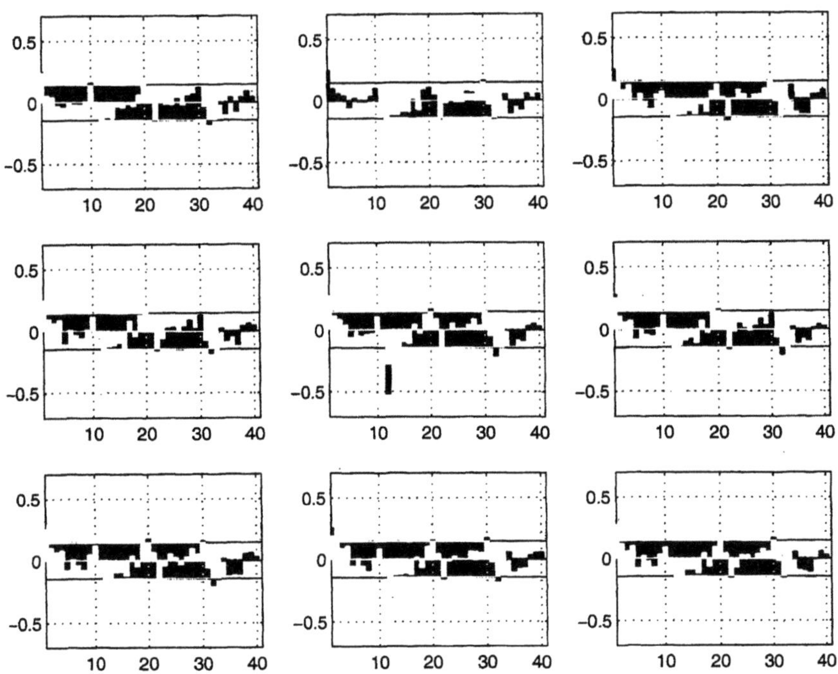

FIG. 9. *Autocorrelation function of each differenced series.*

TABLE 7
Modelling results from fitting separate SARMA models to the nine sites.

site	ϕ_1	Θ_1
1	-0.3111	-0.8157
2	-0.2835	-0.8212
3	-0.2851	-0.8192
4	-0.3156	-0.8372
5	-0.3097	-0.8338
6	-0.3299	-0.8188
7	-0.3179	-0.8302
8	-0.2730	-0.8122
9	-0.3029	-0.8327

The STARMA model gives slightly better forecasts than the separate ARMA models and it is a parsimonious model, therefore we conclude that the STARMA model is overall a better model.

More general forms of the weighting matrix, including various parameterized functions of the correlation between sites, functions of the variogram

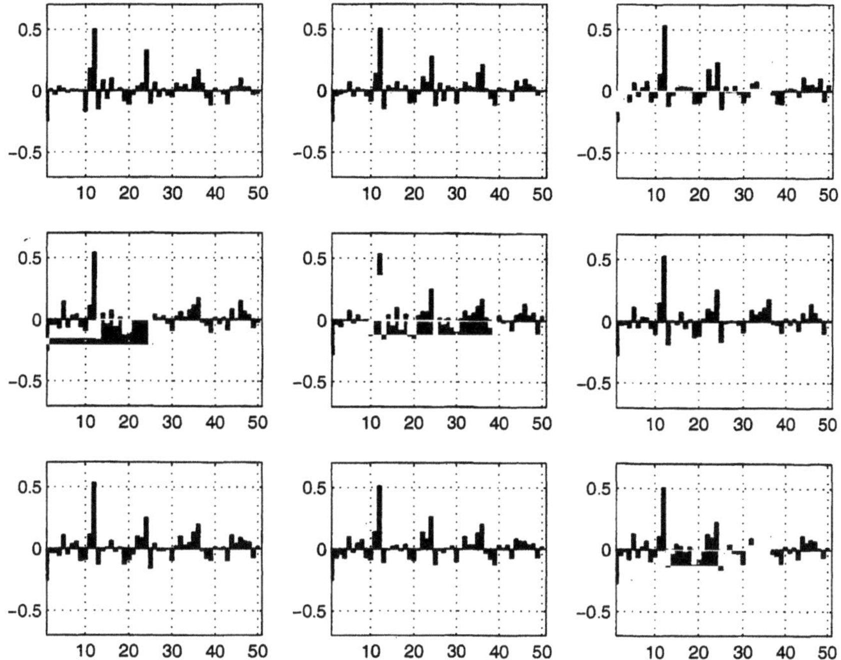

FIG. 10. *Partial autocorrelation function of each differenced series.*

TABLE 8
Covariance matrix of the residuals from the Seasonal STARMA model.

1.8476	1.7409	1.4233	1.6665	1.7708	1.8467	1.7904	1.8099	1.7648
1.7409	1.9214	1.6263	1.7261	1.7947	1.8796	1.9284	1.9398	1.9203
1.4233	1.6263	1.5211	1.4813	1.5035	1.6077	1.6885	1.6869	1.6854
1.6665	1.7261	1.4813	1.8125	1.8014	1.9360	1.8860	1.8488	1.8927
1.7708	1.7947	1.5035	1.8014	1.9083	1.9845	1.9158	1.9071	1.9129
1.8467	1.8796	1.6077	1.9360	1.9845	2.1577	2.0507	2.0254	2.0678
1.7904	1.9284	1.6885	1.8860	1.9158	2.0507	2.0948	2.0374	2.0696
1.8099	1.9398	1.6869	1.8488	1.9071	2.0254	2.0374	2.0964	2.0631
1.7648	1.9203	1.6854	1.8927	1.9129	2.0678	2.0696	2.0631	2.1516

and parameterized functions of the distance between sites have been tried.
However, for this particular data set they resulted in no improvement or
little improvement, in terms of fitting or forecasts, over the results obtained
with the above form of the weights.

6. Discussion. In this paper the STARMA model was described as
well as the estimation procedure developed by Pfeifer and Deutsch [8] to fit

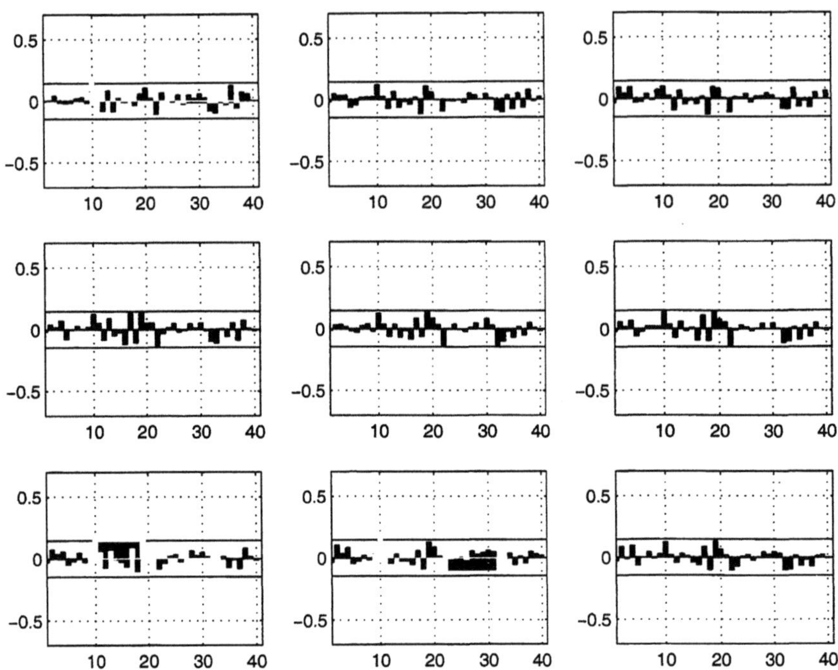

FIG. 11. *Autocorrelation function of the residuals from the SARMA models fitted to each series.*

TABLE 9
Covariance matrix of the residuals from the SARMA model.

1.8542	1.7404	1.4165	1.6645	1.7708	1.8416	1.7805	1.8094	1.7604
1.7404	1.9272	1.6351	1.7220	1.7917	1.8709	1.9247	1.9462	1.9163
1.4165	1.6351	1.5352	1.4754	1.4983	1.5968	1.6890	1.6924	1.6834
1.6645	1.7220	1.4754	1.8049	1.7975	1.9225	1.8768	1.8515	1.8813
1.7708	1.7917	1.4983	1.7975	1.9099	1.9759	1.9054	1.9098	1.9033
1.8416	1.8709	1.5968	1.9225	1.9759	2.1451	2.0355	2.0257	2.0532
1.7805	1.9247	1.6890	1.8768	1.9054	2.0355	2.0859	2.0374	2.0584
1.8094	1.9462	1.6924	1.8515	1.9098	2.0257	2.0374	2.1057	2.0663
1.7604	1.9163	1.6834	1.8813	1.9033	2.0532	2.0584	2.0663	2.1427

the model to a given vector of time series . An approach for obtaining the
initial estimates of the parameters of the STARMA model was developed
and incorporated in the present modelling procedure. The whole procedure was implemented in MATLAB and when tested on simulated data
gave very satisfactory results. After convincing ourselves that the procedure was working satisfactorily, we tested on the real monthly temperature

TABLE 10
Covariance matrix of the 31 forecast errors: STARMA approach.

1.3671	1.5269	1.2872	1.3848	1.1095	1.3767	1.5315	1.4292	1.5477
1.5269	1.9678	1.8137	1.6840	1.3368	1.7469	1.9912	1.8602	2.0217
1.2872	1.8137	1.9837	1.5605	1.2223	1.6746	1.9171	1.7949	1.9791
1.3848	1.6840	1.5605	1.5889	1.2352	1.5918	1.7654	1.6617	1.8403
1.1095	1.3368	1.2223	1.2352	1.1778	1.2665	1.4613	1.3485	1.5038
1.3767	1.7469	1.6746	1.5918	1.2665	1.7320	1.8554	1.7205	1.9393
1.5315	1.9912	1.9171	1.7654	1.4613	1.8554	2.1756	1.9843	2.2010
1.4292	1.8602	1.7949	1.6617	1.3485	1.7205	1.9843	1.9304	2.0357
1.5477	2.0217	1.9791	1.8403	1.5038	1.9393	2.2010	2.0357	2.3321

TABLE 11
Covariance matrix of the 31 forecast errors: Univariate ARMA approach.

1.3808	1.5264	1.2808	1.3744	1.0777	1.3817	1.5233	1.4396	1.5456
1.5264	1.9448	1.7907	1.6701	1.3167	1.7530	1.9778	1.8645	2.0232
1.2808	1.7907	1.9752	1.5554	1.2263	1.6843	1.8971	1.8051	1.9943
1.3744	1.6701	1.5554	1.5845	1.2017	1.6122	1.7654	1.6652	1.8464
1.0777	1.3167	1.2263	1.2017	1.1368	1.2682	1.4379	1.3445	1.4915
1.3817	1.7530	1.6843	1.6122	1.2682	1.7868	1.8791	1.7432	1.9813
1.5233	1.9778	1.8971	1.7654	1.4379	1.8791	2.1747	1.9975	2.2126
1.4396	1.8645	1.8051	1.6652	1.3445	1.7432	1.9975	1.9583	2.0623
1.5456	2.0232	1.9943	1.8464	1.4915	1.9813	2.2126	2.0623	2.3639

data considered in the paper. The STARMA model found to be suitable was the seasonal STARMA $(1_1, 0, 0) \times (0, 1, 1_1)_{12}$. We have compared the results of this fit with separate univariate ARMA models. The univariate ARMA models fitted separately to each time series all have the form SARMA$(1, 0, 0) \times (0, 1, 1)_{12}$. The STARMA model provided a better fit to five of the nine series when compared with the fit from the separate ARMA models. One-step ahead forecasts obtained from the two models and applied to the last 31 points from each of the nine time series were compared and the conclusion is that the STARMA model provided overall better fit to the data as well as providing better forecasts.The comparison was made in terms of mean squared errors of the residuals from the forecasts. The space-time ARMA model fits the nine spatial time series only using only four parameters while separate ARMA models fitted to each site required a total of eighteen parameters. Although the space-time ARMA model is harder to implement and slower to fit to the data than the ARMA model, it provides a much more parsimonious model.

Acknowledgment. We would like to thank the referee for his useful and constructive comments and suggestions.

REFERENCES

[1] AKAIKE H., *Time Series analysis and control through parametric models, in Applied Time Series Analysis* , D.F. Findley (ed.) Academic Press. New York, 1978.

[2] CHATFIELD C., *Time Series Forecasting*, Boca Raton, Chapman & Hall/CRC, 2001.

[3] D'AGOSTINO R.B., STEVENS M.A. EDS, *Goodness-of-fit techniques*, Marcel Dekker, 1986.

[4] DEUTSCH S.J. AND PFEIFER P.E., *Space-time ARMA modelling with contemporaneously correlated innovations*, Technometrics (1981), **23**, 401–9.

[5] HANNAN E.J. AND RISSANEN J., *Recursive estimation of mixed autoregressive-moving average order*, Biometrika (1982), **69**(1), 81–94.

[6] HASLETT J. AND RAFTERY A.E., *Space-time Modelling with Long-memory Dependence: Assessing Ireland's Wind Power Resource*, Appl. Statist. (1989), **38**, 1–50.

[7] MARQUARDT D.W., *An Algorithm for least squares estimation of non-linear parameters*, Jour. Soc. Ind. Appl. Math. (1963), **11**, p. 413.

[8] PFEIFER P.E. AND DEUTSCH S.J., *A three stage interactive procedure for space-time modeling*, Technometrics (1980a), **22**, 35–47.

[9] PFEIFER P.E. AND DEUTSCH S.J., *Identification and interpretation of first order space-time ARMA models*, Technometrics (1980b), **22**, 397–408.

[10] PFEIFER P.E. AND DEUTSCH S.J. , *Independence and sphericity tests for the residuals of space-time ARIMA Models*, Communications in Statistics. B — Simulation and Computation (1980d), **9**(5), 533–549.

[11] PFEIFER P.E. AND DEUTSCH S.J., *Seasonal space-time modelling*, Geographical Analysis (1981a), **13**, 117–33.

[12] PFEIFER P.E. AND DEUTSCH S.J., *Variance of the sample space-time Autocorrelation Function*, J. Roy. Statist. Soc. Ser. B (1981b), **43**(1), 28–33.

[13] PFEIFER P.E. AND BODILY S.E., *A test of Space-time ARMA Modelling and Forecasting of Hotel data*, J. Forecasting (1990), **9**, 255–272.

[14] SMITH R.L., *Spatial Statistics in Environmental Science. In Nonlinear and Nonstationary Signal Processing*, W.J. Fitzgerald, R.L. Smith, A. Warden, and P.C. Young, eds, 2000.

MODELING NORTH PACIFIC CLIMATE TIME SERIES

DONALD B. PERCIVAL*, JAMES E. OVERLAND†, AND
HAROLD O. MOFJELD†

Abstract. The North Pacific (NP) index is a time series related to atmospheric pressure variations at sea level and is an important indicator of the NP climate. We consider three statistical models for the NP index, namely, a Gaussian stationary autoregressive process, a Gaussian stationary fractionally differenced (FD) process, and a 'signal plus noise' process consisting of a square wave oscillation with a pentadecadal period embedded in Gaussian white noise. Each model depends upon three parameters, so all three models are equally simple. Statistically each model fits the NP index equally well. The fact that this index consists of just a hundred observations makes it unrealistic to expect to be able to clearly prefer one model over the other. Although the models fit equally well, their implications for the long term behavior of the NP index can be quite different in terms of, e.g., generating regimes of characteristic lengths (i.e., stretches of years over which the NP index is predominantly either above or below its long term average value). Because we cannot determine a preferred model statistically, we are faced with either entertaining multiple models when considering what the long term behavior of the NP index is likely to be or using physical arguments to select one model. The latter approach would arguably favor the FD process because it has an interpretation as the synthesis of first order differential equations involving many different damping constants.

Key words. Autoregressive process, North Pacific index, fractionally differenced process, time series analysis.

AMS(MOS) subject classifications. 62P12, 60G10, 60G15.

1. Introduction. There has been considerable interest in recent years in understanding how the climate in the North Pacific (NP) influences fish and mammals in the region. For example, the dramatic drop in the abundance of Stellar sea lions in the Gulf of Alaska over the last fifty years has been hypothesized to be due in part to NP climate variability. One particular measure of this climate variability is the Aleutian low sea level pressure field averaged over the months November to March [17, 13]. This series is known as the NP index and is plotted in Figure 1 (thin curve) for the years 1900 to 1999. Our goal is to investigate the nature of interdecadal variability in this climate time series, but the shortness of this series (one hundred points in all) presents major difficulties in our ability to say with some degree of confidence what the NP index will look like in the future. Faced with a lack of data, one approach is to investigate various models. The idea is to see what models are reasonable fits to the observed series and then to investigate what each model implies about the future behavior of the NP index. We cannot expect to identity one 'true' model for the index

*Applied Physics Laboratory, University of Washington, Box 355640, Seattle, WA 98195–5640 (dbp@apl.washington.edu).
†Pacific Marine Environmental Laboratory, National Oceanic and Atmospheric Administration, 7600 Sand Point Way NE, Seattle, WA 98115–6349.

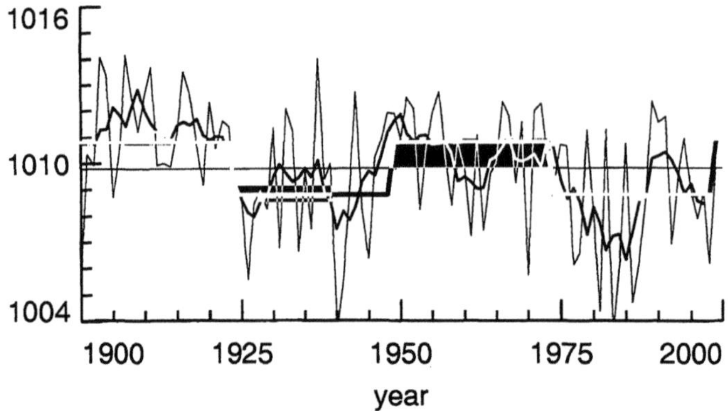

FIG. 1. *Plot of the NP index (thin curve), a five year running average of the index (thicker) and a pentadecadal square wave oscillation used as a model for the index (thickest). There is also a thin horizontal line on the plot. The height of this line depicts the sample mean (1009.8 millibars) for the index.*

based upon the small amount of available data, but we can investigate what implications the different models have on interesting aspects of the index.

The question then arises as to which models we should consider for the NP index. There are many viable candidates. In [14], we compared two purely stochastic models, namely, a 'short memory' model specified by a Gaussian first order autoregressive (AR(1)) process, and a 'long memory' model given by a Gaussian fractionally differenced (FD) process. In [15], we used the matching pursuit algorithm to investigate evidence for a hypothesis by Minobe [12] that the NP index is subject to pentadecadal and bidecadal oscillations in the form of rapid transitions (or 'regime shifts') that 'cannot be attributed to a single sinusoidal-wavelike variability.' This work leads to us to consider a simple 'signal plus noise' model consisting of a square wave oscillation (SWO) observed in the presence of Gaussian white noise. Our intent here is to study the relative merits of this model and the two purely stochastic models.

The remainder of this article is organized as follows. We briefly describe our three models in §2, after which we discuss their parameter estimation in §3. We assess how well each of the models fits the NP index in §4. Given the small amount of available data, we cannot hope to definitively discriminate amongst the three models on statistical grounds, a fact that we quantify in §5. We discuss in §6 the different implications these three models have for extrapolations and for the notion of regime shifts. We state our conclusions in §7.

2. Statistical models for the NP index. In the subsections below, we discuss two Gaussian stationary processes and a 'signal plus Gaussian noise' process as models for the NP index. The stationary processes are an AR(1) process and an FD process, while the 'signal plus noise' process consists of an SWO observed in the presence of Gaussian white noise. All three processes are fully specified by three parameters and hence can be considered to be 'equally simple.' One of the parameters for each process is related to the overall mean value of the NP index, while a second is critical in setting its overall variance. The final parameter controls either (i) the shapes of both the autocovariance sequence (ACVS) and the spectral density function (SDF) for the stationary processes or (ii) the amplitude of the SWO. An important distinguishing characteristic of each process is expressed in their ACVSs. The ACVS for an AR(1) process dies down quickly (exponentially) to zero, so this process is said to have 'short memory;' on the other hand, for an FD process, the ACVS dies down slowly (hyperbolically) and is said to be associated with 'long memory.' For the 'signal plus noise' process, if we force it into the mold of a stationary process by introducing a random phase for the SWO (i.e., location of first zero crossing), the resulting ACVS is periodic and hence never dies down to zero.

2.1. Short memory model. For this model we regard the NP index as a realization of portion $X_0, X_1, \ldots, X_{N-1}$ of a stationary Gaussian AR(1) process $\{X_t\}$; i.e.,

$$(2.1) \qquad X_t - \mu_X = \phi(X_{t-1} - \mu_X) + \epsilon_t = \sum_{k=0}^{\infty} \phi^k \epsilon_{t-k},$$

where $\mu_X = E\{X_t\}$ is the process mean; $\{\epsilon_t\}$ is a Gaussian white noise process with mean zero and variance σ_ϵ^2; and $|\phi| < 1$. We note that, if $\phi = 0$, then $\{X_t\}$ reduces to Gaussian white noise. The ACVS for an AR(1) process is given by

$$(2.2) \qquad s_{X,\tau} \equiv \mathrm{cov}\{X_t, X_{t+\tau}\} = \sigma_\epsilon^2 \frac{\phi^{|\tau|}}{1 - \phi^2},$$

where $\tau \in \mathbb{Z}$ (the set of all integers). Its SDF is

$$(2.3) \qquad S_X(f; \theta_X) = \frac{\sigma_\epsilon^2}{1 + \phi^2 - 2\phi \cos(2\pi f)},$$

where $\theta_X \equiv \{\phi, \sigma_\epsilon^2\}$, and $|f| \leq 1/2$. An AR(1) process is widely used in climatology as a default model for correlated time series [18]. It is suggested as a viable model for the NP index by the fact that, when presented with all possible autoregressive/moving average models of orders ranging from zero to ten, the default order selection criterion used by the ITSM software package [5] picked the AR(1) process as the best model [8]. An AR(1) process is related to a discretized first order stochastic differential equation with a single damping constant that is related to ϕ. In [18], a measure of the decorrelation time (or integral time scale) of an AR(1) process is given by

$$(2.4) \qquad \tau_D \equiv 1 + 2 \sum_{\tau=1}^{\infty} \frac{s_{X,\tau}}{s_{X,0}} = \frac{1+\phi}{1-\phi};$$

i.e., the subseries $X_{n\lceil \tau_D \rceil}$, $n \in \mathbb{Z}$, is approximately a white noise process, where $\lceil x \rceil$ is the smallest integer no less than x.

2.2. Long memory model. For this model we regard the NP index as a realization of a portion $Y_0, Y_1, \ldots, Y_{N-1}$ of a stationary Gaussian FD process $\{Y_t\}$; i.e,

$$
\begin{aligned}
(2.5) \qquad Y_t - \mu_Y &= \sum_{k=0}^{\infty} \frac{\Gamma(1+\delta)}{\Gamma(k+1)\Gamma(1+\delta-k)} (-1)^k (Y_{t-k} - \mu_Y) + \varepsilon_t \\
&= \sum_{k=0}^{\infty} \frac{\Gamma(1-\delta)}{\Gamma(k+1)\Gamma(1-\delta-k)} (-1)^k \varepsilon_{t-k},
\end{aligned}
$$

where $\mu_Y = E\{Y_t\}$ is the process mean; $\{\varepsilon_t\}$ is a Gaussian white noise process with mean zero and variance σ_ε^2; and $|\delta| < 1/2$. If $\delta = 0$, then $\{Y_t\}$ reduces to a white noise process. The process exhibits long memory if $\delta > 0$. The ACVS for an FD process is given by

$$(2.6) \quad s_{Y,0} = \frac{\sigma_\varepsilon^2 \Gamma(1-2\delta)}{\Gamma^2(1-\delta)} \quad \text{and} \quad s_{Y,\tau} = s_{Y,\tau-1} \frac{\tau + \delta - 1}{\tau - \delta}, \qquad \tau = 1, 2, \ldots$$

(for $\tau < 0$, recall that $s_{Y,\tau} = s_{Y,-\tau}$). Its SDF is given by

$$(2.7) \qquad S_Y(f; \theta_Y) = \frac{\sigma_\varepsilon^2}{|2\sin(\pi f)|^{2\delta}},$$

where $\theta_Y \equiv \{\delta, \sigma_\varepsilon^2\}$. For large τ, we have $s_{Y,\tau} \propto |\tau|^{2\delta-1}$ approximately, while, for small f, we have $S_Y(f) \propto 1/|f|^{2\delta}$ approximately. An FD model is suggested for the NP index by the fact that, when presented with all autoregressive/fractionally differenced models with the autoregressive order ranging from zero to ten, the ITSM order selection criterion [5] picked the FD process as the best model [8]. An FD process can be synthesized using an aggregation of first order differential equations involving many different damping constants [1]. The measure of the decorrelation time given by Equation (2.4) is infinite for long memory FD processes.

2.3. Square wave oscillation plus noise model. Recently Minobe [12] analyzed the NP index and found evidence for 'regime' shifts (see his paper for citations to prior work on this hypothesis). Loosely speaking, a regime is a stretch of years over which the index is essentially either above or below the sample mean of the entire record. Consider, for example, the index from 1901 to 1923 (thin curve in Figure 1), over which it is essentially greater than its sample mean (thin horizontal line), with the only exceptions being the years 1905 and 1919. This positive regime with a duration of 23 years is more clearly seen in a five year running mean (thicker curve), which lies strictly above the sample mean up until 1924. Continuing on, we find this running mean to be essentially (but not strictly) below the sample mean from 1924 to 1946, indicating a negative regime of 23 years. This pattern arguably persists up to the end of the recorded data, providing evidence for a pentadecadal oscillation [12].

A regime shift is the transition between regimes of different polarities. Part of the current interest in the notion of regimes is that the shifts occur rapidly, signaling a change in conditions that can influence the relative abundances of fish and mammal populations. To quote from Minobe [12],

> 'Although the regime shifts in the present century are characterized by a pentadecadal timescale, rapid transitions from one regime to another cannot be attributed to a single sinusoidal-wavelike variability. The rapid-transition nature of 20th century regime shifts suggests that the pentadecadal variability is characterized by a non-sinusoidal variation such as a rectangular wave ....'

In [15], we informally assessed Minobe's hypothesis using matching pursuit. The basic idea behind matching pursuit is to construct a 'dictionary' of vectors with characteristics that might be useful in describing a time series. We constructed a dictionary consisting of (i) sinusoidal vectors from a discrete Fourier transform, (ii) SWOs with periods ranging from two up to one hundred years in combination with all possible alignments (i.e., phase shifts) and (iii) wavelet and scaling vectors from a Haar wavelet transform spanning two up to one hundred years (again with all possible alignments). When presented with this dictionary, matching pursuit found the best single approximating vector to be a square wave oscillation with a pentadecadal period; i.e., given the choice between square wave and sinusoidal oscillations, matching pursuit preferentially favored the former over the latter, in agreement with Minobe's hypothesis. In addition, the alignment of the selected pentadecadal oscillation was such that its regime shifts are in quite good agreement with those given in Minobe [12].

In view of this result from matching pursuit, we will also consider the NP index to be a realization of a portion $Z_0, Z_1, \ldots, Z_{N-1}$ of an SWO process defined by

$$(2.8) \qquad Z_t = \mu_Z + \beta D_{t+1} + e_t,$$

where μ_Z and β are unknown parameters; $\{e_t\}$ is a Gaussian white noise process with mean zero and variance σ_e^2; and

$$(2.9) \qquad D_t \equiv \begin{cases} +0.1, & \text{when } t = 0, \ldots, 24; \\ -0.1, & \text{when } t = 25, \ldots, 49; \text{ and} \\ D_{t \bmod 50}, & \text{otherwise} \end{cases}$$

(by definition, '$t \bmod 50$' is equal to $t + 50n$, where n is the unique integer such that $0 \le t + 50n \le 49$). Note that, if $\beta = 0$, then $\{Z_t\}$ becomes a white noise process. When $\beta \neq 0$, an intuitive measure for the decorrelation time is fifty since the subseries Z_{j+50k}, $k \in \mathbb{Z}$, for any fixed $j \in \mathbb{Z}$ is a white noise process whose components all have the same mean value (the same is not true for Z_{j+lk} when $l = 1, \ldots, 49$).

3. Estimation of model parameters. Each of the statistical models for the NP index described in §2 has three parameters: μ_X, ϕ and σ_ϵ^2 for the AR(1) process $\{X_t\}$; μ_Y, δ and σ_ε^2 for the FD process $\{Y_t\}$; and μ_Z, β and σ_e^2 for the SWO process $\{Z_t\}$. The parameters μ_X, μ_Y and μ_Z capture the overall level of the index and can be estimated via the sample mean; i.e.,

$$\hat{\mu}_X = \frac{1}{N} \sum_{t=0}^{N-1} X_t, \qquad \hat{\mu}_Y = \frac{1}{N} \sum_{t=0}^{N-1} Y_t \quad \text{and} \quad \hat{\mu}_Z = \frac{1}{N} \sum_{t=0}^{N-1} Z_t.$$

For the AR(1) and FD models, the sample mean is a suboptimal estimator, but is commonly used in time series analysis because there is little practical loss of efficiency.

After computing the sample mean, we use it to recenter the NP index, yielding

$$\widetilde{X}_t \equiv X_t - \hat{\mu}_X, \quad \widetilde{Y}_t \equiv Y_t - \hat{\mu}_Y \quad \text{and} \quad \widetilde{Z}_t \equiv Z_t - \hat{\mu}_Z.$$

As an approximation, we consider $\widetilde{X}_t$, $\widetilde{Y}_t$ and $\widetilde{Z}_t$ to be AR(1), FD and SWO processes with $\mu_X = \mu_Y = \mu_Z = 0$. There are now two parameters left to estimate for each process. We estimate these using the maximum likelihood (ML) method ([14] has details about this approach for the AR(1) and FD models). Let $\hat{\phi}$, $\hat{\sigma}_\epsilon^2$, $\hat{\delta}$, $\hat{\sigma}_\varepsilon^2$, $\hat{\beta}$ and $\hat{\sigma}_e^2$ denote the ML estimators. Large sample theory says that these estimators are all (at least) approximately normally distributed and unbiased (both statements are exact for $\hat{\beta}$, and $\hat{\sigma}_e^2$ is also exactly unbiased). The variances of $\hat{\phi}$ and $\hat{\delta}$ are approximately given by, respectively, $(1 - \phi^2)/N$ and $6/\pi^2 N$, whereas the variance of $\hat{\beta}$ is given exactly by σ_e^2 (the dependence on N is implicit here because the SWO of Equation (2.9) is defined to have unit norm over N observations). The variances of $\hat{\sigma}_\epsilon^2$, $\hat{\sigma}_\varepsilon^2$ and $\hat{\sigma}_e^2$ all have a similar form, namely, $2\sigma_\epsilon^4/N$, $2\sigma_\varepsilon^4/N$ and $2\sigma_e^4/N$, respectively (these are approximations for the AR(1) and FD processes but exact for the SWO process). Monte Carlo experiments indicate that the large sample approximations we have quoted for the AR(1) and FD processes are in fact very accurate for $N \geq 100$ [14].

Table 1 shows the parameter estimates for the three models, along with associated 95% confidence intervals (CIs) for the unknown parameters based upon ML theory. Based upon the CIs, we can reject the white noise hypotheses $\phi = 0$, $\delta = 0$ and $\beta = 0$ at the 0.05 level of significance. The estimated residual standard deviations are remarkably similar for the three fitted models, with the one for the SWO model being the smallest. The fitted SWO model $\hat{\mu}_Z + \hat{\beta}D_{t+1}$ is shown in Figure 1 as the thickest curve.

TABLE 1

Autoregressive (AR), fractionally differenced (FD) and square wave oscillator (SWO) process parameter estimates for the NP index, along with associated 95% confidence intervals (CIs).

model	parameter	95% CI	σ	95% CI
AR	$\hat{\phi} \doteq 0.21$	$[0.02, 0.40]$	$\hat{\sigma}_\epsilon \doteq 2.37$	$[2.01, 2.67]$
FD	$\hat{\delta} \doteq 0.17$	$[0.02, 0.32]$	$\hat{\sigma}_\epsilon \doteq 2.35$	$[2.00, 2.66]$
SWO	$\hat{\beta} \doteq -10.09$	$[-14.51, -5.67]$	$\hat{\sigma}_e \doteq 2.21$	$[1.88, 2.50]$

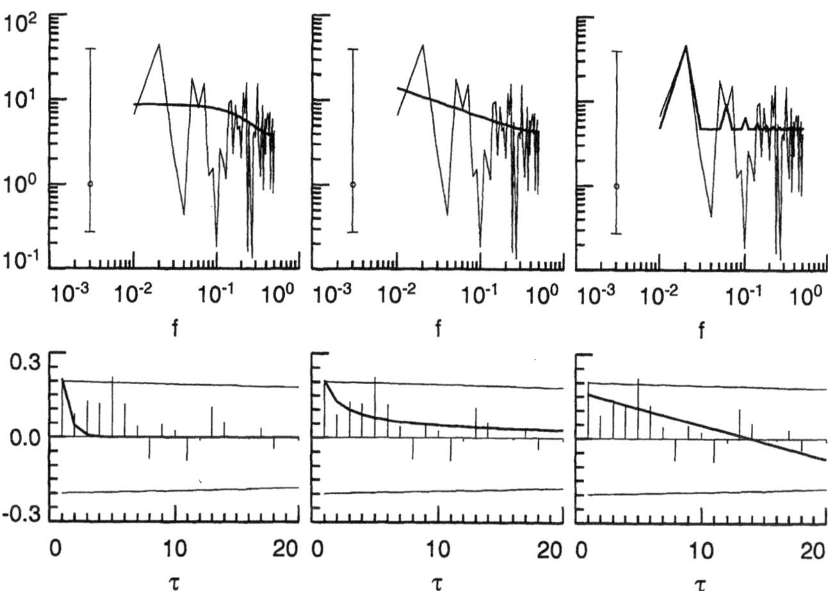

FIG. 2. *Spectral density functions (thick curves, top row) for the fitted AR, FD and SWO models (left-hand, middle and right-hand columns, respectively) and the corresponding periodogram for the NP index (thin curves, top row), along with theoretical autocorrelation sequences (ACSs, thick curves, bottom row) for the models and the corresponding sample ACS (depicted as deviations from zero in each plot in the bottom row). The two thin curves above and below the zero line in each lower plot indicate limits within which individual sample ACS estimates should fall 95% of the time when the true process is white noise. The error bar in each of the upper plots gives a similar assessment of the inherent variability in the periodogram.*

4. Assessment of fitted models. As a first step in assessing how well the three fitted models capture the characteristics of the NP index, let us compare plots of the theoretical SDFs corresponding to the fitted models with the periodogram for the index (top row of Figure 2). By definition the SDFs for the AR(1) and FD processes are given by Equations (2.3) and (2.7), respectively. With θ_X set equal to $\hat{\theta}_X \equiv \{\hat{\phi}, \hat{\sigma}_\epsilon^2\}$ and θ_Y set to $\hat{\theta}_Y \equiv \{\hat{\delta}, \hat{\sigma}_\epsilon^2\}$, the theoretical SDFs for the fitted models are given by the solid curves in the upper left-hand and middle plots. For the SWO process, we can define a theoretical SDF to be the expected value of the periodogram for $\widehat{Z}_t \equiv \hat{\beta} D_{t+1} + e_t$ conditional on regarding $\hat{\beta}$ as fixed. By definition, this periodogram is given by

$$(4.1) \qquad \hat{S}_{\widehat{Z}}(f_k) \equiv \frac{1}{N} \left| \sum_{t=0}^{N-1} \widehat{Z}_t e^{-i2\pi f_k t} \right|^2, \quad f_k \equiv k/N, \ k = 1, \ldots, N/2,$$

and its expected value is given by

$$E\{\hat{S}_{\widehat{Z}}(f_k)\} = \frac{1}{N} \left| \sum_{t=0}^{N-1} \hat{\beta} D_{t+1} e^{-i2\pi f_k t} \right|^2 + \sigma_e^2.$$

With σ_e^2 set to its ML estimate, the above is plotted as the thick curve in the upper right-hand plot. The periodogram for the NP index is given by Equation (4.1) with $\widetilde{X}_t$ substituted for $\widehat{Z}_t$ and is shown on all three upper plots as the thin jagged curves. An assessment of the sampling variability in the periodogram is given by the CIs shown in the left-hand portion of each plot in the top row. These CIs emanate from a small circle. If we move the CI so that the circle is centered on a particular value of $\hat{S}_{\widetilde{X}}(f_k)$, then we have a 95% CI for the true SDF at frequency f_k.

If we take into account the sampling variability inherent in the periodogram, all three theoretical SDFs are viable summaries of the spectral content in the NP index. While the fitted AR(1) and FD SDFs are broadband summaries, the SDF for the SWO model locks onto the largest value in the periodogram, associated with a period of $1/f_2 = 50$ years. There are additional peaks in the SWO SDF visible at certain harmonics of f_2. The most prominent is the second harmonic, i.e., $3f_2 = f_6$, which is associated with a period of $1/f_6 \doteq 16.7$ years. The fact that there appears to be some spectral content surrounding this second harmonic in the periodogram lends some additional (but statistically questionable) support for Minobe's hypothesis that a pure sinusoidal oscillation is inadequate to describe the NP index. Note that the background level from which these harmonics arise is determined by the estimated variance $\hat{\sigma}_e^2$ of the white noise process $\{e_t\}$ in the model.

Let us now compare plots of the theoretical autocorrelation sequences (ACSs) with the sample ACS for the NP index (bottom row of Figure 2). The ACSs for the AR(1) and FD processes are given by, respectively, $\rho_{X,\tau} \equiv s_{X,\tau}/s_{X,0}$ and $\rho_{Y,\tau} \equiv s_{Y,\tau}/s_{Y,0}$, where $s_{X,\tau}$ and $s_{Y,\tau}$ are given in Equations (2.2) and (2.6). The inverse Fourier transform of $E\{\hat{S}_{\widehat{Z}}(f)\}$, $|f| \leq 1/2$, can be used to form a theoretical ACVS for the SWO process, from which the corresponding ACS follows. The theoretical ACSs for the fitted AR(1), FD and SWO models are the solid curves in, respectively, the left-hand, middle and right-hand plots. Letting $\widetilde{X}_t$ stand for recentered NP index, the sample ACS is given by

$$\hat{\rho}_{\widetilde{X},\tau} \equiv \frac{\sum_{t=0}^{N-\tau-1} \widetilde{X}_t \widetilde{X}_{t+\tau}}{\sum_{t=0}^{N-1} \widetilde{X}_t^2}, \quad \tau = 0, 1, \ldots, N-1.$$

The sample ACS is shown in each plot as deviations from the zero line. An assessment of the sampling variability in the sample ACS is given by the upper and lower thin curves in each plot, which depict 95% CIs under the assumption that the NP index is a realization of a white noise process (see Corollary 6.3.6.2 of [9] for details).

Each model captures certain aspects of the sample ACS for the NP index. The AR(1) ACS captures $\hat{\rho}_{\widetilde{X},1}$ almost perfectly, but then damps down to zero very quickly. The FD ACS also matches $\hat{\rho}_{\widetilde{X},1}$ very well, but then decays to zero slowly and hence visually captures the sample ACS a bit better. The SWO ACS underestimates $\hat{\rho}_{\widetilde{X},1}$ somewhat and, like the FD ACS, has much larger values than the AR(1) ACS for lags below about $\tau = 10$. When we take into account the sampling variability in the sample ACS, however, all three models seem qualitatively reasonable.

Let us now look at three test statistics that give us quantitative assessments of the fitted models. The first statistic T_1 compares the periodogram $\hat{S}_{\widetilde{X}}(f_k)$ for the NP index to the fitted $S(f_k; \hat{\theta})$ from a particular model [11, 1]:

$$T_1 \equiv \frac{NA}{4\pi B^2}, \text{ with } A \equiv \sum_{k=1}^{\lfloor (N-1)/2 \rfloor} \left(\frac{\hat{S}_{\widetilde{X}}(f_k)}{S(f_k; \hat{\theta})} \right)^2 \text{ and } B \equiv \sum_{k=1}^{\lfloor (N-1)/2 \rfloor} \frac{\hat{S}_{\widetilde{X}}(f_k)}{S(f_k; \hat{\theta})},$$

where $\lfloor x \rfloor$ is the largest integer no more than x. Here $S(f_k; \hat{\theta})$ is taken to be either $S_X(f_k; \hat{\theta}_X)$ of Equation (2.3) for the AR(1) model or $S_Y(f_k; \hat{\theta}_Y)$ of Equation (2.7) for the FD model (the theory behind T_1 developed in [11] does not extend to the SDF we defined for the SWO model). Under the null hypothesis that the model corresponding to $S(f_k; \hat{\theta})$ is correct, T_1 is asymptotically normal with mean $1/\pi$ and variance $2/(\pi^2 N)$. We reject the null hypothesis at a level of significance of α when $\sqrt{N/2}(\pi T_1 - 1)$ exceeds $Q_1(1 - \alpha)$, which is the upper $(1 - \alpha) \times 100\%$ percentage point for the standard normal distribution.

The other two test statistics make use of residuals from each model. For the AR(1), FD and SWO models, we denote the residuals as, respectively, $\hat{\epsilon}_t$, $\hat{\varepsilon}_t$ and $\hat{e}_t$ This notation emphasizes the fact that the residuals can be regarded as estimates of the white noise processes ϵ_t, ε_t and e_t involved in each model (see Equations (2.1), (2.6) and (2.8)). There are details in [14] about how to compute ϵ_t and ε_t for the AR(1) and FD models. For the SWO model, we have

$$\hat{e}_t \equiv \widetilde{Z}_t - \hat{\beta} D_{t+1}.$$

If a particular model is adequate for the NP index, then the residuals from the fitted model should resemble a sample from a white noise process. The cumulative periodogram test statistic assesses this resemblance by determining if the periodogram for the residuals is consistent with the white noise assumption [2, 4]. This test statistic is defined as

$$T_2 = \max\left\{ \max_l \left(\frac{l}{\lfloor (N-1)/2 \rfloor - 1} - \mathcal{P}_l \right), \ \max_l \left(\mathcal{P}_l - \frac{l-1}{\lfloor (N-1)/2 \rfloor - 1} \right) \right\},$$

where $\mathcal{P}_l$ is the normalized cumulative periodogram for, say, $\hat{\epsilon}_t$:

$$\mathcal{P}_l \equiv \frac{\sum_{k=1}^{l} \hat{S}_{\hat{\epsilon}}(f_k)}{\sum_{k=1}^{\lfloor (N-1)/2 \rfloor} \hat{S}_{\hat{\epsilon}}(f_k)}$$

(analogous expressions hold for $\bar{\epsilon}_t$ and $\hat{e}_t$). We reject that null hypothesis of white noise at the α level of significance if T_2 exceeds

$$Q_2(1-\alpha) \equiv \frac{C(1-\alpha)}{(M-1)^{1/2} + 0.12 + \frac{0.11}{(M-1)^{1/2}}},$$

where $C(0.9) = 1.224$, $C(0.95) = 1.358$ and $C(0.99) = 1.628$ [16].

The last test statistic determines if the residuals are consistent with the white noise hypothesis by examining their sample ACS. Given a positive integer K (taken to be small compared the sample size N), the Box–Pierce portmanteau test statistic [3] is defined for, e.g., the AR(1) residuals $\hat{\epsilon}$ as

$$T_3 = N \sum_{\tau=1}^{K} \hat{\rho}_{\hat{\epsilon},\tau}^2,$$

where $\rho_{\hat{\epsilon}_t,\tau}$ is the sample ACS for $\hat{\epsilon}_t$ (similar expressions hold for $\bar{\epsilon}_t$ and $\hat{e}_t$). We reject the null hypothesis of white noise at a level of significance α if T_3 exceeds $Q_3(1-\alpha)$, which is the $(1-\alpha) \times 100\%$ percentage point for the chi-square distribution with $K-1$ degrees of freedom. In keeping with recommendations in the literature, we set $K = N/20 = 5$, but we also looked at $K = 10$ and obtained virtually the same results. (We note that there is a variation on T_3 known as the Ljung–Box–Pierce portmanteau test statistic [10], which takes the form

$$T_4 = N(N+2) \sum_{\tau=1}^{K} \frac{\hat{\rho}_{\hat{\epsilon}_t,\tau}^2}{N-\tau}.$$

The results that we got using T_3 and T_4 were virtually identical.)

For all three test statistics T_j, we reject the 'model is adequate' hypothesis when T_j is 'too big' as quantified by a percentage point from a distribution under the null hypothesis. Table 2 shows the results of these

TABLE 2

Model goodness of fit tests for the NP index. In the column reporting the result of using an $\alpha = .05$ level test, 'accept' should be interpreted as shorthand for 'fail to reject the null hypothesis.'

j	model	T_j	$Q'_j(0.90)$	$Q_j(0.95)$	$Q_j(0.99)$	$\alpha = .05$ test	$\hat{\alpha}$
1	AR	0.30	0.38	0.39	0.42	accept	0.67
	FD	0.28	"	"	"	accept	0.78
	WN	0.39	"	"	"	reject	0.05
2	AR	0.10	0.17	0.19	0.23	accept	$\gg 0.10$
	FD	0.07	"	"	"	accept	$\gg 0.10$
	SWO	0.10	"	"	"	accept	$\gg 0.10$
	WN	0.21	"	"	"	reject	≈ 0.03
3	AR	4.65	7.74	9.45	13.31	accept	0.32
	FD	3.12	"	"	"	accept	0.54
	SWO	2.83	"	"	"	accept	0.59
	WN	12.63	"	"	"	reject	0.01

goodness of fit tests for the AR(1), FD and SWO models, along with an additional model (denoted as 'WN') that regards the NP index as a realization of a white noise process and has 'residuals' that are taken to be the NP index itself. At the 0.05 level of significance, all the test statistics reject the hypothesis that the NP index is white noise, but all fail to reject the adequacy of the AR(1), FD and SWO models; i.e., statistically, all three models are viable. The table also gives an indication of the observed level of significant $\hat{\alpha}$ (i.e., the smallest α for which we would end up rejecting the null hypothesis). For all three test statistics and for all three models, $\hat{\alpha}$ is so large that we cannot reject the null hypothesis at any reasonable level of significance.

Finally, let us comment upon the Gaussian assumption that we have made for each model. Quantile-quantile plots [6] of the residuals from the three models indicate some possible departures from Gaussianity. These departures are not severe, but are a topic for future research since it is unclear how they impact the analysis presented here.

5. Model discrimination. The fact that the AR(1), FD and SWO models are all viable for the NP index from a statistical point of view raises the question as to whether or not we could reasonably hope to distinguish amongst these models given the fact that we only have one hundred values for the NP index. To address this question, we consider the following experiment. For the sake of argument, let us assume that the fitted FD model is in fact exactly correct for the NP index. Using procedures outlined in [14], we can generate simulated time series of a desired length $N' \geq N$ from this fitted model. We can then fit an AR(1) model to each simulated

FD series and evaluate the fitted AR(1) model using each of our three test statistics T_j. By repeating the above a large number of times (where 'large' is here taken to be 2500), we can estimate the probability that T_j will (correctly) reject the null hypothesis that the AR(1) model is the correct one. This experiment will tell us how much power T_j has in being able to say that the AR(1) model is incorrect. We can also try to fit an SWO model to the same simulated FD series to get an idea as to how much power each test statistic has in this case. We can repeat the above for a variety of sample sizes N'. We can also repeat all of the above by generating simulation from either the fitted AR(1) model or the fitted SWO model and then fitting the other two models to the simulated time series.

Figure 3 shows plots of the estimated probabilities versus sample size N' that, using a level of significance $\alpha = 0.05$, the three test statistics T_j will reject a fitted model A when the true model is B. For completeness, we also estimated these probabilities when A and B were the same model – theoretically these probabilities should be close to $\alpha = 0.05$. From this figure, we can see that, at best, we have only a 30% chance of rejecting the null hypothesis when we have only one hundred observations. The best scenario is when we use the T_1 test statistic on an AR(1) fit to a time series generated by an SWO process. If the SWO model is indeed correct for the NP index, we should have a reasonable chance of rejecting the AR(1) and FD models by the end of the current century. We would need two to three hundred years worth of data in order to have a 50% chance of rejecting the SWO model when either of the other two models is correct. The test statistics are less successful at discriminating between the AR(1) and FD models, where we would need about five hundred years of the NP index in order to have even a 50% chance of correctly rejecting the null hypothesis. Finally, it is of interest to note that no one test statistic performs uniformly better than the other two, which emphasizes the need to use them all.

6. Model implications. We have seen that, based upon our three test statistics T_j, there is no statistical reason to prefer either an AR(1), FD or SWO model for the NP index. In addition, all three models depend on just three parameters and hence can be regarded as equally simple, so the principle of parsimony would not prefer any of the models (this ignores the fact that the SWO model was picked out by matching pursuit rather than *a priori*). Even though all three models seem to match the NP index equally well, the models can have implications that are quite different and potentially important, as the following demonstrates.

As a starting point for our discussion, the left-hand plots in the bottom three rows of Figure 4 show simulated series from each of the three models. From the second down to fourth rows, the series are from, respectively, the AR(1), FD and SWO models. These time series were all created using the exact simulation method described in [7] and [19], which transforms realizations of independent standard Gaussian random variables into a time

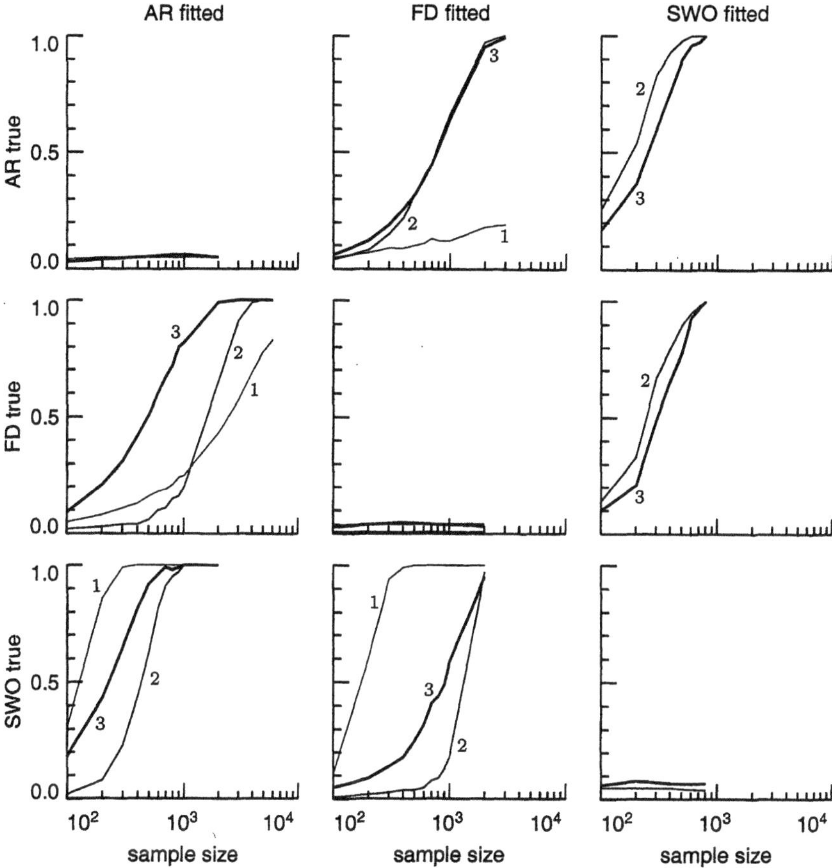

FIG. 3. *Probability (as a function of sample size) of rejecting the null hypothesis (using a test with a 0.05 level of significance) that a fitted model A is adequate for a realization of a process B when using the test statistics T_1, T_2 and T_3. For the plots in the left- to right-hand columns, the fitted models A is, respectively, an FD, AR(1) and SWO model. The same ordering is used for the process B in the top to bottom rows.*

series. Here all three time series were formed from the *same* set of realizations to make it easier to appreciate the qualitative differences between the three models. For comparison, the top row shows the recentered NP index itself. Each simulated series is a thousand years long, and the three series are visually rather similar. The right-hand plots show the periodograms corresponding to each series. With the passage of time, the square wave oscillation in the SWO model determines the dominant feature in the periodogram, but note that there is also considerable power at low frequencies in the simulated AR(1) and FD processes. The periodic nature of the SWO and the low frequency power in the other two models can all contribute to

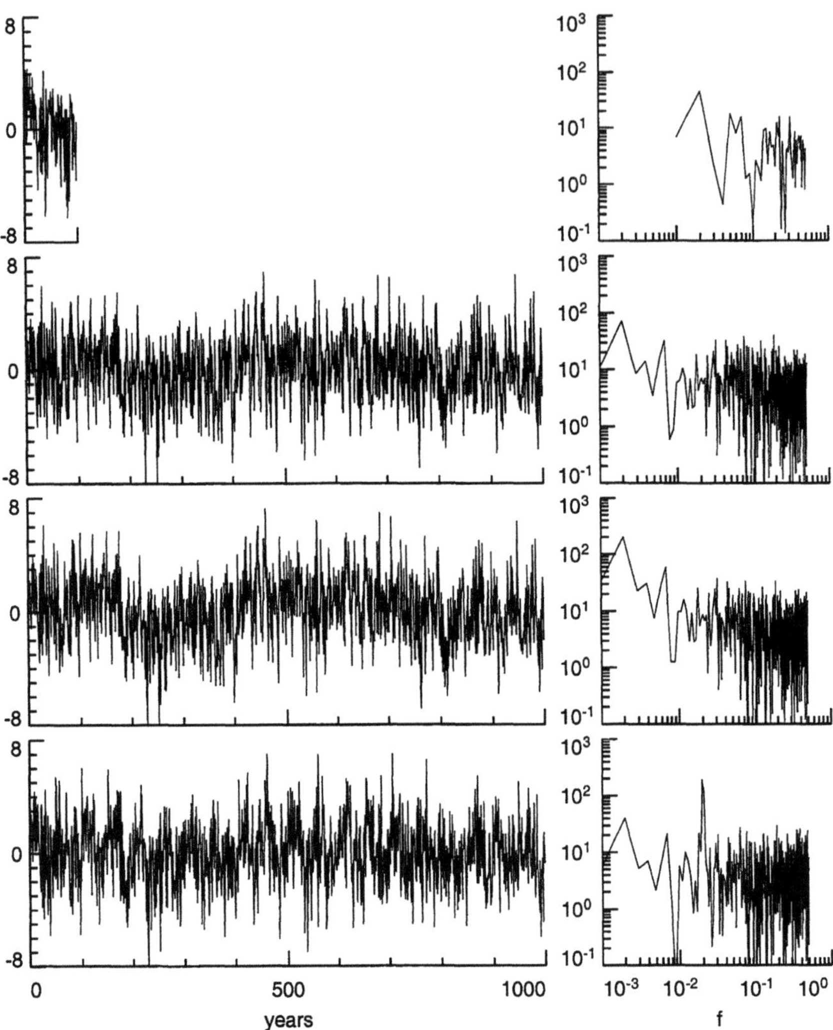

FIG. 4. *Comparison of recentered NP index (left-hand plot in top row) and its periodogram (right-hand) with thousand year simulations of this index and their periodograms using the fitted AR(1) model (second row), FD model (third row) and the SWO model (bottom row).*

the generation of regimes. We thus pose the following two questions. First, how well do these three models support the notion of regimes? Second, are there any differences in the regimes generated by these models? Answers to these questions are important in understanding the implications of using any one of these models to say something about expected future patterns of the NP index.

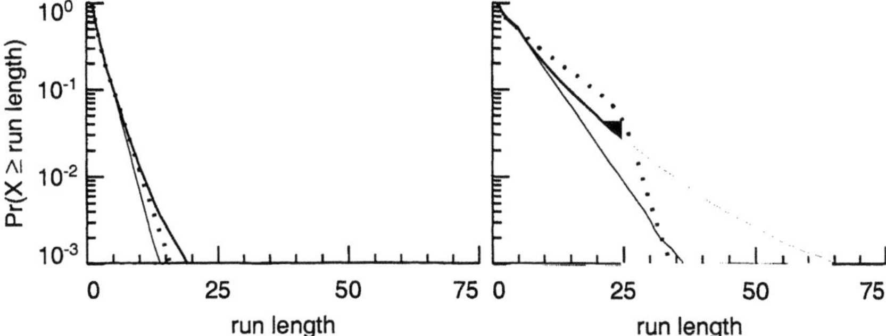

FIG. 5. *Probability of observing a run that is greater than or equal to a specified run length. The thin, thick and dotted curves correspond to the AR, FD and SWO models. The left-hand plot is for processes without smoothing, whereas the right-hand plot is for processes subjected to a five year running average.*

To address these questions, we consider the following computer experiment. For the sake of argument, suppose that the NP index is actually a realization of an FD process. To account for the fact that our parameter estimate $\hat{\delta}$ is not perfect but is subject to sampling error, we start by generating a deviate $\tilde{\delta}$ from the presumed distribution for $\hat{\delta}$, namely, a Gaussian distribution with mean $\hat{\delta}$ and variance $6/(\pi^2 N) = 6/(\pi^2 100)$. We next create a simulated series of length 1024 from an FD process with zero mean and parameter $\tilde{\delta}$ (the value we use for σ_ε^2 doesn't matter). We then tabulate the sizes of the observed regimes in the series, where the size of a regime is taken to be the number of contiguous values that are either entirely above or entirely below zero. We also do the same for a five year running average of the simulated series. If we repeat the above procedure a thousand times and if we do the same for the fitted AR(1) and SWO models (making use of the large sample distributions for $\hat{\phi}$ and $\hat{\beta}$), we can empirically determine the probability that a regime has a size greater than or equal to a specified run length.

The results of this experiment are shown in Figure 5. The left- and right-hand plots show the estimated probabilities for, respectively, the original and the five year averaged series. The AR(1), FD and SWO models are indicated by the thin, thick and dotted curves. The more interesting of the two plots is the right-hand one, for which a regime is defined in terms of run lengths of five year running averages. We see that intermediate regime sizes are most likely to occur under the SWO model and that quite large regime sizes are most likely under the FD model. The AR(1) model is the least supportive of the idea of extensive regimes. For example, in comparison to the AR(1) model, a regime whose size is greater than or equal to 23 years (i.e., the 'typical' size indicated by our inspection of Figure 1) is four times more likely under the FD model and is eight times more likely under the SWO model. Thus, even though all three models are viable from a statistical point of view, these models can lead to quite different statements about what the NP index will look like in the long term.

7. Conclusions. We have shown that, from a statistical point of view, the AR(1), FD and SWO models are all adequate descriptions of the NP index despite the fact that these models are quite different from one another. Our inability to identify a best model is directly related to the fact that we only have one hundred observations of the index. We would need several hundred more observations before we would have a reasonable chance of selecting a single best model. All three models include a white noise process as a special case, and one of the parameters in each of the models can be used to test the hypothesis of white noise. In all three cases, we can reject the white noise hypothesis. In terms of their ACSs, the ACS for the AR(1) model has the most rapid decay toward zero. The one for the FD model has a long tail of small positive correlations, while the SWO model has an oscillating ACS. The FD model is more supportive of the notion of extensive regimes than the other two models, but the SWO model supports regimes with sizes dictated by the half-period of its oscillation (25 years). In the absence of a viable physical explanation for a pentadecadal oscillation in the NP index, loose physical considerations might favor the FD model as the best single choice because this model corresponds to an aggregation of first order differential equations, each presumably associated with geophysical phenomena with different time scales. In lieu of picking a single best model, another strategy (to be explored in future research) would be to use a 'model averaging' approach to combine all three statistically feasible models in making statements about the future behavior of the NP index.

Acknowledgments. The work of the first author was supported in part by grants from the Cooperative Institute for Arctic Research (CIFAR), Stellar Sea Lion Research, and from the Pacific Marine Environmental Laboratory, National Oceanic and Atmospheric Administration. We would like to thank the organizers of the IMS Workshop on 'Time Series Analysis and Applications to Geophysical Systems' for the invitation to participate and to thank the other participants for their many helpful comments about our work. The paper is Contribution No. 2527 from NOAA/Pacific Marine Environmental Laboratory.

REFERENCES

[1] J. BERAN, *Statistics for Long-Memory Processes*, Chapman and Hall, NY, 1994.

[2] G.E.P. BOX, G.M. JENKINS, AND G.C. REINSEL, *Time Series Analysis: Forecasting and Control (Third Edition)*, Prentice Hall, Englewood Cliffs, NJ, 1994.

[3] G.E.P. BOX AND D.A. PIERCE, *Distribution of residual autocorrelations in autoregressive-integrated moving average time series models*, J. Amer. Stat. Assoc. **65** (1970), pp. 1509–1526.

[4] P.J. BROCKWELL AND R.A. DAVIS, *Time Series: Theory and Methods (Second Edition)*, Springer, NY, 1991.

[5] P.J. BROCKWELL AND R.A. DAVIS, *Introduction to Time Series and Forecasting (Second Edition)*, Springer, NY, 2002.

[6] J.M. CHAMBERS, W.S. CLEVELAND, B. KLEINER, AND P.A. TUKEY, *Graphical Methods for Data Analysis*, Duxbury Press, Boston, 1983.

[7] R.B. DAVIES AND D.S. HARTE, *Tests for Hurst effect*. Biometrika **74** (1987), pp. 95–101.

[8] R.A. DAVIS, private communication, 2001.

[9] W.A. FULLER, *Introduction to Statistical Time Series (Second Edition)*, Wiley–Interscience, NY, 1996.

[10] G.M. LJUNG AND G.E.P. BOX, *On a measure of lack of fit in time series models*, Biometrika **65** (1978), pp. 297–303.

[11] A. MILHØJ, *A test of fit in time series models*, Biometrika **68** (1981), pp. 177–187.

[12] S. MINOBE, *Resonance in bidecadal and pentadecadal climate oscillations over the North Pacific: Role in climate regime shifts*, Geophys. Res. Lett. **26** (2001), pp. 855–858.

[13] J.E. OVERLAND, J.M. ADAMS, AND N.A. BOND, *Decadal variability of the Aleutian low and its relation to high-latitude circulation*, J. Climate **12** (1999), pp. 1542–1548.

[14] D.B. PERCIVAL, J.E. OVERLAND, AND H.O. MOFJELD, *Interpretation of North Pacific variability as a short- and long-memory process*, J. Climate **14** (2001), pp. 4545–4559.

[15] ———, *Using matching pursuit to assess atmospheric circulation changes over the North Pacific* (2002), submitted to J. Climate.

[16] M.A. STEPHENS, *EDF statistics for goodness of fit and some comparisons*, J. Amer. Stat. Assoc. **69** (1974), pp. 730–737.

[17] K.E. TRENBERTH AND D.A. PAOLINO, JR., *The Northern Hemisphere sea-level pressure data set: Trends, errors and discontinuities*, Mon. Weather Rev. **108** (1980), pp. 855–872.

[18] H. VON STORCH AND F.W. ZWIERS, *Statistical Analysis in Climate Research*, Cambridge University Press, Cambridge, UK, 1999.

[19] A.T.A. WOOD AND G. CHAN, *Simulation of stationary Gaussian processes in* $[0, 1]^d$, J. Comp. Graph. Stat. **3** (1994), pp. 409–432.

SKEW-ELLIPTICAL TIME SERIES WITH APPLICATION TO FLOODING RISK

MARC G. GENTON* AND KEITH R. THOMPSON†

Abstract. In this article, skew-elliptical time series are defined in order to account for both skewness and kurtosis, with particular emphasis on the skew-normal and skew-t distributions. The bivariate skew-t distribution is then used to describe a 63 year time series of hourly sea levels measured at Charlottetown, Atlantic Canada. It is shown that the skew-t fits the data better than the normal distribution and it can be used to recover return periods of extreme levels based on a standard analysis of 63 annual maxima. Preliminary results are presented to show how the skew-t distribution may be used to estimate changes in flooding risk resulting from changes in sea level rise, storminess, and other climatic factors.

Key words. Extremes, flooding risk, kurtosis, nonstationarity, skewness, storm surge.

AMS(MOS) subject classifications. Primary 62M10, 62H05, 62P12.

1. Introduction. This article presents time series modeling with a class of continuous multivariate distributions that can simultaneously account for both skewness and heavy tails. To date, most time series modeling has focused on symmetric distributions, with particular emphasis on the normal distribution. Nevertheless, there are many natural phenomena that do not follow the normal law including the example discussed in this paper: the nontidal changes in sea level caused by variations in air pressure and wind acting at the surface of the ocean. Non-normal distributions are needed to model such phenomena and departures from normality can be achieved by varying both the skewness and kurtosis in the distribution. For this purpose, two main approaches are available.

The first approach consists of modifying a random variable, and thus also its quantiles, through an appropriate transformation. It was suggested by John W. Tukey in 1977 and discussed by Hoaglin et al. (1985) in the univariate setting. Basically, a standard normal random variable Z is transformed to $Y = \tau_{g,h}(Z)$, where:

$$(1.1) \qquad \tau_{g,h}(Z) = \left(\frac{\exp(gZ) - 1}{g} \right) \exp\left(\frac{h}{2} Z^2 \right).$$

The resulting random variable Y is said to have a g-and-h distribution, where g is a real constant controlling the skewness and h is a nonnegative real constant controlling the kurtosis, or elongation. The quantiles of

---

*Department of Statistics, North Carolina State University, Raleigh, NC 27695-8203, USA (genton@stat.ncsu.edu).

†Department of Mathematics and Statistics, and Department of Oceanography, Dalhousie University, Halifax, Nova Scotia, Canada B3H 4J1 (keith@phys.ocean.dal.ca).

$Y = \tau_{g,h}(Z)$ can easily be computed since $\tau_{g,h}$ increases monotonically in Z and is therefore a bijective transformation. Thus, the p-th quantile of the distribution of Y is simply $\tau_{g,h}(z_p)$ where z_p is the p-th quantile of the standard normal distribution. The estimation of the parameters g and h from data is carried out by computing and fitting empirical quantiles. However, the generalization to the multivariate setting is not straightforward, mainly because multivariate quantiles need to be defined appropriately. Research is currently underway on this topic.

The second approach consists of modifying the probability density of a random variable instead of the random variable itself. This approach was first developed by Azzalini (1985) for the univariate normal distribution and by Azzalini and Dalla Valle (1996) for the multivariate normal distribution, yielding the so-called skew-normal distribution. Extensions to skew-elliptical distributions were proposed by Azzalini and Capitanio (1999), Branco and Dey (2001), Sahu et al. (2003), and include for instance skew-t and skew-Cauchy distributions. All these distributions are particular types of generalized skew-elliptical distributions recently introduced by Genton and Loperfido (2002), i.e. they are defined as the product of a multivariate elliptical density g with a skewing function π:

$$(1.2) \quad h_n(\boldsymbol{x}) = 2|\Omega|^{-1/2} \cdot g\big[\Omega^{-1/2}(\boldsymbol{x}-\boldsymbol{\xi})\big] \cdot \pi\big[\Omega^{-1/2}(\boldsymbol{x}-\boldsymbol{\xi})\big], \quad \boldsymbol{x} \in \mathbb{R}^n,$$

where $\boldsymbol{\xi} \in \mathbb{R}^n$ and $\Omega \in \mathbb{R}^{n \times n}$ are location and scale parameters respectively, $0 \leq \pi(\boldsymbol{x}) \leq 1$, and $\pi(-\boldsymbol{x}) = 1 - \pi(\boldsymbol{x})$. The skew-elliptical distributions are attractive because their properties are very similar to those of the normal distribution, and include the normal distribution as a particular case.

In this paper we will provide a practical application of the skew-elliptical distribution (1.2) by modeling the distribution of a long time series of sea level and then using the distribution to predict changes in flooding risk associated with rising sea level. It will be shown that the skew-t distribution leads to an effective and parsimonious description of the sea level process and can be used to take into account its strong seasonality and other forms of nonstationarity.

The paper is structured as follows. In Section 2, we define skew-elliptical random processes and describe in detail the skew-normal and skew-t distributions. Section 3 presents the sea level analysis and conclusions are presented in Section 4.

2. Skew-elliptical distributions and stochastic processes. We say that a stochastic process $\{X_t, t \in T\}$ is a skew-elliptical time series if and only if its distribution functions are all multivariate skew-elliptical. In particular, Gaussian times series are skew-elliptical time series. Skew-elliptical time series are very flexible and allow us to model a wide variety of natural phenomena. In this section we focus on two simple examples, namely skew-normal time series and skew-t time series.

2.1. Skew-normal time series. A skew-normal time series is a skew-elliptical time series defined by the multivariate pdf (1.2) where g is a multivariate normal pdf ϕ_n with correlation matrix Ω and π is the standard normal univariate cdf Φ, i.e.

$$(2.1) \qquad h_n(x) = 2\phi_n(x - \xi; \Omega)\ \Phi\big(\alpha^T(x - \xi)\big).$$

The parameter $\alpha \in \mathbb{R}^n$ controls the skewness and $\alpha = 0$ reduces to the multivariate normal case. A random vector $x \in \mathbb{R}^n$ with pdf given by (2.1) is said to be multivariate skew-normal and was first introduced by Azzalini and Dalla Valle (1996). Its first two moments are given by

$$(2.2) \qquad \mathrm{E}(x) = \xi + \sqrt{\frac{2}{\pi}}\delta,$$

$$(2.3) \qquad \mathrm{Var}(x) = \Omega - \frac{2}{\pi}\delta\delta^T,$$

where

$$(2.4) \qquad \delta = \frac{\Omega\alpha}{(1 + \alpha^T\Omega\alpha)^{1/2}}.$$

Note that both the expectation and the variance of x depend on the skewness parameter α (or equivalently δ). In order to describe a stationary skew-normal time series with constant mean, we require that $\xi = \xi\mathbf{1}_n$ and $\delta = \delta\mathbf{1}_n$, where $\mathbf{1}_n = (1 \ldots 1)^T \in \mathbb{R}^n$. Figure 1 depicts the contours of the pdf (2.1) for $n = 2$, $\xi = (0, 0)^T$, Ω the correlation matrix with correlation 0.5, with $\alpha = (0, 0)^T$ (left panel) and $\alpha = (2, 2)^T$ (right panel). Note that the left panel corresponds to the bivariate standard normal distribution with correlation 0.5.

It is rather straightforward to simulate realizations of skew-normal time series. Indeed, Azzalini and Capitanio (1999) showed that if

$$\begin{pmatrix} Z_0 \\ z \end{pmatrix} \sim N_{n+1}(0, \Omega^*), \qquad \Omega^* = \begin{pmatrix} 1 & \delta^T \\ \delta & \Omega \end{pmatrix}$$

where Z_0 is a scalar component, Ω^* is a correlation matrix, then

$$x = \begin{cases} z & \text{if } Z_0 > 0, \\ -z & \text{otherwise,} \end{cases}$$

has a skew-normal distribution with parameters $\xi = 0$, Ω, and α, where

$$\alpha = \frac{\Omega^{-1}\delta}{(1 - \delta^T\Omega^{-1}\delta)^{1/2}}.$$

The multivariate skew-normal distribution enjoys many pleasant properties. For instance, marginal distributions of a multivariate skew-normal

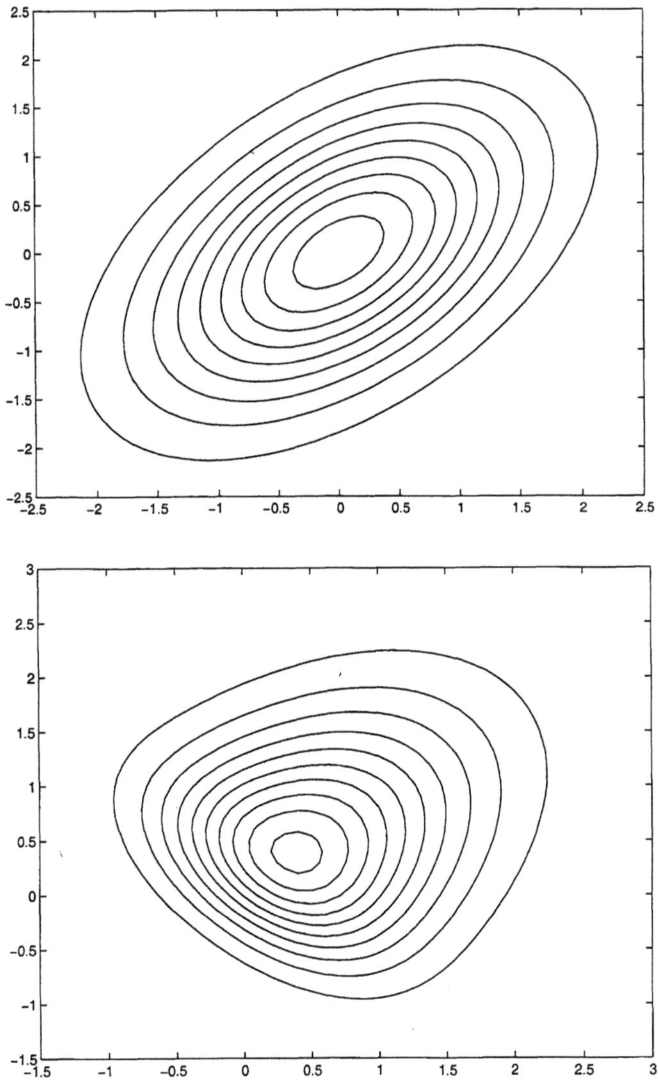

FIG. 1. *Contours of the multivariate skew-normal pdf (2.1) for $n = 2$, $\boldsymbol{\xi} = (0,0)^T$, Ω the correlation matrix with correlation 0.5, with $\boldsymbol{\alpha} = (0,0)^T$ (top panel) and $\boldsymbol{\alpha} = (2,2)^T$ (bottom panel).*

distribution are still skew-normal. Further properties and applications of skew-normal distributions can be found in Azzalini and Dalla Valle (1996), Azzalini and Capitanio (1999), and Genton et al. (2001). One drawback of skew-normal time series is that only skewness is accounted for, and not kurtosis. The next section deals with this issue.

2.2. Skew-t time series. A skew-t time series is a skew-elliptical time series defined with the multivariate pdf (1.2) where g is a multivariate t pdf t_n with correlation matrix Ω and π is the standard t univariate cdf T, i.e.

$$h_n(x) = 2t_n(x - \xi; \Omega)T(\alpha^T(x - \xi)), \tag{2.5}$$

where

$$t_n(x; \Omega) = \frac{\Gamma((\nu + n)/2)}{(\pi\nu)^{n/2}\Gamma(\nu/2)}|\Omega|^{-1/2}\left(1 + \nu^{-1}x^T\Omega^{-1}x\right)^{-(\nu+n)/2}. \tag{2.6}$$

The parameter $\alpha \in \mathbb{R}^n$ controls the skewness and $\alpha = 0$ reduces to the multivariate t case. The parameter ν controls the kurtosis and $\nu = 1$ reduces to the multivariate skew-Cauchy distribution, whereas $\nu \to \infty$ reduces to the multivariate skew-normal distribution. The multivariate skew-t distribution was introduced by Branco and Dey (2001) and its first two moments are given by

$$\mathrm{E}(x) = \xi + \frac{\Gamma((\nu - 1)/2)}{\Gamma(\nu/2)}\sqrt{\frac{\nu}{\pi}}\delta, \qquad \text{if } \nu > 1, \tag{2.7}$$

$$\mathrm{Var}(x) = \frac{\nu}{\nu - 2}\Omega - \left(\frac{\Gamma((\nu - 1)/2)}{\Gamma(\nu/2)}\right)^2\frac{\nu}{\pi}\delta\delta^T, \qquad \text{if } \nu > 2, \tag{2.8}$$

where

$$\delta = \frac{\Omega\alpha}{(1 + \alpha^T\Omega\alpha)^{1/2}}. \tag{2.9}$$

Note that both the expectation and the variance of x depend on the skewness parameter α (or equivalently δ). In order to describe a stationary skew-t time series with constant mean, we require that $\xi = \xi 1_n$ and $\delta = \delta 1_n$, where $1_n = (1\ldots 1)^T \in \mathbb{R}^n$. Here again, the marginal distributions of a multivariate skew-t distribution are still skew-t, see Branco and Dey (2001).

3. Modeling the distribution of sea level. To illustrate the usefulness of multivariate skew-t time series we will now model variations in sea level recorded at Charlottetown, a coastal city on the south shore of Prince Edward Island in the Gulf of St Lawrence (the tide gauge is located at 46.23°N, 63.12° W). Charlottetown was chosen for two reasons. First, it is a low-lying coastal city and so there is considerable interest in the possible effect of climate change on flooding risk. Second, Charlottetown has one of the longest sea level records in Canada and this has allowed us to examine the usefulness of the multivariate skew-t distribution in modeling both seasonality and other forms of nonstationarity.

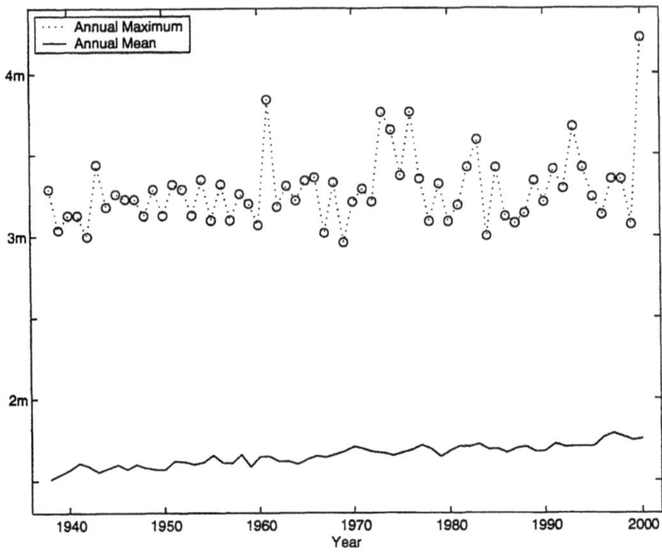

FIG. 2. *Annual means and annual maxima of Charlottetown hourly sea levels, 1938 to 2000 inclusive. The levels are specified with respect to local Chart Datum.*

3.1. Annual means, annual maxima, and residuals. The annual means of sea level at Charlottetown have been increasing almost linearly at a rate of about 3 mm per year from 1938 to the present day (Figure 2, lower trace). This linear trend is due in part to a rise in global sea level of about 1 mm per year; the remainder is believed to be due to subsidence of the Earth's crust in this region.

The annual maxima of the hourly sea levels are, not surprisingly, more variable than the annual means (Figure 2, upper trace). Variations in the annual maxima can exceed 1 m (compare the maxima for 1999 and 2000). From Figure 3 we can conclude that the distribution of the annual maxima for each half of the record are reasonably consistent with a Type I extreme value distribution, i.e. with a cdf of the form $\exp(-\exp(-(y-\alpha)/\beta))$ where α and β are location and scale parameters (see for example Leadbetter et al., 1982). From Figure 3 it can also be seen that the probability of an annual maximum not exceeding a specified level is higher for the first half of the record, pointing to an increase in the annual maxima in recent decades.

Note the 1.5 m offset of the two traces in Figure 2. A significant part of this difference between annual means and annual maxima is due to the tide, the dominant component of sea level variability at Charlottetown. The decomposition of sea level into its tidal and aperiodic component is illustrated in Figure 4 for the latter part of January, 2000. During this period an intense storm piled up water in the southern Gulf of St Lawrence and caused extensive flooding of downtown Charlottetown. In fact this

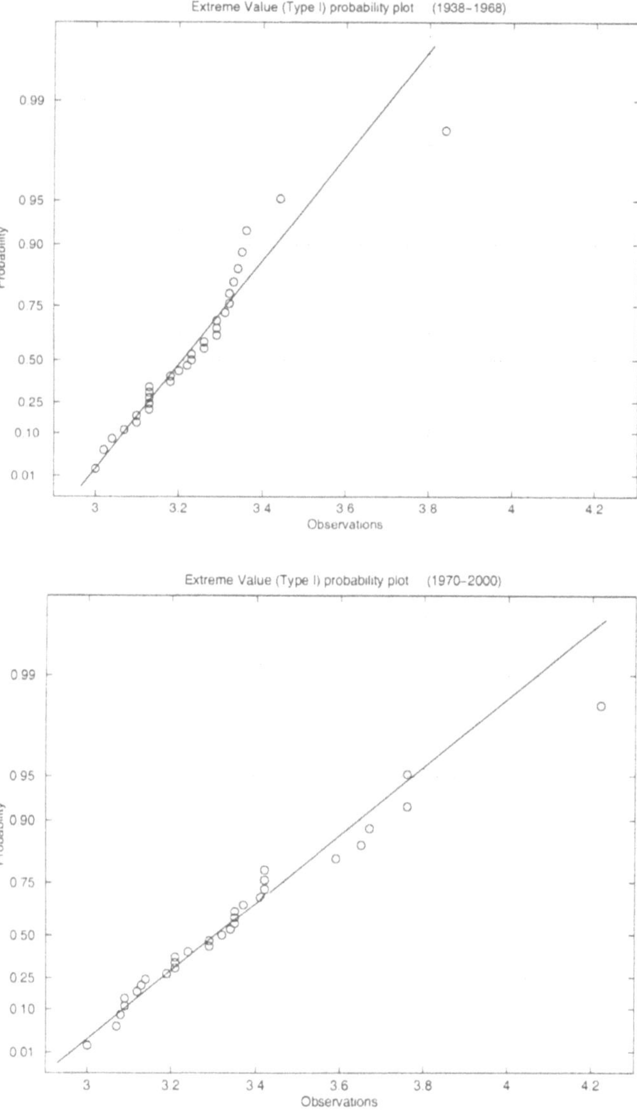

FIG. 3. *Extreme value (Type I) probability plots based on the annual maxima of Charlottetown hourly sea levels. The top and bottom panels are for the first and last halves of the record respectively. The straight lines are the maximum likelihood fits.*

storm caused the highest sea level observed at Charlottetown between 1938 and 2000 (compare Figures 2 and 4).

The tidal component of sea level is forced by the gravitational pull of the sun and moon and, for the purposes of this study, can be treated as deterministic. The difference between the observed sea level and the

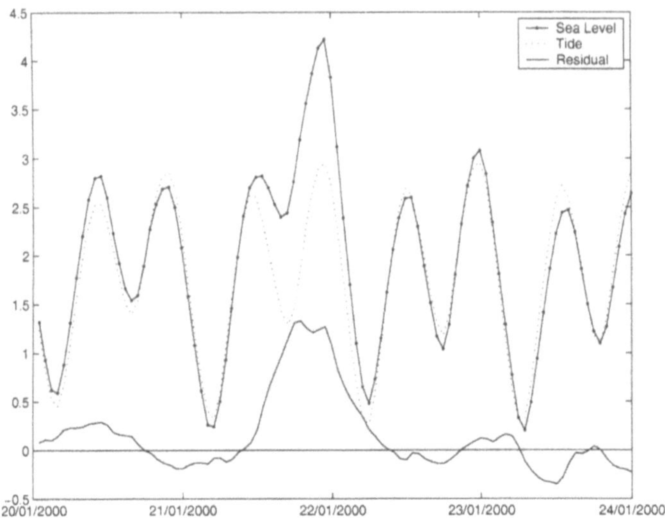

FIG. 4. *Decomposition of sea level into tidal and residual components. The solid line with dots shows the observed hourly sea level variation at Charlottetown from 20 to 23 January, 2000. The dotted line is the predicted tide. The difference between the observed sea level and predicted tide is termed the residual. It is shown by the solid line that fluctuates about zero.*

tide is caused by oceanographic and meteorological factors and is usually called the "residual" (i.e. the part that remains after the predicted tide is subtracted from the observed level). The most important cause of residual variability at Charlottetown is the passage of intense storms and the associated variations in air pressure and wind. It should therefore not be surprising to learn that the residuals exhibit strong seasonal dependence with the largest residuals occurring in winter (Figure 5).

In this study we have treated the residuals as stochastic and modeled their bivariate distribution with a bivariate skew-t distribution. From the fitted bivariate skew-t distribution we were then able to estimate flooding risk as explained in the following section.

Before fitting the bivariate skew-t distribution to the residuals the sea level record was split into two equal halves (1938-1968, 1970-2000 inclusive) and the residuals were stratified by calendar month. This allowed us to model secular and seasonal changes in the residual distribution. Twelve bivariate skew-t distributions were then fit to pairs of adjacent hourly residuals from each half of the record: one for January, one for February and so on through to December.

The method of maximum likelihood was used to fit the bivariate skew-t distribution to pairs of adjacent residuals. Let $\mathbf{x}_t = (x_t, x_{t+1})^T$ denote a residual pair starting at time t. To reduce the dependence amongst

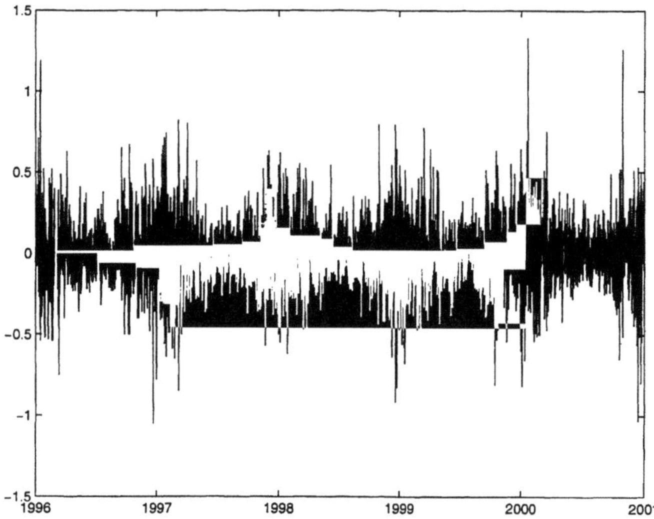

FIG. 5. *Variations in the hourly sea level residuals for the last 5 years of the Charlottetown record. Note the strong seasonality in the record with the largest residuals always occurring in winter. The ticks mark the first day of January for each year.*

pairs, they were subsampled every τ hours and the following likelihood function, which assumes independence, was maximized with respect to ξ, Ω_{11}, Ω_{12}/Ω_{11}, ν and α:

$$h(\mathbf{x}_1, \mathbf{x}_{1+\tau}, \mathbf{x}_{1+2\tau}, \ldots) = h_2(\mathbf{x}_1)h_2(\mathbf{x}_{1+\tau})h_2(\mathbf{x}_{1+2\tau}) \cdots$$

where $h_2(\mathbf{x})$ is the bivariate skew-t distribution (2.5). It is possible to obtain another $\tau - 1$ estimates by starting the subsampling at times 2 through τ. Using the fact that maximum likelihood estimators are asymptotically unbiased we averaged the τ estimates to provide a more reliable estimate of the 5 parameters.

Table 1 summarizes the statistics of residual variability and the estimated parameters of the bivariate skew-t distribution for the second half of the record. (We assumed a subsampling rate of $\tau = 24$ hours which is reasonable given the strength of serial correlation in the residual record.) We see that the sample standard deviation (s) exhibits a pronounced seasonal variation with winter values that are more than double the summer values. This is consistent with Figure 5. The autocorrelations at lag 1, $\hat{\rho}(1)$, are close to unity but slightly smaller in summer. If the residual process is AR(1) the e-folding time in February is $- [\log(0.97)]^{-1} = 32$ hours; in August the e-folding time is $= [\log(0.94)]^{-1} = 16$ hours. This means the residual process has a shorter "memory" in summer and this is another aspect of the seasonality of the process. The kurtosis indicates heavier tails

TABLE 1

Statistics of residual variability and the estimated parameters of the bivariate skew-t distribution for the period 1970 to 2000. The first column gives the month of the year for which the residuals were analyzed. The next four columns give the sample standard deviation, autocorrelation at a lag of 1 hour, kurtosis (ku) and skewness (sk) calculated directly from the residuals. The remaining 5 columns list the estimated parameters of the bivariate skew-t distribution as defined in the text.

Month	s	$\hat{\rho}(1)$	ku	sk	$\hat{\xi}$	$\hat{\Omega}_{11}^{1/2}$	$\dfrac{\hat{\Omega}_{12}}{\hat{\Omega}_{11}}$	$\hat{\nu}$	$\hat{\alpha}$
1	0.23	0.96	4.9	0.13	0.005	0.19	0.97	6.2	0.01
2	0.21	0.97	4.1	0.15	-0.036	0.18	0.97	7.5	0.38
3	0.20	0.97	4.4	0.29	-0.015	0.17	0.97	7.2	0.29
4	0.15	0.96	4.3	0.49	-0.046	0.13	0.97	6.6	2.03
5	0.11	0.95	3.9	0.25	-0.056	0.11	0.96	9.9	2.11
6	0.10	0.95	4.0	0.33	-0.035	0.10	0.96	9.6	2.26
7	0.09	0.95	3.6	0.32	-0.039	0.10	0.96	11.8	3.67
8	0.10	0.94	3.6	0.36	-0.056	0.10	0.96	10.4	3.91
9	0.13	0.95	4.4	0.37	-0.032	0.11	0.96	7.5	1.45
10	0.16	0.96	5.4	0.41	-0.033	0.13	0.96	5.8	0.79
11	0.18	0.96	4.2	0.30	-0.024	0.16	0.97	6.9	0.60
12	0.23	0.96	4.8	0.13	-0.010	0.19	0.97	5.4	0.14

than the normal, with lightest tails in summer, and the skewness is positive throughout the year. For the estimated parameters of the bivariate skew-t distribution, we see that estimates of ν and α tend to be highest in summer.

Typical probability plots are shown in Figure 6. It is clear that the skew-t distribution fits the residual histograms better than the normal distribution. Of particular note is the ability of the skew-t distribution to describe the heavy tails of the January residual histogram. Similar fits were found for the other 11 months. Figure 7 shows probability density contours for the bivariate normal and skew-t distributions fit to the residuals from April, 1970 to 2000. This figure clearly shows the ability of the skew-t distribution to model the positive skewness evident in the residuals.

3.2. Flooding risk. The extreme value probability plot shown in Figure 3 gives some indication of flooding risk. For example, the probability an annual maximum will not exceed the critical level $\eta_c = 3.8$ m is estimated to be 0.957 (corresponding to a "return time" of $1/(1-0.957) = 23.2$

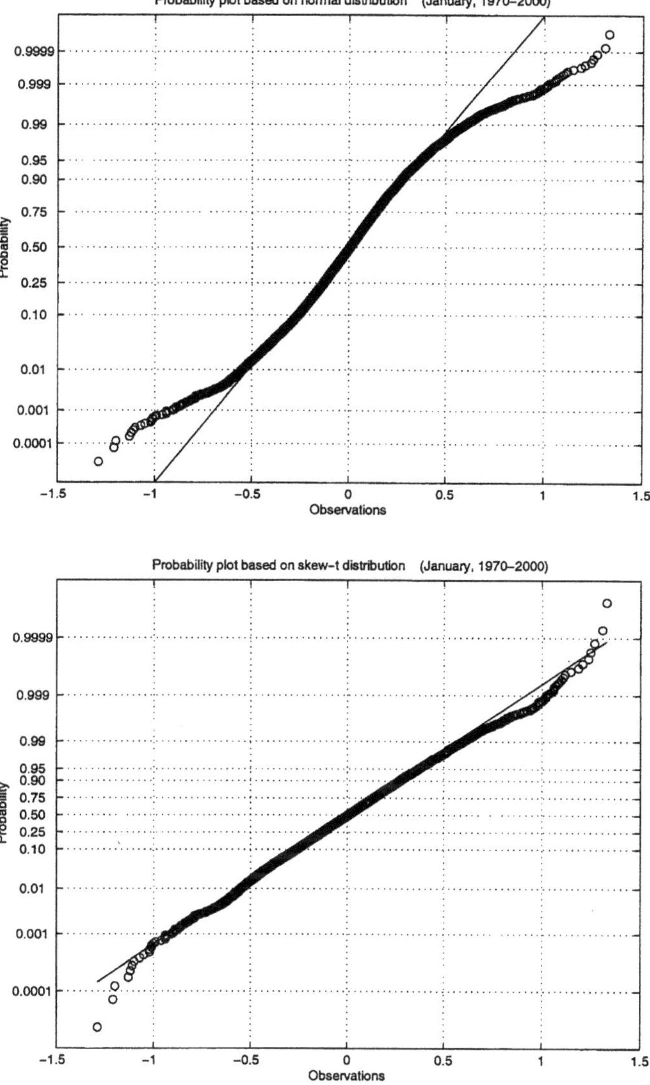

FIG. 6. *Probability plots for the January residuals, 1970 to 2000 inclusive. The top-hand panel is based on the normal distribution and the bottom-hand panel is based on the skew-t distribution.*

years). Assuming the annual maxima are independent, the probability that no hourly sea level will exceed η_c over a 31 year period (the time span of 1970 to 2000) is $0.957^{31} = 0.256$.

To show how the skew-t distributions can be used to calculate flooding probabilities, such as those described in the preceding paragraph, we

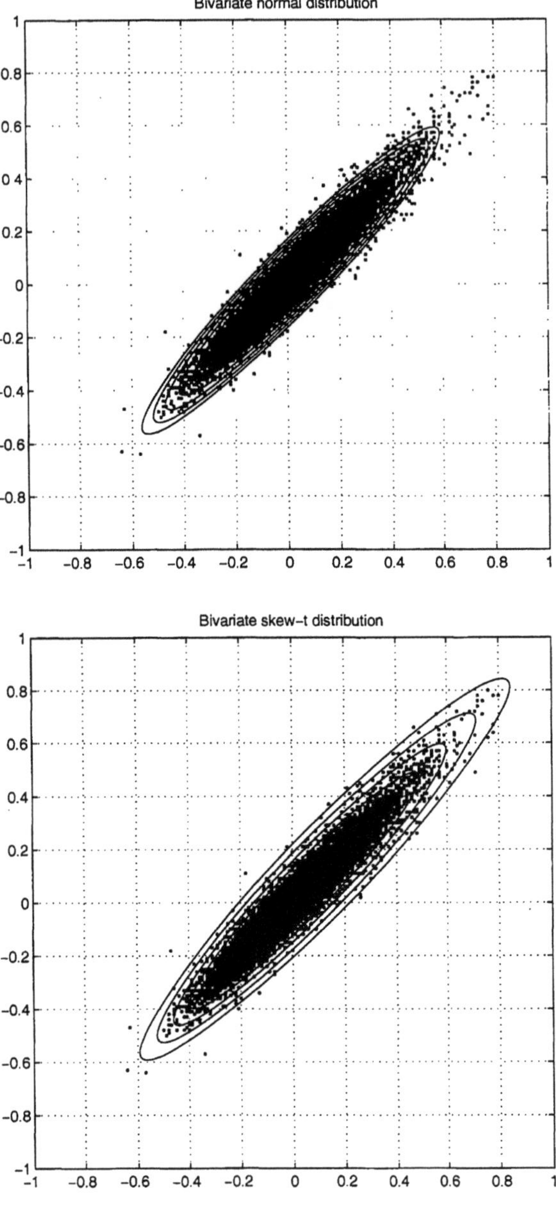

FIG. 7. *Contours of the bivariate normal pdf (top panel) and the bivariate skew-t pdf (bottom panel) fitted to all pairs of adjacent April residuals from 1970 to 2000 (using a subsampling rate of $\tau = 24$ hours). The contour values are logarithmically spaced and equal to 10^{-2}, $10^{-1.5}$, 10^{-1}, $10^{-0.5}$, 10^{0}, $10^{0.5}$ and 10. The dots are the observed residual pairs.*

expand as follows the probability that all hourly sea levels between hour 1 and n are less than a specified critical level η_c:

$$(3.1) \qquad p_{12...n} = p_1 \, p_{2|1} \, p_{3|2,1} \, \cdots \, p_{n-1|n-2,...,1} \, p_{n|n-1,...,1}$$

where $p_{t|t-1,t-2,...,1}$ is the probability the sea level is below η_c at time t given it was below for the earlier times $t-1$ through 1. To simplify (3.1) we assume that for large values of η_c it is only necessary to condition on the preceding M hours. This leads to the approximation

$$(3.2) \qquad p_{12...n} \approx p_1^{(M)} \, p_2^{(M)} \, p_3^{(M)} \, \cdots \, p_{n-1}^{(M)} \, p_n^{(M)}$$

where $p_t^{(M)} = p_{t|t-1,t-2,...,t-M}$.

To evaluate $p_t^{(M)}$ we write the sea level at hour t as follows:

$$\eta_t = \eta_t^T + \eta_t^R$$

where superscript T and R denote the tidal and residual components respectively. The event $\eta_t < \eta_c$ is equivalent to $\eta_t^R < \eta_c - \eta_t^T$ i.e. the residual at time t is not large enough to "jump the gap" between the predicted tide and the critical level. Thus $p_t^{(M)}$ can be expressed in terms of conditional probabilities involving η_n^R which, in turn, can be estimated by fitting an $M+1$ dimensional skew-t distribution to the observed residuals. This has been done for the case $M = 1$ (i.e. conditioning on the previous sea level) after stratifying the residuals by month to allow for seasonality as explained earlier. It was then a straight forward calculation to compound the conditional probabilities according to (3.2) and estimate the probability the sea level would not exceed η_c.

The result of a typical calculation of $p_{12...n}$ using the skew-t and taking $\eta_c = 3.8$ m is shown in Figure 8. The small steps in the trace correspond to the seasonal transitions from summer to winter when the residuals tend to be larger and the probability of an exceedance of η_c is greatest. (This is consistent with the seasonal variation in variance shown in Table 1.) The probability η_c will not have been exceeded during this 31 year period is 0.126 according to Figure 8. This is equal to the probability that all annual maximum over this 31 year period are below $\eta_c = 3.8$ m. If we assume the annual maxima are independent, and denote by p the probability that an annual maximum is below η_c, it follows that $p^{31} = 0.126$. This implies $p = 0.126^{1/31} = 0.934$. This calculation allows us to take the probabilities calculated using (3.2) and plot them on the extreme value probability plot of the annual maxima. This has been done in Figure 9 for η_c increasing from 3.40 m to 4.25 m in steps of 0.05 m. The agreement between return periods calculated by the two approaches is encouraging and suggests we may be able to use the skew-t distribution and conditional probability approach to calculate flooding risks in a strongly nonstationary setting (e.g. changing

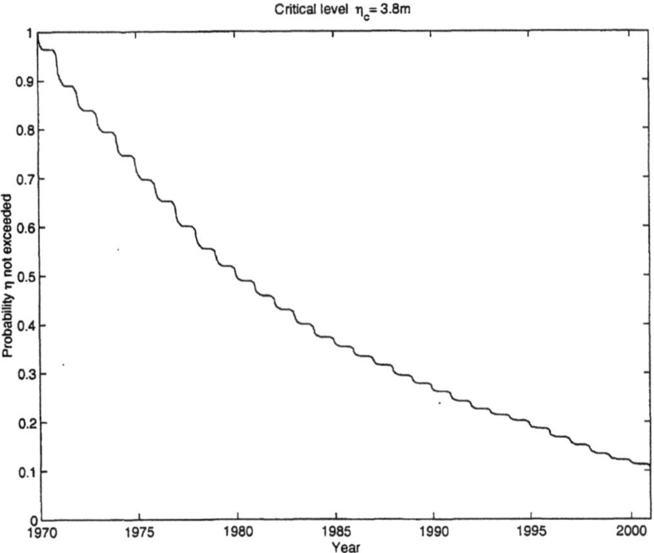

FIG. 8.   *Probability that sea level will not exceed $\eta_c = 3.8$ m as a function of time starting January 1, 1970. The trace was obtained by compounding conditional probabilities calculated from the skew-t distributions fit to the residuals from 1970 to 2000. See text for details.*

the mean, or perhaps slowly increasing the standard deviation of the skew-t distribution through time to reflect changes in storminess under various climate change scenarios).

To conclude this section we present in Figure 10 a calculation showing the probability of Charlottetown sea level not exceeding $\eta_c = 4.5$ m over the next 100 years. The predicted tides for this period were calculated in the standard way and the seasonally stratified, bivariate skew-t distributions based on 1970-2000 residuals were used to calculate $p_t^{(1)}$. To gauge the effect of rising sea level on flooding risk we added two linear sea level trends to the predicted tide with different slopes: 3 mm per year is the present value and 7 mm per year is the value proposed by the International Panel on Climate Change as a plausible rate for the coming century. Superimposed on the gradual drop and seasonal steps in probability can be seen a small oscillation with a period of about 20 years. This is due to the nodal tide which is forced by moon's gravitational pull; it has an amplitude of several centimeters and a period of 18.6 years. The effect of the increased rate of sea level rise on flooding risk is clearly evident in Figure 10: the probability of at least one exceedance of 4.5 m during the next century is about 0.3 if sea level continues to rise at its present rate, and about 0.8 if it increases to 7 mm per year.

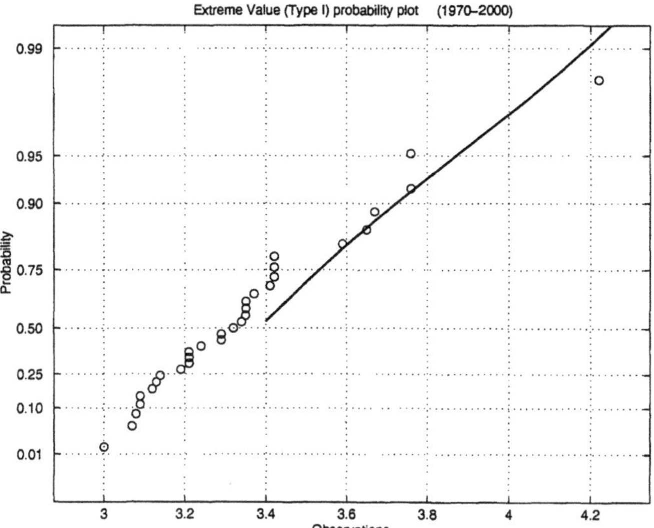

FIG. 9. *Extreme value probability plot based on annual maxima of hourly sea levels, 1970-2000 inclusive. The line shows the probabilities calculated using the skew-t distribution and (3.2) as explained in the text.*

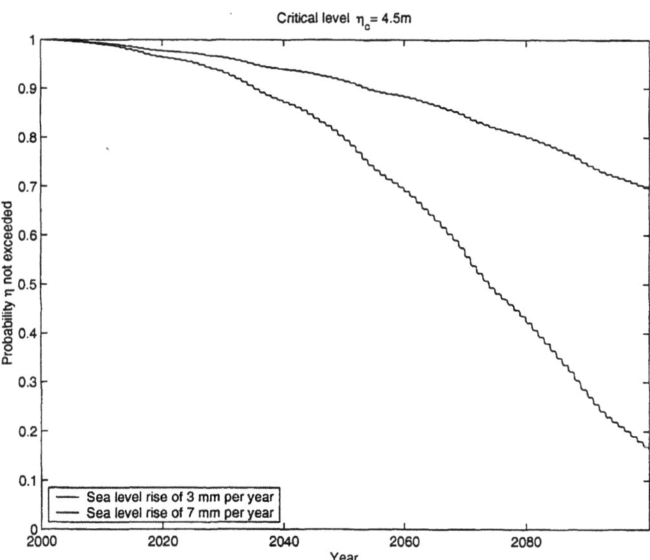

FIG. 10. *Probability of not exceeding the critical level $\eta_c = 4.5$ m over the next century assuming a rate of sea level rise of 3 mm per year (upper trace) and 7 mm per year (lower trace). The results are based on compounding conditional probabilities calculated from the skew-t distribution fit to the seasonally stratified residuals, 1970 to 2000 inclusive.*

4. Conclusions. In this article, the multivariate skew-t distribution has been shown to fit well the distribution of residual sea level at Charlottetown. In particular it captured the heavy tails and skewness in the residuals, features not reproducible by the normal distribution. The bivariate skew-t distribution was used to estimate conditional probabilities of sea level not exceeding a given level, given it was below on the previous time step. By compounding these conditional probabilities it was possible to recover the return periods of extreme sea level calculated in the standard way using extreme value theory and observed annual maxima. This encouraged us to use the conditional probability approach to calculate the risk of flooding a specified critical level as a function of time over the next century. The calculated probabilities reflect the strong seasonality in the sea level process and also long period tidal effects. An increase in the rate of rise of sea level from 3 to 7 mm per year was shown to have a dramatic effect on flooding risk.

The reason simple compounding of conditional probabilities can give reasonable return period of extreme levels is that sea level at Charlottetown is dominated by the tide. This means that even though the highest residuals in the record may not have occurred at the highest tides, which we treat at deterministic, the probability of this coincidence occurring can be estimated without extrapolating into the tails of the residual distribution. This idea has been used by a number of authors following the lead of Pugh and Vassie (1980).

The present approach was designed specifically to quantify flooding risk in a strongly nonstationary setting and the preliminary results presented here are encouraging. There are however a number of issues that need to be addressed before the method is used for practical purposes including, most importantly, the number of residual levels on which to condition i.e. the choice of M and its dependence on η_c in strongly nonstationary situations.

Acknowledgments. Part of this research was carried out while the first author was visiting the second at Dalhousie University, whose hospitality is gratefully acknowledged. The authors thank an anonymous referee for valuable and constructive comments on the article. We would also like to thank Bruce Smith and Chris Field for useful discussions.

REFERENCES

[1] A. AZZALINI, *A class of distributions which includes the normal ones*, Scand. J. Stat., **12** (1985), pp. 171–178.

[2] A. AZZALINI AND A. DALLA VALLE, *The multivariate skew-normal distribution*, Biometrika, **83** (1996), pp. 715–726.

[3] A. AZZALINI AND A. CAPITANIO, *Statistical applications of the multivariate skew normal distribution*, J. R. Statist. Soc., B **61** (1999), pp. 579–602.

[4] M.D. BRANCO AND D.K. DEY, *A general class of multivariate skew-elliptical distributions*, Journal of Multivariate Analysis, **79** (2001), pp. 99–113.

[5] M.G. GENTON, L. HE, AND X. LIU, *Moments of skew-normal random vectors and their quadratic forms*, Statistics and Probability Letters, **51** (2001), pp. 319–325.

[6] M.G. GENTON AND N.M. LOPERFIDO, *Generalized skew-elliptical distributions and their quadratic forms* (2002), Institute of Statistics Mimeo - Series #2539, under review.

[7] D.C. HOAGLIN, F. MOSTELLER AND J.W. TUKEY, *Exploring Data Tables, Trends, and Shapes*, Wiley, 1985.

[8] M.R. LEADBETTER, G. LINDGREN, AND H. ROOTZEN, *Extremes and Related Properties of Random Sequences and Processes*, Springer Series in Statistics, Springer-Verlag, 1982.

[9] D.T. PUGH AND I.M. VASSIE, *Applications of the Joint Probability Method for Extreme Sea Level Computations*, Proceedings of the Institute of Civil Engineers, Part 2, 69, 959–975, 1980.

[10] S.K. SAHU, D.K. DEY, AND M.D. BRANCO, *A new class of multivariate skew distributions with applications to Bayesian regression models*, The Canadian Journal of Statistics, **31** (2003).

HIDDEN PERIODICITIES ANALYSIS AND ITS APPLICATION IN GEOPHYSICS

ZHONGJIE XIE*

Abstract. This paper describes the use of spatial hidden periodicity analysis (SHPA) for the determination of the number of the harmonic components and hidden frequencies. All the estimators are strongly consistent. The method is used for the modeling and forecasting of spatial data of permeability in oil fields of China.

Key words. Spatial time series, hidden periodicities analysis.

1. Introduction. In oil field development, the determination of the permeability values of the rock strata is one of the most important tasks required for the evaluation of the oil reserves in the field. The determination of the permeability of the rock strata is usually based on data obtained in the drilling of wells. The cost of drilling wells, particularly deep wells in desert areas, is a great outlay for an oil company. As a result there is a strong desire to reduce the number of the wells that have to be drilled in the oil field, while still obtaining comparatively accurate permeability values. Permeability forecasting by statistical means represents an interesting and challenging approach to the estimation of oil reserves without the need to drill additional wells.

For the sparse data obtained from real measurements, geologists and statisticians generally use regression methods for interpolation and forecasting. Unfortunately, many researchers have shown that regression methods sometimes give very poor performance on model fitting and forecasting (Makridakis et al., 1982).

There is a large and important literature on two-dimensional periodic models, as exemplified by Priestley (1964, 1997), Chen, Wu, and Dahlhaus (2000), and Zhang and Mandrekar (2001), Harbauch and Merriam (1968) suggest harmonic trend analysis for the modeling of geological data and showed that good results can be obtained for many real applications in geosciences. However, in the use of the two-dimensional harmonic model, one of the difficulties is the determination of the number of harmonic components, i.e., the orders M and N (see (6–10) in Harbauch and Merriam, 1968). These authors give several suggestions on how to determine the integers M and N, but their approach lacks of mathematical basis, e.g. asymptotic statistical analysis of the estimates of the model.

In recent years, many papers discuss the spectral analysis of spatial data (He, 1999; Yaglom, 1987). In this paper, we will briefly introduce the theory and method of He (1999), and will show how to apply Spatial Hidden Frequencies Analysis (SHFA) to permeability modeling in the S3

*School of Mathematical Sciences, Peking University, Beijing 100871, China (zjxie@pku.edu.cn).

rock stratum in Northwest China. Compare with the regression method, SHFA shows better performance. In Section 2, we will summarize the theory and method of SHFA. In Section 3, we demonstrate the practical modeling and forecasting of the permeability.

2. Main results of SHFA. For hidden-frequency analysis in the one-dimensional case, many results may be found in the literature (Brillinger, 1981, Priestly, 1981, Grenander and Rosenblatt, 1957; Xie, 1993). In the case of spatial time series, He (1999) introduced the following:

Suppose that the model of the spatial time series is

$$y(t, s) = \sum_{k=1}^{P} \eta(k) \exp\{i\lambda_k t + i\mu_k s\} + \xi(t, s)$$
$$t = 1, 2, \cdots, N; \qquad s = 1, 2, \cdots, M .$$

In the above equation,$P, \{\lambda_k, \mu_j\}, \{\eta(k)\}$ are unknown (or random) parameters, $\xi(t, s)$ is a stationary random field defined on $Z \times Z$. The determination of the order P and $\{(\lambda_j, \mu_j, \eta_j), j = 1, 2, \ldots, p\}$ is an interesting problem in modern time series analysis. Suppose that the frequencies $\{(\lambda_k, \mu_k)\}_1^P$ have been arranged in alphabetic order. Then

Step 1. Put

$$J(\lambda, \mu, N, M) = \sum_{n=1}^{N} \sum_{m=1}^{M} y(n, m) \exp\{-i(n\lambda + m\mu)\};$$
$$C(k, j) = |J(d_k, e_j, N, M)|, \qquad 1 \le k \le 2N; \qquad 1 \le j \le 2M$$
$$\text{Where} \qquad d_k = \frac{k\pi}{N} - \pi, \qquad e_j = \frac{j\pi}{M} - \pi.$$

Step 2. Select an appropriate constant (threshold)

$$\gamma_0 = o(N \times M) = A(N^2 M\sqrt{M} + M^2 N\sqrt{N})^{1/2}; \quad A = 1 \sim 4.$$

If $\lambda_k \ne \lambda_j$, $\mu_k \ne \mu_j$, $k \ne j$ then $\gamma_0 = A(N \times M)^{3/4}$.

Step 3. Put

$$\Sigma = \{(d_k, e_j); \ C(k, j) > \gamma_0; \ 1 \le k \le 2N, \ 1 \le j \le 2M\} .$$

Step 4. Decompose the set Σ into several subsets, where any two subsets should satisfy the distance criterion:

$$\Sigma = \Sigma_1 + \Sigma_2 + \cdots + \Sigma_{\hat{P}(N,M)} .$$

We say $(\lambda_k, \mu_j)$and $(\lambda_{k'}, \mu_j')$ belong to the same subset if

$$\left(|\lambda_k - \lambda_{k'}| \le \frac{2\pi}{\sqrt{N}}\right) \cup \left(|\lambda_k - \lambda_{k'}| \ge 2\pi - \frac{2\pi}{\sqrt{N}}\right);$$
$$\left(|\mu_j - \mu_{j'}| \le \frac{2\pi}{\sqrt{N}}\right) \cup \left(|\mu_j - \mu_{j'}| \ge 2\pi - \frac{2\pi}{\sqrt{N}}\right) .$$

Step 5. Suppose that the $C(j, k)$ achieves its maximal value at $(\hat{d}_i, \hat{e}_i)$ within Σ_i. Then put

$$\hat{\lambda}_i(N, M) = \hat{d}_i \; ; \quad \hat{\mu}_i(N, M) = \hat{e}_i \; .$$

We arrange all the two-dimensional indices $\{(\hat{d}_i, \hat{e}_i)\}$ in alphabetic order, and denote the frequency estimators as

$$\{\hat{\lambda}_k(N, M), \; \hat{\mu}_k(N, M); \; k = 1, 2, \cdots, \hat{P}(N, M)\} \; .$$

Then we have the following result (He, 1999):

THEOREM 1. *If the noise model $\xi(t, s)$ is the linear process*

$$\xi(t, s) = \sum_{k \geq 0} d_k w_{n-k}, \quad n \in \mathbf{Z}^2; \qquad \sum_{k \geq 0} (k_1 + k_2)|d_k| < +\infty$$

where $\{w_k\}$ is a two-dimensional strictly stationary ergodic random field, $\{w_k, k \in Z_+^2\}$ is a $1/4$ martingale difference white noise

$$Ew_k = 0; \quad Ew_0^2 \log |w_0| < +\infty \; .$$

Then

$$\underset{N,M \to \infty}{Lim} \; \hat{P}(N, M) = P, \; \text{a.s.}$$

Similar results

$$\hat{\lambda}_k(N, M) \to \lambda_k, \; a.s. \quad \text{when } N, M \to +\infty;$$
$$\hat{\mu}_k(N, M) \to \mu_k, \; a.s. \quad \text{when } N, M \to +\infty$$

also hold.

Figure 1 shows a decomposition of the set Σ.

In practical applications, one of the following two real models is often used

$$y(t, s) = \xi(t, s) + \sum_{k=1}^{P} \big[A_k \cos(\lambda_k t) \cos(\mu_k s) + B_k \cos(\lambda_k t) \sin(\mu_k s)$$
$$+ \, C_k \sin(\lambda_k t) \cos(\mu_k s) + D_k \sin(\lambda_k t) \sin(\mu_k s) \big]$$

or

$$y(t, s) = \xi(t, s) + \sum_{k=1}^{P} [A_k \cos(\lambda_k t + \mu_k s) + B_k \sin(\lambda_k t + \mu_k s)].$$

If we put

$$p = 2q \text{ and } j = 1, 2, \cdots, q$$
$$\lambda_{q+j} = -\lambda_j, \; \mu_{q+j} = -\mu_j; \; \beta_j = (a_j - ib_j)/2, \; \beta_{q+j} = \bar{\beta}_j$$

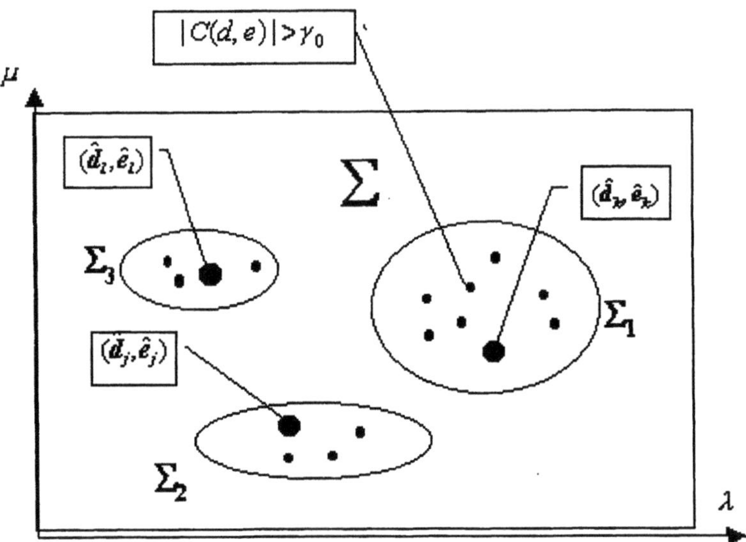

FIG. 1. *Decomposition of the set Σ and estimation of frequencies.*

then

$$\sum_{j=1}^{q} [A_j \cos(\lambda_j t + \mu_j s) + B_j \sin(\lambda_j t + \mu_j s)] = \sum_{j=1}^{p} \beta_j \exp[i(\lambda_j t + \mu_j s)] .$$

3. The permeability modeling of sandstone S3 at Turfan-Hami basin. Permeability is one of the most important parameters needed for the oil reserve evaluation. Permeability forecasting is a very new research topic in oil field development. In the following, we want to introduce our preliminary research work on the modeling and forecasting of the permeability of sandstone, Jurassic strata S3, at the Turfan-Hami basin, North-West China.

Some of the real sparse permeability samples are as in the following

(01, 06, 1222), (03, 06, 823), (06, 04, 1020), (13, 04, 1195), (17, 06, 769),
(18, 06, 1161), (24, 06, 1007), (27, 08, 1153), (28, 08, 1563), (29, 08, 961),
(29, 07, 1384), (30, 08, 1192), (30, 07, 1393), (31, 07, 1746), (31, 09, 1211),
(32, 07, 1012), (32, 09, 943), (33, 08, 968), (33, 09, 1580), (34, 08, 1204),
(36, 03, 873), (35, 06, 1026), (38, 06, 830), (39, 03, 844), (40, 04, 1466),
(41, 03, 1130), (44, 05, 753), (50, 05, 1020), (51, 05, 646), (52, 07, 1758),
(54, 06, 956), (56, 06, 1036), (57, 07, 997), (58, 09, 1711), (61, 09, 1355),
(60, 10, 793), (62, 09, 1068), (63, 08, 1046), (64, 09, 1119), (65, 09, 1131),
(66, 09, 653), (67, 10, 739), (68, 11, 801), (69, 13, 991), (70, 13, 776).

Each vector value denotes $(x_k, y_k, p(x_k, y_k))$, where p is the permeability value (in units of $10^{-3} \mu m^2$) at the location (x_k, y_k). For the purpose of forecasting how does one construct a good model based upon sparse data?

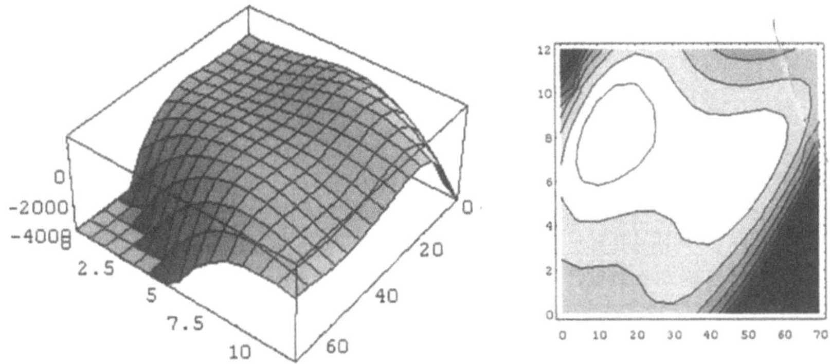

FIG. 2. *Spatial polynomial fitting: three-dimensional plot (left) and the corresponding contour plot (right).*

Generally geologists use the polynomial regression, but practical experience shows that good results are not obtained in many cases. For example, using a polynomial regression to fit a data set, we obtain the optimal fitting model given by the following function:

$$\begin{aligned}
\varsigma(x,y) = {} & 0.35858 + 2.426x - 2.033x^2 + 0.173x^3 \\
& - 0.0038x^4 + 7.37 \times 10^{-6}x^5 + 1.595y + 11.064xy \\
& - 0.906x^2y + 0.0324x^3y + 6.0353y^2 - 0.143x^2y^2 \\
& + 15.524y^3 + 0.383xy^3 - 2.636y^4 + 0.0849y^5 \; .
\end{aligned}$$

The 3-dimensional plot and the corresponding counter plot are depicted in Fig. 2.

From the figure we see that the model fits comparatively well in the medium portion, but fits badly at the edges. Particularly, it is unacceptable that some permeability values are negative.

Now, we introduce the hidden frequencies model. For the purpose of easy comparison and identification from the geoscience point of view, the stratum is divided into 12 layers of equal thickness. In each layer, the sparse permeability data are distributed on a two-dimensional field. Then in each layer independently, the data are fitted by the SHFA model.

Consider the real model

$$\begin{aligned}
y(t,s) = \xi(t,s) + \sum_{k=1}^{P} \big[& A_k \cos(\lambda_k t)\cos(\mu_k s) + B_k \cos(\lambda_k t)\sin(\mu_k s) \\
& + C_k \sin(\lambda_k t)\cos(\mu_k s) + D_k \sin(\lambda_k t)\sin(\mu_k s) \big] \; .
\end{aligned}$$

The sample size in two-dimensions is $N = 35$, $M = 26$. Select

$$\gamma_0 = 3(N \times M)^{0.75} = 497$$

as the threshold.

For the data of the 1st layer, we have the following frequencies

$$\{(\lambda_k, \mu_k); \ k = 1, 2, 3, 4\},$$

where

$$|C(\lambda_k, \mu_k)| > \gamma_0 = 497$$

$|C|$ value	680	578	584	1777
λ_k	0.0897	0.3590	0.1795	0.2693
μ_k	0.4488	1.7952	0.8976	1.3464

Put

$$\hat{y}(t, s) = C + \sum_{k=1}^{3} \Big[A_k \cos(\lambda_k t) \cos(\mu_k s) + B_k \cos(\lambda_k t) \sin(\mu_k s)$$
$$+ \, C_k \sin(\lambda_k t) \cos(\mu_k s) + D_k \sin(\lambda_k t) \sin(\mu_k s) \Big]$$

and select parameters $\{(A_k, B_k, C_k, D_k)\}$ by minimizing the S.S.E.

$$\min_{\{A_k, B_k, C_k, D_k\}} \sum_{t=1}^{N} \sum_{s=1}^{M} |y(t, s) - \hat{y}(t, s)|^2 \ .$$

By SHFA, the final fitted model of the 1st layer is

$$
\begin{aligned}
\hat{y}(t, s) = \ & 1098.95 \\
& - 70.19 \cos(.08976t) \cos(.4488s) + 191.34 \cos(.359t) \cos(1.7952s) \\
& + 256.57 \cos(.17952t) \cos(.8976s) - 479.81 \cos(.2693t) \cos(1.3464s) \\
& + 258.92 \cos(.4488s) \sin(.08976t) - 348.99 \cos(.8976s) \sin(.1795t) \\
& - 196.86 \cos(1.3464s) \sin(.2693t) - 30.42 \cos(1.7952s) \sin(.359t) \\
& + 35.03 \cos(.08976t) \sin(.4488s) - 130.93 \sin(.08976t) \sin(.4488s) \\
& + 36.57 \cos(.1759t) \sin(.8976s) + 257.56 \sin(.1795t) \sin(.8976s) \\
& + 48.82 \cos(.2693t) \sin(1.3464s) - 178.78 \sin(.2693t) \sin(1.3464s) \\
& - 21.34 \cos(.359t) \sin(1.7952s) + 209 \sin(.359t) \sin(1.7952s) \ .
\end{aligned}
$$

The 3D plot of the fitted model (for the 1st layer) is shown in Fig. 3.

The following shows the comparison of fitted permeability values (the first number in each couplet) vs. real samples (the second number in each couplet)

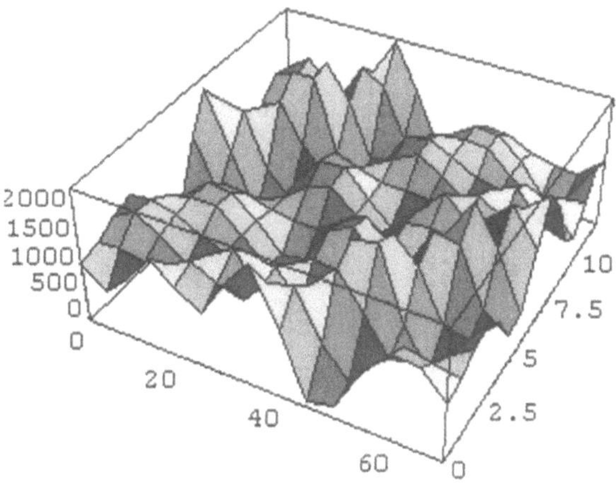

FIG. 3. *SHFA modeling of the first layer of permeability value at the Jurassic strata S3 in Turfan-Harmi.*

(968, 1222),   (1045, 822),   (1160, 1020),   (968, 830),   (973, 1195),
(982, 769),   (962, 1016),   (952, 1561),   (968, 1161),   (1097, 1007),
(1147, 1153), (1189, 1563), (1178, 961),   (1228, 1384), (1152, 1192),
(1358, 1393), (1384, 1746), (1100, 1211),   (1285, 1021), (1100, 943),
(1237, 968),   (1190, 1580), (1232, 1204),   (1243, 991),   (1058, 873),
(1048, 1026), (961, 830),   (1178, 844),   (1025, 1466), (1155, 1130),
(1027, 953),   (1038, 1020), (975, 646),   (1285, 1758), (1090, 956),
(1012, 1036), (972, 997),   (1077, 1711),   (1107, 1355), (1038, 793),
(1043, 1068), (973, 1046),   (947, 1119),   (988, 1131),   (1067, 653),
(915, 739),   (961, 801),   (1020, 991),   (1083, 776).

The bias of the difference is $|E[y - \hat{y}]| = 10$, and the S.D. error is R.M.S. $= \sigma(y - \hat{y}) = 251$. This result is not seen to be bad since the error of our instrument is R.M.S. $= 176$. Contour plots of the first layer and the second layer are shown in Fig. 4

Since our modeling procedure is based upon independent models for each layer, the comparable and reasonable permeability distributions of different layers show that our modeling coincides with the geological background, such as the direction of the ancient river flow and the rock stratum structure etc.

4. Conclusion. It has been shown in this paper that the harmonic model sometime is better than the polynomial regression. Particularly, when we use the spatial hidden periodicities analysis (SHFA), the theory may show us how to determine the number of the harmonic components. The method introduced in this paper is also easy to use in practical computation.

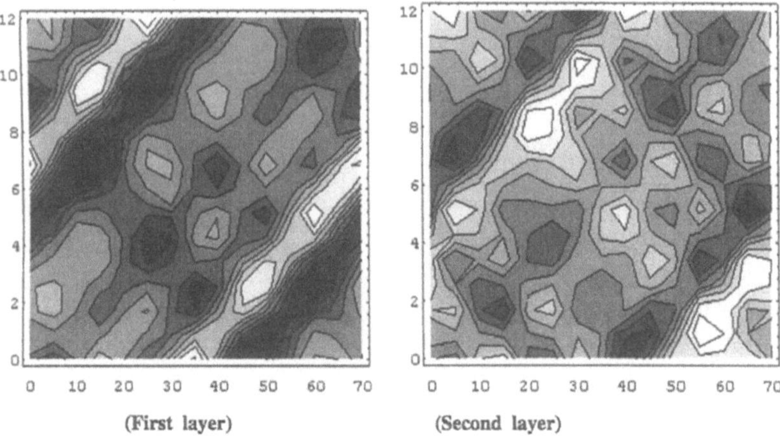

(First layer)   (Second layer)

FIG. 4. *The contour plot of the fitted permeability of the first and second layers.*

Acknowledgements. This research work was supported by the National Natural Science Foundation of China (Grant No. 10171005).

REFERENCES

[1] AN H. AND Z. XIE (1992). Time series analysis in China. In: The development of statistics: recent contributions from China. *Longman Scientific & Technical,* UK, 7–40.

[2] BRILLINGER DAVID R. (1981). Time Series: Data Analysis and Theory. Holden-Day.

[3] CHEN Z.H., WU K.H., AND DAHLHAUS R. (2000). Hidden frequency estimation with data taper. *Journal of Time Series,* **21**, 113–142.

[4] GRENANDER U. AND M. ROSENBLATT (1957). Statistical analysis of stationary time series. Wiley, New York.

[5] HARBAUGH J.W. AND G.B. CARTER (1980). Computer simulation in geology. Wiley-Interscience, N.Y.

[6] HE S. (1999). Parameters estimation for the spatial hidden periodicities model. *Science in China,* Series A, **42**(3), 238–245.

[7] MAKRIDAKIS S. et al. (1982). The accuracy of extrapolation (time series) methods: results of a forecasting competition. *Journ. of Forec.* 1, 111–153.

[8] PRIESTLEY M.B. (1964). The analysis of two dimensional stationary processes with discontinuous spectra, *Biometrica,* **51**, 192–217.

[9] PRIESTLY M.B. (1981). Spectral analysis of time series, Academic Press, London.

[10] PRIESTLEY M.B. (1997). Detection of periodicities, Applications of Time Series in Astronomy and Meteorology, ed. by T. Subba Rao, M.B. Priestley, and O. Lesse, Chapman and Hall, London.

[11] XIE Z. (1993). Case studies in time series analysis, *World Scientific,* Singapore.

[12] YAGLOM A.M. (1987), Correlation theory of stationary and related random functions. Springer-Verlag, New York.

THE INNOVATION APPROACH TO THE IDENTIFICATION OF NONLINEAR CAUSAL MODELS IN TIME SERIES ANALYSIS

T. OZAKI[*], J.C. JIMENEZ[†], H. PENG[‡], AND V.H. OZAKI[§]

Abstract. This paper shows how the innovation approach developed by Wiener (1949), Kalman (1960) and Box and Jenkins (1970) has found wide application in modern nonlinear time series analysis. Nonlinear models, such as the chaos, stochastic or deterministic differential equation models, neural network models and nonlinear AR models developed in the last two decades are reviewed as useful causal models in time series analysis for nonlinear dynamic phenomena in many scientific fields. The merit of the use of the innovation approach in conjunction with these new models is pointed out. Further, the computational efficiency and advantage of RBF-AR models over RBF neural network models is demonstrated in real data analysis of EEG time series of subjects with epilepsy. The advantage of multivariate RBF-ARX models in the modeling of thermal power plants is also shown using numerical results.

Key words. Innovation approach, maximum likelihood method, nonlinear Kalman filter, Markov diffusion process, Pearson system, Gamma distributed processes, stochastic differential equations, chaos, Local Linearization scheme, ExpAR models, RBF neural network, RBF-AR models, RBF-ARX models.

1. Introduction. The innovation approach in time series analysis tries to find a causal model that "whitens" the observed time series data into Gaussian white noise. By doing so it provides us with a method for predicting the series and characterizing the dynamics behind the time series using the identified causal model. At the same time, the optimal predictor, optimal smoother and the optimal filter of the variables used in the model for the time series are obtained from the identified causal model. This technique was developed by Wold (1938), Kolmogorov (1941), Wiener (1949), Kalman (1960), Kailath (1968), Akaike (1969), Box and Jenkins (1970), among others. The full exploitation of Wiener's innovation approach led Kalman to his celebrated dual theory for the optimal linear prediction filter and optimal linear control filter. When these theories were applied to practical control problems, optimal controllers were designed on the basis of identified causal models. Here the causal model for the real data is, of course, unknown and we have to identify an appropriate model from several candidates. The success of the approach in real applications is closely dependent on the "causal model" used in the innovation approach. Naturally, the models employed by the approach were limited to linear models at the early stage of development from the 1940s to 1960s. However there

---

[*]The Institute of Statistical Mathematics, 4-6-7 Minami Azabu, Minato-ku, Tokyo 106-8569, Japan (ozaki@ism.ac.jp).

[†]Institute of Cybernetics, Mathematics and Physics, Cuba.

[‡]Central South University, Changsha 410083, P.R. China. Currently a visiting researcher at the Institute of Statistical Mathematics (peng@ism.ac.jp).

[§]Sophia University, Japan (v_ozaki@sophia.ac.jp).

is no reason why we should restrict ourselves only to linear causal models. Unfortunately the problem of developing useful applicable nonlinear causal models did not attract the attention of time series analysts as much as it attracted applied scientists in neuroscience, financial engineering, meteorology, and biology. The emergence of chaos studies in the 1970s may be regarded as evidence of the shift of scientists' concern from general linear models to more specific nonlinear causal models. The surge of chaos studies was a challenge to the traditional approach of time series analysis. May (1976) made the idea of deterministic chaos widely known among scientists working in time series analysis. In spite of the much-emphasized claim of the determinism of chaos dynamics, many scientists were led to pay more attention to stochastic models with alternative nonlinear dynamics capable of reducing uncertainty, i.e. the prediction errors, with the aim of ultimately making the errors zero, and the model deterministic. In recent time series analysis, as several linear models were overtaken by newly developed nonlinear chaos models and neural network models, the original idea of the innovation approach developed and used by Wiener (1949), Kalman (1960) and Box and Jenkins (1970) seemed to have been put aside as well.

The innovation approach, however, remained the focus of some time series analysts. Studies of nonlinear dynamic models, including those new animals such as deterministic and stochastic differential equation models and neural network models, have been pursued also from the perspective of the traditional innovation approach. In the 1960s, it was widely known, in the traditional time series circles, that nonlinear polynomial AR models are not appropriate for simulating a nonlinear process although they improve prediction performance significantly compared with linear models. They are computationally easy to estimate but they almost always lead to computational explosions if the models are simulated with Gaussian white noise. Because of this, nonlinear time series models that can be used for simulation as well as prediction started attracting the attention of researchers in traditional time series analysis in the 1970s (see the discussions of Campbell and Walker (1977), Tong (1977), Ozaki and Oda (1978)). Here time series models demonstrating some kind of nonlinear dynamic structure were the analysts' main concern.

In the present paper we will see that the innovation approach can play an essential role in the assessment of chaos, neural network models and nonlinear differential equation models as well as linear and nonlinear time series models. All these models belong to the class of Markov process models. One of the important features of Markov process models is that prediction errors are Gaussian when the trajectory of the process defined by the Markov model is continuous (Doob, 1953). This is the reason why the innovation approach can still play an important role in nonlinear time series modeling, where the data are usually sampled from continuous processes with a sufficiently small time interval. We will see, in Section 2, that classic theories developed for Markov processes are useful in directing nonlinear time series modeling. Examples of nonlinear stochastic differential equation models successfully applied to real time series data and artificial data are also shown in Section 2.

The most useful, in terms of prediction and simulation, newly developed nonlinear dynamic models are neural network rather than chaos models. However, a difficulty for neural network modeling is that although the multi-layer network models are very flexible and general, it is very difficult to take advantage of the model flexibility, because the identification method is so complicated and time consuming. Among the many families of neural network models, Radial Basis Function (RBF) neural network models have been recognized as among the most useful in applications (Dyn, 1989a, 1989b). In Section 3, we introduce a recently developed model, called the RBF-AR model, and its performance is compared with the RBF neural network model, by applying both model families to the EEG data of subjects suffering from epilepsy. The study shows the superiority of RBF-AR models over RBF neural network models. Another set of numerical results shows the advantage of using RBF-ARX models for the modeling of power plant controls. Finally a strategy for the future development of time series analysis is discussed in Section 5. The revitalization of the classic innovation approach with more emphasis on modern nonlinear causal modeling, including RBF-AR modeling and stochastic differential equation modeling, is suggested as essential in the future.

2. Markov processes and the innovation approach. By the innovation of a time series, we mean the one step-ahead prediction error, i.e. $\nu_t = x_t - E[x_t | x_{t-1}, x_{t-2}, ...]$, and the innovation approach searches for a model which minimizes the innovation variance. The conditional distribution of x_t given the past observations up to time $t - 1$ is obtained when the model is specified. When the conditional distribution is specified its innovations and the likelihood can be calculated. Many mathematical theories, including the least squares estimation theory of Gauss and Kalman, support this idea. One of the most important and useful theorems came from Frost and Kailath (1971) for continuous time processes, where the prediction error, $\nu(t) = x(t) - E[x(t) | x(\tau), \ \tau \le t - \Delta t]$ is said to be approximately Gaussian white noise even though the process x(t) itself is non-Gaussian. (Here we assume x(t) is Markov and its trajectory is continuous, i.e. $[x(t + \Delta t) - x(t)] \to 0$ in probability for $\Delta t \to 0$).

This theorem justifies the use of Gaussian likelihood for the estimation of the parameters of a nonlinear stochastic differential equation model,

$$\dot{x} = f(x) + n(t)$$

using a discretized model,

$$x_t = A_{t-1} x_{t-1} + B_{t-1} n_t$$

when the sampling interval is reasonably small. Here the Gaussian likelihood is written down using the Gaussian innovation,

$$\nu_t = x_t - A_{t-1}x_{t-1}$$
$$= B_{t-1}n_t.$$

Then (-2)log-likelihood function is given by

$$(-2)\log p(x_1, x_2, ..., x_N) = (-2)\log p(x_2, ..., x_N|x_1) + (-2)\log p(x_1)$$
$$= (-2)\sum_{t=1}^{N-1}\log p(\nu_{t+1}|x_t, ...x_1) + (-2)\log p(x_1)$$
$$= \log\sum_{t=1}^{N-1}\frac{\{B_t^{-1}(x_{t+1}-A_tx_t)\}^2}{\sigma_n^2} + (N-1)\log\sigma_n^2$$
$$+ (N-1)\log 2\pi + 2\sum_{t=1}^{N-1}\log|B_t| + (-2)\log p(x_1).$$

The theorem also suggests that Gaussian likelihood may be used for the estimation of parameters by using prediction errors obtained through the discrete time nonlinear time series model obtained when the time series data are sampled from continuous time dynamic phenomena with a reasonably short time interval.

2.1. Markov processes and their distributions. The need for a time series model which is useful for both prediction and simulation in engineering applications is one of the reasons why some time series analysts (Ozaki, 1985a, 1985b) started looking into continuous time models, where Markov process theories play an important role. In fact to the founders of time series analysts such as Kolmogorov and Wiener, no essential difference existed between continuous time processes and discrete time processes. Both were as important and useful for getting to the solution of prediction and control problems in applications. Wiener meant by "time series" much more than Markov processes and ARMA processes in his book (Wiener, 1961). Kalman also did not bring in any essential discrimination between the continuous time model and the discrete time model. He introduced his celebrated filtering theory for both the discrete and continuous time cases. The separation between time series analysis and continuous time processes is only a recent phenomenon dating back to the 1960s. Looking back, the recent split between stochastic process researchers and statistical time series analysts seems like a transient phenomenon.

One of the advantages of Markov process models is that they are well equipped with strong theories about the distributional characteristics of the processes. Kolmogorov (1931) and later Wong (1963) pointed out an interesting relationship between nonlinear Markov diffusion processes and their marginal distributions. They showed that for any distribution $V(x)$ which belongs to the Pearson system,

$$(2.1) \qquad \frac{dV(x)}{dx} = \frac{c_0 + c_1 x}{d_0 + d_1 x + d_2 x^2} V(x)$$

it is possible to find a diffusion process whose marginal distribution is $V(x)$. Ozaki (1985a) showed that this can be extended to any distribution $V(x)$ defined by proper analytic functions $c(x)$ and $d(x)$ in a distribution system given by,

$$(2.2) \qquad \frac{dV(x)}{dx} = \frac{c(x)}{d(x)} V(x)$$

and that the corresponding Markov diffusion process is given, with $c(x)$ and $d(x)$, by the following Fokker-Planck equation,

$$(2.3) \qquad \frac{\partial p}{\partial t} = -\frac{\partial}{\partial x}[\{c(x) + d'(x)\}p] + \frac{1}{2}\frac{\partial^2}{\partial x^2}[2d(x)p] \ .$$

The class of distributions defined by (2.2) includes not only the distribution family of Pearson systems but also all the distributions in the exponential family. Here we note that $c(x)$ and $d(x)$ of the extended Pearson system (2.2) need not be mutually irreducible. There could be infinitely many pairs of $c(x)$ and $d(x)$ that define one and the same distribution $V(x)$. For example we can think of a diffusion process corresponding to

$$(2.4) \qquad \frac{dV(x)}{dx} = \frac{c(x)x}{d(x)x} V(x),$$

instead of (2.1). From (2.4) we have a diffusion process defined by the following Fokker-Planck equation,

$$\frac{\partial p}{\partial t} = -\frac{\partial}{\partial x}\big[\{c(x)x + d(x) + d'(x)x\}p\big] + \frac{1}{2}\frac{\partial^2}{\partial x^2}[2d(x)xp] \ .$$

This diffusion process has the same marginal distribution $V(x)$ as the diffusion process of (2.3). Each diffusion process defined by the Fokker-Planck equation has a stochastic differential equation representation of the process such as

$$dx = a(x)dt + b(x)dw(t)$$

with a different pair of $a(x)$ and $b(x)$.

By using the variable transformation, $y = h(x)$ satisfying the relation,

$$\frac{\partial h(x)}{\partial x} b(x) = 1$$

each stochastic differential equation may be transformed into a constant variance noise driven stochastic dynamical system,

$$dy = f(y)dt + dw(t).$$

This means that the mechanism generating a certain non-Gaussian distribution from Gaussian white noise $dw(t)$ is not unique. We will see this in the following examples for the Gamma distributed processes.

Example 1. Type I Gamma distributed process.

The Pearson system for the Gamma distribution

$$V(x) = \frac{x^{\alpha-1}\exp(-x/\beta)}{\Gamma(\alpha)\beta^{\alpha}}$$

is given by

$$\frac{dV(x)}{dx} = \frac{(\alpha-1)\beta - x}{\beta x}V(x).$$

Wong (1963) showed a systematic way to introduce a diffusion process whose stationary distribution is $V(x)$ specified by the Pearson system. For the above $V(x)$ we have a diffusion process defined by the following Fokker-Planck equation,

$$\frac{\partial p}{\partial t} = -\frac{\partial}{\partial x}[\alpha\beta - x)p] + \frac{1}{2}\frac{\partial^2}{\partial x^2}[2\beta x p].$$

The diffusion process is called a type I Gamma distributed process and has the following stochastic differential equation representation,

$$dx = (\alpha\beta - x)dt + \sqrt{2\beta x}dw(t).$$

From the stochastic differential equation we are lead to the following state space representation in continuous time,

$$dy = \left\{\frac{(\alpha - 1/2)}{y} - \frac{y}{2}\right\}dt + dw(t)$$

$$x(t) = \frac{\beta y^2(t)}{2}.$$

This shows that the Gamma distributed process $x(t)$ is generated from a Gaussian white noise $dw(t)$ by the instantaneous square transformation and the nonlinear dynamical system,

$$dy = \left\{\frac{(\alpha - 1/2)}{y} - \frac{y}{2}\right\}dt.$$

Example 2. Type II Gamma distributed process.
Since $c(x)$ and $d(x)$ of system (2.1) for the distribution $V(x)$ need not be mutually irreducible, we can think of defining the same Gamma distribution by the following system,

$$\frac{dV(x)}{dx} = \frac{(\alpha - 1)\beta x - x^2}{\beta x^2} V(x).$$

Following the scheme of Ozaki (1985a, 1992) we are led to another diffusion process defined by

$$\frac{\partial p}{\partial t} = -\frac{\partial}{\partial x}[\{(\alpha + 1)\beta x - x^2\}p] + \frac{1}{2}\frac{\partial^2}{\partial x^2}[2\beta x^2 p].$$

The process is called a type II Gamma distributed process and is defined by the stochastic differential equation,

$$dx = \{(\alpha + 1)\beta x - x^2\}dt + \sqrt{2\beta}\,x dw(t).$$

Then the process $x(t)$ has the following continuous time state space representation.

$$x(t) = e^{\sqrt{2\beta}\,y(t)}$$
$$dy = \left\{\frac{\alpha\beta}{\sqrt{2\beta}} - e^{\sqrt{2\beta}\,y}\right\}dt + dw(t).$$

This shows that the Gamma distributed process $x(t)$ is generated from Gaussian white noise $dw(t)$ by the instantaneous exponential transformation and the nonlinear dynamical system,

$$dy = \left\{\frac{\alpha\beta}{\sqrt{2\beta}} - e^{\sqrt{2\beta}\,y}\right\}dt.$$

These examples show that even though the original time series has a non-Gaussian distribution, it may be transformed into Gaussian white noise by using an instantaneous variable transformation and a nonlinear dynamic model. This is also true for processes with fat-tailed distributions. It means that there are infinitely many ways of whitening the non-Gaussian distributed dependent process into independent Gaussian noise. Which model to choose from those possible candidates is a very important problem in time series analysis, and a practical solution is given by the maximum likelihood method and AIC (Akaike Information Criterion).

2.2. Discretization of Markov diffusion processes and nonlinear AR models. We know that Markov diffusion models are useful causal models for stochastic dynamic phenomena. They have stochastic differential equation forms, providing useful ideas about the dynamic causal relations between the variables. However when we perform a computer simulation we need to use a discrete time model derived from the continuous time model by a time discretization scheme. The discretized model used is a kind of nonlinear AR model, and it is possible that the simulation will explode computationally as was often experienced by nonlinear time series analysts in the 1960s. For example if we discretize the stochastic differential equation model,

$$\frac{dx}{dt} = f(x) + dw(t)$$

by the well known Euler scheme, we have

$$x_{t+\Delta t} = x_t + \Delta t f(x_t) + \sqrt{\Delta t}\, n_{t+\Delta t}.$$

For $f(x) = -x^3$, this gives a polynomial AR model and we know that its simulation is explosive in general. To prevent the explosion, we have to use either a very small time interval Δt or an implicit scheme. Both of these approaches are computationally expensive and time consuming. In addition, making Δt smaller is not a complete solution to this problem. Explosions will be less frequent and so it will take longer for an explosion to occur, but it will certainly explode when we simulate the model for a long time. This computational explosion problem became more and more important for numerical analysts as the need for nonlinear dynamic model simulation became stronger in many application fields. A scheme which is known to preserve the stability of a continuous time process as it is transformed into a discrete time process is called the A(absolute)-stable scheme and has become a major concern of recent numerical studies of stochastic differential equations (Mil'stein (1995), Kloeden and Platen (1995)).

It took a while before the A-stable scheme presented by time series analysts (Ozaki, 1985a, 1985b, 1986), started to be recognized by numerical analysts for stochastic differential equations. One of the first models introduced by Ozaki was the ExpAR model, which was developed from a linear AR(2) model for modeling ship rolling data (Ozaki and Oda, 1978). Here the key idea was to implement a mechanism to stop an AR model with amplitude-dependent coefficients diverging in simulations. Later he found that the LL scheme (Ozaki, 1985a) for discretizing the original stochastic differential equation gives essentially the same model as the ExpAR model. From the beginning, the main concern of the LL scheme was the computational stability of the discrete time model. The original presentation of the LL scheme for the model $dx = f(x)dt + dW(t)$ was

$$x_{t+\Delta t} = x_t + J_t^{-1}\{Exp(J_t\Delta t) - I\}f(x_t) + \sqrt{\Delta t}\, n_{t+\Delta t}$$
$$x_{t+\Delta t} = A_t x_t + B_t n_{t+\Delta t}$$
$$A_t = I + J_t^{-1}\{Exp(J_t^\Delta t) - I\}F_t$$
$$B_t = \sqrt{\Delta t}\, I.$$

where $J_t = \frac{\partial f(x)}{\partial x}\big|_{x=x_t}$, and F_t is such that $F_t x_t = f(x_t)$.

Several modifications have been introduced for the improvement of the scheme, and its theoretical characteristics, such as A-stability, consistency and speed of convergence of the scheme were all clarified later by Biscay et al. (1996), Shoji and Ozaki (1998), Jimenez et al. (1999), Jimenez and Ozaki (2000a, 2000b). The numerical performance of the maximum likelihood method through the local linearized model has also been studied by Ozaki (1992, 1994) and Shoji and Ozaki (1997).

2.3. The Frost-Kailath theorem and the nonlinear filter. It is well known that a scalar linear ARMA$(k, k-1)$ model, for some $k > 1$, is sufficient to characterize the dynamics of time series x_t even though the series itself is generated from a multi-dimensional linear dynamical system driven by Gaussian white noise. Here the k-dimensional state vector contains k unobservable variables and the observed variable x_t is formally related to other variables through the state dynamic model (this require-ment is called the observability condition). When the dynamics are non-linear, however, scalar linear ARMA models are no longer sufficient for characterizing the high-dimensional dynamics of the time series.

The Frost-Kailath theorem (1971) became useful in such situations. The theorem guarantees that the innovation likelihood can be used even if the dynamics of the multi-dimensional stochastic differential equation model are nonlinear and only scalar time series observations are available. We can write down the (-2)log-likelihood as

$$(-2) \log p(x_1, x_2, ..., x_N | z_0, V_0, \theta) = (-2) \log p(\nu_1, \nu_2, ..., \nu_N | z_0, V_0, \theta)$$

$$= \sum_{t=1}^{N} \left(\log \sigma_\nu^2 + \frac{\nu_t^2}{\sigma_\nu^2} \right) + N \log 2\pi$$

using the innovations ν_t of the state space representation,

$$z_{t+1} = A(z_t)z_t + B(z_t)n_{t+1}$$
$$x_t = h(z_t) + \varepsilon_t$$

where the state model is derived from the original continuous time stochas-tic or deterministic differential equation by the LL scheme, the dimension of $x_t = 1$ and z_t has dimension $k > 1$. The innovations ν_t of the state space model may be calculated recursively using a nonlinear Kalman filter (Ozaki (1993a), Shoji (1998)). Here the use of the stable LL discretization scheme is essential for the validity of the innovation approach. When the discretization scheme is not stable and if the observation error is large, the filtering scheme is often computationally explosive, and the filtering scheme does not produce appropriate innovations. Therefore the use of a stable discretization scheme is essential not only for simulation but also for the estimation of the continuous time model with observation errors. The method is very powerful indeed in many application fields.

2.4. Applications. We will see interesting examples of applications in three areas; chaos modeling in geophysics, dynamical system modeling for alpha rhythms in neuroscience and dynamical system modeling for the foreign currency exchange market in finance.

Example 3. Rikitake chaos.

$$d\xi = (\theta_1\xi + \varsigma\eta)dt + \sigma_w dw$$

$$d\eta = (\theta_2\xi - \theta_1\eta + \varsigma\xi)dt$$

$$d\varsigma = (1 - \xi\eta)dt$$

$$x_t = (0,0,1)z_t + \varepsilon_t$$

$$z_t = (\xi_t, \eta_t, \varsigma_t)' \ .$$

The model was first introduced by Rikitake (1958) as a dynamic model for a disc dynamo in geophysics. Later this model was analyzed by Ito (1980) as an example of a continuous time chaos model. Since the stochastic differential equation model includes deterministic chaos as a special case, where the variance σ_w^2 of the driving noise is zero, the estimation of the parameters of chaos and estimation of initial values of the process may be realized by the present innovation approach (Ozaki et al. (2000)). For example, deterministic chaos develops if we set $\sigma_w^2 = 0$. If we simulate the model with $\sigma_w^2 > 0$, noise driven stochastic chaos develops. The present method can actually estimate the parameter σ_w^2 at the same time as θ_1 and θ_2. This means that the present method may be used for checking the validity of the model as a deterministic chaos model for a data set. The top graph in Fig. 1 shows the simulated time series $x_1, x_2, ..., x_n$ for a Rikitake model with parameters, $\theta_1 = 5.0, \theta_2 = 124.8, \sigma_w^2 = 0.05$ and $\sigma_e^2 = 0.001$. Table 1 shows the estimated parameters found by the present method. The

TABLE 1

Maximum likelihood estimates for Rikitake data.

	θ_1	θ_2	σ_w^2	ξ_0	η_0	ς_0
Theoretical	5.00	124.8	0.05	1.99	0.0	0.0
Initial values	5.00	124.8	0.05	0.015	0.31	0.0
Estimated	4.95	123.1	0.078	1.002	0.08	0.003

lower graph in Fig. 1 shows the time series plot of the innovations obtained by the identified model. The histogram of the data and the histogram of the innovations are shown in Fig. 2, where we can see that the data has clearly been transformed into Gaussian distributed white noise. The results show how the maximum likelihood method provides useful information by reducing the prediction errors. Fig. 3 shows the filtered estimate of the state variables ξ_t, η_t, and ς_t as well as the prediction errors when the initial state variables, ξ_0, η_0, and ς_0, are not optimized in terms of the log-likelihood.

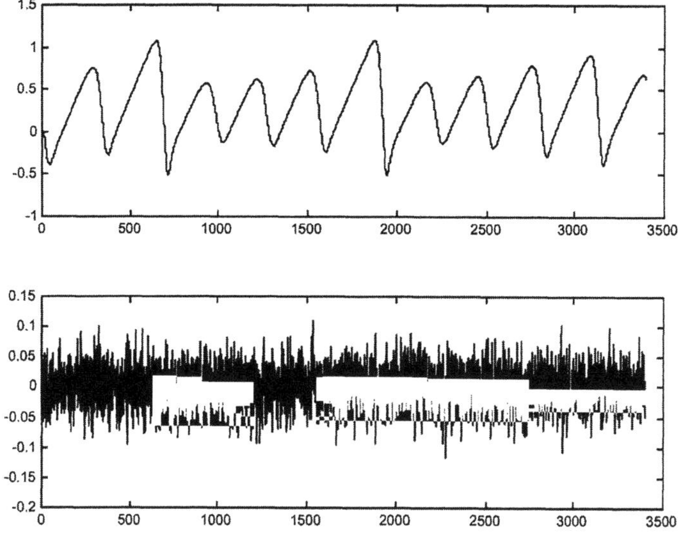

FIG. 1. *Simulated Rikitake chaos data (above) and innovations (below).*

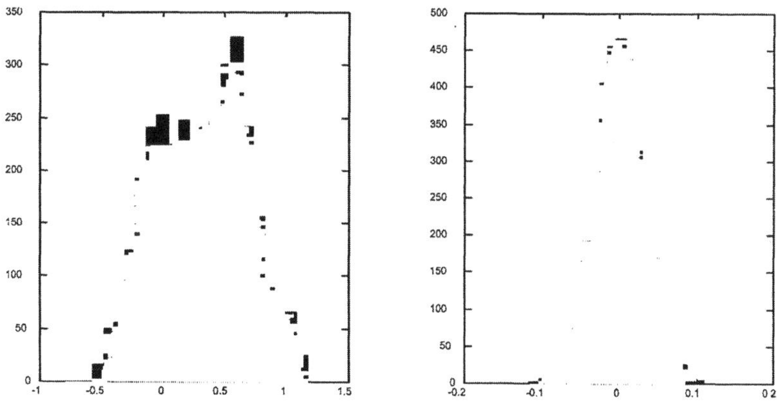

FIG. 2. *Histograms of the Rikitake data (left) and innovations (right).*

Fig. 4, on the other hand, shows the results when the initial state variables, ξ_0, η_0, and ζ_0, are optimized in terms of the log-likelihood. The two figures show that we can extract useful information about the initial variables of the chaos model by the present method. It also shows that by estimating the initial values properly, we can reliably estimate the unobserved variables ξ_t and η_t. It is easy to see that the "initial value sensitivity" of chaos is no reason to ignore the statistical method for the identification of deterministic processes.

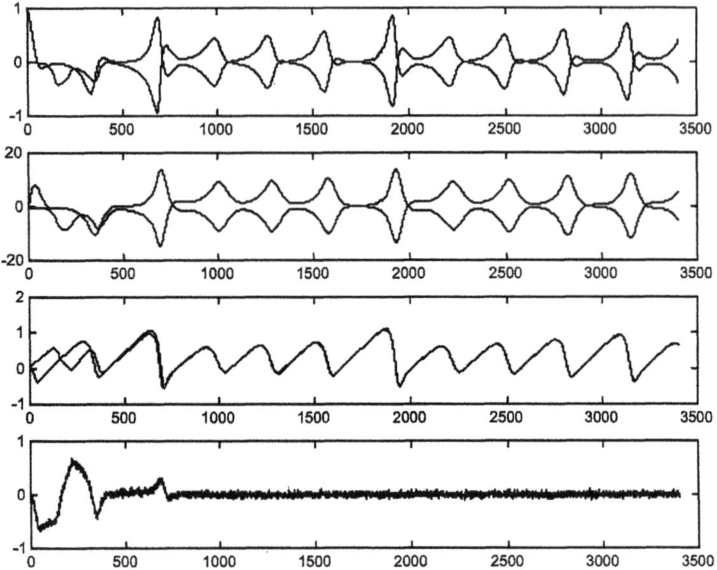

FIG. 3. *The true state variables, ξ_t, η_t and ζ_t, and the filtered state variables, $\xi_{t|t}$, $\eta_{t|t}$ and $\zeta_{t|t}$ (top three graphs) and the innovations ν_t (bottom graph) when the initial state variables, ξ_0, η_0, and ζ_0, are not optimized.*

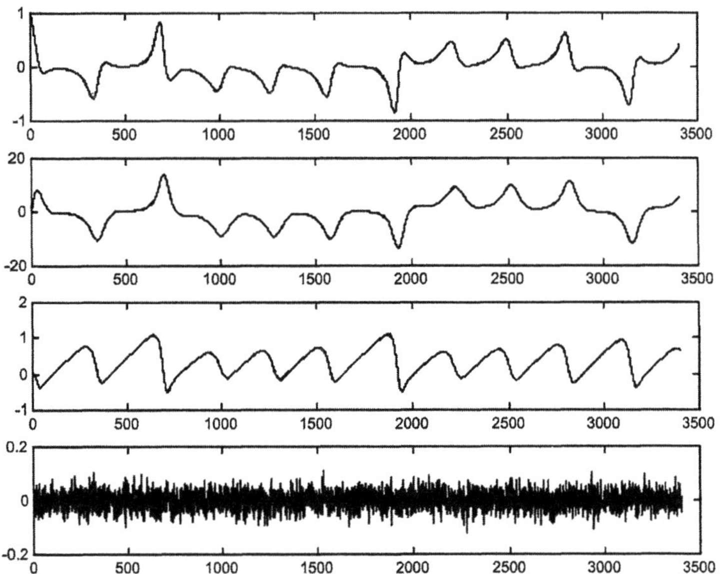

FIG. 4. *The true state variables, ξ_t, η_t and ζ_t, and the filtered state variables, $\xi_{t|t}$, $\eta_{t|t}$ and $\zeta_{t|t}$ (top three graphs) and innovations ν_t (bottom graph) when the initial state variables, ξ_0, η_0, and ζ_0, are optimized.*

Example 4. The Zetterberg model for alpha rhythms.

This model was developed through the work of Lopes da Silva et al. (1976) and Zetterberg et al. (1978). It provides a neural mass model for the alpha rhythms in EEG time series data. The model is a combination of three excitatory compartments, one inhibitory compartment and one compartment for the measurement. The compartment models are described in detail by Valdes Sosa et al. (1999a) and are specified by the following dynamical system with a constant input $\delta_e \Pi$ and a white noise input $\omega(t)$.

$$(2.5)\quad \begin{aligned} \frac{d^2 V_{1e}(t)}{dt^2} &= \alpha_e \frac{dV_{1e}(t)}{dt} + \beta_e V_{1e}(t) + \delta_e c_4 g(V_{2e}(t)) \\ &\quad + c_2 \alpha \frac{dI_f(t)}{dt} + c_2 \beta\, I_f(t) + c_2 \delta_i g(V_i(t)) \\ &\quad + \delta_e \Pi + \omega(t) \end{aligned}$$

$$(2.6)\quad \frac{d^2 V_{2e}(t)}{dt^2} = \alpha_e \frac{dV_{2e}(t)}{dt} + \beta_e V_{2e}(t) + \delta_e c_3 g(V_{1e}(t))$$

$$(2.7)\quad \frac{d^2 V_i(t)}{dt^2} = \alpha_e \frac{dV_i(t)}{dt} + \beta_e V_i(t) + \delta_e c_1 g(V_{1e}(t))$$

$$(2.8)\quad \frac{d^2 I_f(t)}{dt^2} = \alpha_e \frac{dI_f(t)}{dt} + \beta_i I_f(t) + \delta_i g(V_i(t))$$

$$(2.9)\quad \frac{d^3 V_{1f}(t)}{dt^3} = \kappa_2 \frac{d^2 V_{1f}(t)}{dt^2} + \kappa_1 \frac{dV_{1f}(t)}{dt} + \kappa_0 V_{1f}(t) + a\delta_n^2 \frac{dV_{1e}(t)}{dt}.$$

The V_s' and I_s' represent voltages and currents in the different compartments, the equations (2.5)–(2.8) describe the dynamics of compartments and (2.9) describes the observation system (amplifier) of the voltage at the skull. Using these equations of the compartments, the whole system may be represented in the following state space form,

$$\frac{dx}{dt} = f(x) + w(t)$$
$$z_t = (0, 0, ..., 0, 1)x_t + \varepsilon_t \ .$$

Here the driving noise $w(t) = (\omega(t), 0, ..., 0)'$ has zero mean and $\sigma_\omega^2 = \delta_e^2 \sigma_p^2$, and the 11-dimensional state space is $x(t) = \left(\frac{dV_{1e}(t)}{dt}, V_{1e}(t), \right.$ $\frac{dV_{2e}(t)}{dt}, V_{2e}(t), \frac{dV_i(t)}{dt}, V_i(t), \frac{dI_f(t)}{dt,} I_f(t), \frac{d^2 V_{1f}(t)}{dt^2}, \frac{dV_{1f}(t)}{dt}, \left. V_{1f}(t) \right)'$.

Most coefficients in the model are known from physiological reasoning, and the parameters that we need to estimate from the data are $(c_1, c_2, c_3, c_4, \Pi, \sigma_p^2, a)'$. The upper graph in Fig. 5 shows the EEG time series $x_1, x_2, ..., x_n$. The lower graph in Fig. 5 shows the innovations obtained by the estimated model. Here the estimated parameters of the model are as follows: $c_1 = 10.03, c_2 = 2.16, c_3 = 42.57, c_4 = 8.95, \Pi = 229, a = 4.7, \sigma_p^2 = 18.81$. Fig. 6 shows the histogram of the EEG data and the histogram of the innovations. It has been pointed out by Valdes Sosa et al. (1999a) that

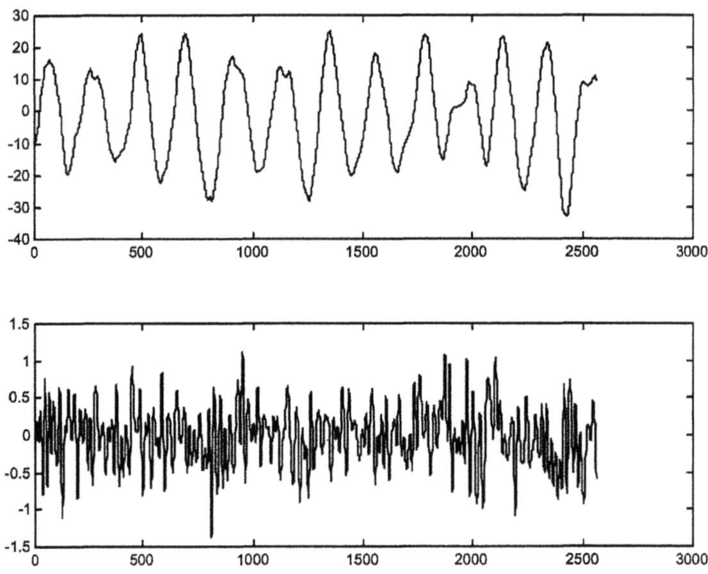

FIG. 5. *(Above) EEG time series (eyes closed), (below) Innovations of the estimated Zetterberg model.*

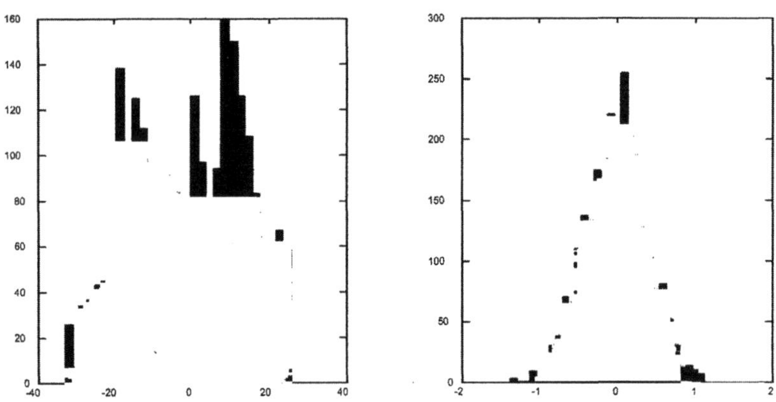

FIG. 6. *Histogram of EEG data (left) and innovations (right) of Fig. 5.*

the estimated Zetterberg model has nonlinear limit cycle properties. This provides neuroscientists with evidence that the alpha rhythm of EEG data is generated by a self-exciting mechanism in the brain.

Example 5. Dynamic micro-market model.

Dynamical system modeling of macroeconomic or financial market data has attracted many applied mathematicians and physicists (Ozaki

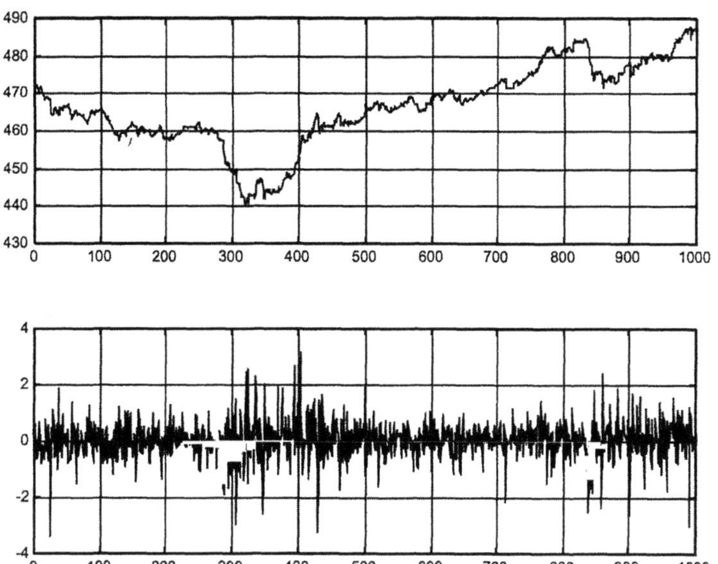

FIG. 7. *(Above) US Dollar-Yen exchange rate time series (daily), (below) Innovations of the estimated micro-market structure model.*

and Ozaki (1989), Bouchaud and Cont (1998)). Iino and Ozaki (2000) extended Bouchard and Cont's idea and introduced a three-dimensional nonlinear stochastic dynamical system model for the identification of foreign currency exchange rate dynamics. For example the dynamics of the logarithm of the price, $S(t) = log P(t)$, of a US Dollar in Japanese Yen is modeled using two unobserved variables $\phi(t)$ and $\lambda(t)$ and the following state space representation with state, $x = (S, \phi, \log \lambda^2)'$. Iino and Ozaki (2000) specified the state dynamics by

$$dS(t) = \phi(t)\lambda^2(t)dt + \gamma_1\lambda(t)dw_1(t)$$
$$d\phi(t) = \beta_1\phi(t)dt + \gamma_2 dw_2(t)$$
$$d\log\lambda^2(t) = \{\alpha_2 + \beta_2\log\lambda^2(t)\}dt + \gamma_3 dw_3(t).$$

The observation equation for the data, $z_t = \log S(t)$, is given by

$$z_t = (1, 0, 0)x_t + \varepsilon_t.$$

Here variable $\phi(t)$ shows whether the Yen is over-valued ($\phi(t) > 0$) or under-valued ($\phi(t) < 0$). Variable $\lambda(t)$ shows the degree of illiquidity. If $\lambda(t)$ is large the market liquidity is low and the market price is volatile. Parameters to be estimated are $(\alpha_2, \beta_1, \beta_2, \gamma_1, \gamma_2, \gamma_3, \sigma_\varepsilon^2)'$. The innovation approach can also be applied to this case. The top graph in Fig. 7 shows the daily data of the US Dollar-Japanese Yen exchange rate.

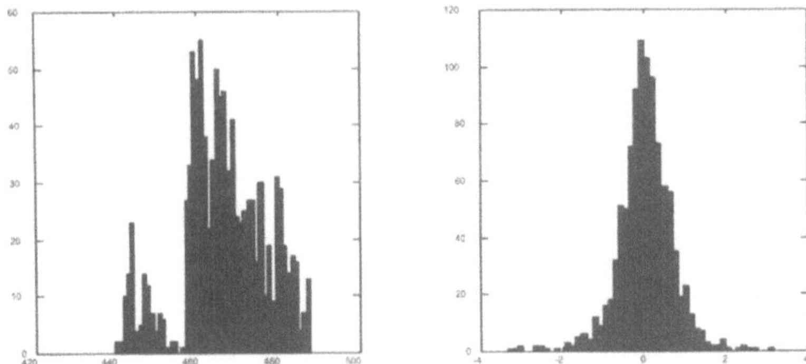

FIG. 8. *Histograms of the log (exchange rate) data (left) and innovations (right).*

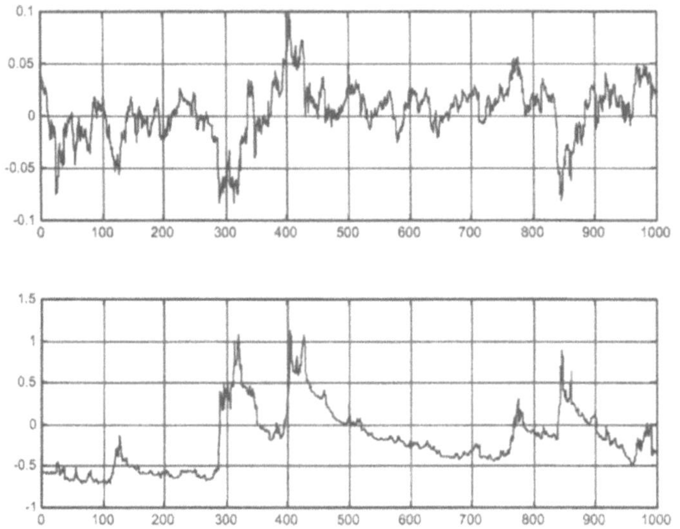

FIG. 9. *(Above)* $\phi_{t|t}$, *the filtered estimates of* $\phi(t)$, *(below)* $\log \lambda^2_{t|t}$, *the filtered estimates of* $\log \lambda^2(t)$.

The estimated parameters of the model are $\alpha_2 = 0.1054$, $\beta_1 = -0.0095$, $\beta_2 = -0.0068$, $\gamma_1 = -0.0351$, $\gamma_2 = 0.0500$, $\gamma_3 = -0.0514$, $\sigma_\varepsilon^2 = 0.7999$. The innovations associated with the estimated model are shown in the lower graph of Fig. 7. The innovation histogram is given in Fig. 8 and shows that the non-Gaussian distributed data is converted to Gaussian-like innovations by the identified dynamic micro-market model. It has been shown that filtered estimates of the unobserved variable $\phi(t)$ and $\log \lambda^2(t)$ (see Fig. 9) hold useful information for real time daily trading in the exchange market. Actually it may be confirmed by numerical experiment that if you shift money between currencies on the basis of the filtered estimates of the $\phi(t)$, capital may be prevented from losing value caused by the fluctuating exchange rate (see Ozaki et al. (2001)).

3. Nonlinear AR models for prediction and simulation. We have seen in the previous section that the discrete time models obtained from continuous time stochastic differential equation models by discretization are nonlinear AR models, useful in simulation and prediction. On the other hand nonlinear AR models developed by time series analysts were not very successful. Masani and Wiener (1959) proved that the conditional expectation of x_t is given by

$$E[x_t|x_{t-1}, x_{t-2}, ...] = \lim Q_n(x_{t-1}, x_{t-2}, ..., x_{t-m_n})$$

where m_n is a non-negative integer which depends on n, Q_n is a real polynomial in $m_n + 1$ variables whose coefficients may be computed from the moments of the process. This suggests the use of the following finite parameter nonlinear autoregressive type model for the prediction of the time series.

$$x_t = Q(x_{t-1}, x_{t-2}, ..., x_{t-m}) + \varepsilon_t$$

Here ε_t is a Gaussian white noise. Unfortunately these nonlinear polynomial AR models are explosive and inappropriate for simulation, although they improve the prediction performance significantly compared with linear models (Ozaki and Oda, 1978). Researchers turned their attention towards models which could provide successful simulations as well as predictions.

In this section, we will see that nonlinear polynomial AR models may be modified, into ExpAR models and RBF-AR models, in such a way that their prediction capability is preserved and at the same time their simulation works well without computational explosion. We will see that these ExpAR models and RBF-AR models have the capability of generating processes similar to the observed time series data no matter how large the variance of the Gaussian white noise inputs becomes. These models in combination with the innovation method are shown to be useful both in simulation and prediction of the time series as well as in controlling highly nonlinear dynamic systems.

3.1. Generalized ExpAR models and RBF-AR models. The ExpAR model is one of the typical nonlinear time series models developed in the statistical time series school during the project of searching for a nonlinear model suitable for both simulation and prediction. The model was introduced to realize nonlinear vibrations such as the amplitude-dependent frequency shift phenomena of the Duffing equation driven by random white noise $n(t)$,

$$\ddot{x} + c\dot{x} + \alpha x + \beta x^3 = n(t)$$

and the perturbed limit cycles of the stochastic Van der Pol equation model,

$$\ddot{x} + c(1 - x^2)\dot{x} + \alpha x = n(t).$$

The following ExpAR(2) model,

$$x_t = \{\phi_{1,0} + \phi_{1,1} \exp(-\gamma\, x_{t-1}^2)\}x_{t-1} + \{\phi_{2,0} + \phi_{2,1} \exp(-\gamma\, x_{t-1}^2)\}x_{t-2} + \varepsilon_t$$

was presented by Ozaki and Oda (1978) for the prediction and simulation of nonlinear random vibration processes. The model has the following Markov representation,

$$\begin{pmatrix} x_t \\ x_{t-1} \end{pmatrix} = \begin{pmatrix} \phi_{1,0} + \phi_{1,1}\exp(-\gamma\,x_{t-1}^2) & \phi_{2,0} + \phi_{2,1}\exp(-\gamma\,x_{t-1}^2) \\ 1 & 0 \end{pmatrix} \begin{pmatrix} x_{t-1} \\ x_{t-2} \end{pmatrix} + \begin{pmatrix} \varepsilon_t \\ 0 \end{pmatrix}$$

and if the eigenvalues of the transition matrix stay inside the unit circle for large $|x_{t-1}|$, the model defines an ergodic process and has a stationary marginal distribution (Ozaki, 1980, 1985a, 1993b). For example the following first order ExpAR(1) models are ergodic Markov chains and have stationary marginal distributions of three kinds; fat-tailed, centered and bimodal.

i) $x_{t+1} = \{\, 1 - 0.2\exp(-\gamma\, x_t^2)\}x_t + \varepsilon_{t+1}$ (fat-tail distributed process)

ii) $x_{t+1} = \{0.8 + 0.2\exp(-\gamma\, x_t^2)\}x_t + \varepsilon_{t+1}$ (center distributed process)

iii) $x_{t+1} = \{0.8 + 0.4\exp(-\gamma\, x_t^2)\}x_t + \varepsilon_{t+1}$ (bi-modal distributed process).

The ExpAR model is useful not only in analyzing random oscillatory time series data, but also for the characterization of nonlinear dynamics of the process such as stable and unstable singular points and chaos (see Ozaki (1985a)).

The General order ExpAR(p) model is defined by

$$x_t = \{\phi_{1,0} + \phi_{1,1}\exp(-\gamma x_{t-1}^2)\}x_{t-1} + ... + \{\phi_{p,0} + \phi_{p,1}\exp(-\gamma x_{t-1}^2)\}x_{t-p} + \varepsilon_t$$

and has been used in many fields for the characterization of nonlinear dynamic structure, such as limit cycles, singular points and bifurcation. It turned out, however, that there are many different types of nonlinear dynamic phenomena in real data analysis in applications. ExpAR models are not really suitable for modeling some nonlinear dynamic processes where the amplitude is not the only variable causing nonlinear dynamics. Several generalizations have been proposed since. Among those generalizations, the following model (Ozaki et al. (1999)) is known to have a much stronger capability in prediction and simulation than an ExpAR model.

$$x(t) = \phi_0(X(t-1)) + \sum_{i=1}^{p} \phi_i(X(t-1))x(t-i) + \varepsilon(t)$$

$$\phi_i(X(t-1)) = c_{i,0} + \sum_{k=1}^{m} c_{i,k} \exp\{-\lambda_k \|X(t-1) - Z_k\|^2\}$$

(3.1)
$$X(t-1) = \{x(t-1), \Delta x(t-1), \Delta^2 x(t-1), ..., \Delta^{d-1}x(t-1)\}'$$

$$Z_k = \{z_{k,1}, ..., z_{k,d}\}' \ .$$

The instantaneous dynamics of the model is dependent not only on the present amplitude of the series but also its velocity Δx_t and/or accelerations $\Delta^2 x_t$. Therefore it could produce, for example, asymmetric nonlinear wave patterns in time series, since the model dynamics may be different when the series is increasing or decreasing. The model turns out to be equivalent to an RBF-AR model (Vesin, 1993),

$$x(t) = \phi_0(X(t-1)) + \sum_{i=1}^{p} \phi_i(X(t-1))x(t-i) + \varepsilon(t)$$

$$\phi_i(X(t-1)) = c_{i,0} + \sum_{k=1}^{m} c_{i,k} \exp\{-\lambda_k \|X(t-1) - Z_k\|^2\}$$

(3.2)
$$X(t-1) = [x(t-1), ..., x(t-d)]'$$

$$Z_k = [z_{k,1}, ..., z_{k,d}]' \ .$$

RBF expansions have good interpolation properties in dealing with scattered data points, and are endowed with the "universal approximation" and "best approximation" capabilities of any continuous function. Universal approximation implies the possibility of approximating a function to any required degree of accuracy. The stronger property of best approximation entails that the approximation error surface always has a unique global minimum for any approximation performance measure (Park and Sandberg, 1991, 1993).

Vesin (1993) did not try to optimize the parameter estimation based on the maximum likelihood method or least squares method. Shi et al. (1999) tried to use the maximum likelihood method, but had to use a genetic algorithm because the conditions for the optimization of the likelihood were very poor. It turned out that the resulting estimate by the genetic algorithm was one of the local maximum values. Peng et al. (2001) found that this ill-conditioned convergence problem is easily solved by using an iterative method where the parameter space is divided into nonlinear parameters and linear parameters.

3.2. Comparison of RBF-AR models and RBF-neural network models. Since the late 1980s neural network models seem to have established a reputation in application fields as being the most capable and useful models for prediction and simulation among the newly developed identifiable models including chaos and stochastic differential equation models. Using the innovation approach, we will see this is not the case.

The special feature of the neural network model is its method of identifying the nonlinear predictors, which are derived as a linear combination of nonlinear basis functions through a few layers (Pogio and Girosi, 1990), (Roberts and Tarassenko, 1995). Here the basis function form at each layer is fixed and weights (coefficients) are estimated for each layer using the sample data. There are several computational methods for the estimation of these coefficients. However the computational speed of these estimation (or learning) procedures is extremely slow compared with the ordinary least squares estimation procedure for polynomial AR models and ExpAR models. Although the nonlinear functional form (called the activation function) is fixed for each layer, the choice of the form is rather arbitrary. There are many possible forms for the activation function (some examples are given in Cichoki and Unbehauen(1993)) and the prediction results will depend on the initial choice of the functional form.

Although the multi-layer neural network models provide us with a wide range of tractable nonlinear prediction models, the above-mentioned computational burden of estimating the weights of each layer in the model generally forces researchers to move from the general multi-layer neural network model to a family of single layer networks with a general nonlinear function family. The shift is commonly seen in many scientific fields where scientists become more and more interested in a specific dynamic structure for their own problem, as their vision becomes clearer in the light of preliminary analysis with a general model such as a multi-layer neural network model.

As a consequence of its high approximation capabilities, the single layer RBF model has been recognized as an alternative to the multi-layer neural network model (Lapedes and Farber, 1987). The single layer RBF model has a clear computational advantage over the multi-layer neural network model, derived from its "linear-in-the-parameter" formulation. The single-layer structure of the RBF model is a feature which can be exploited for parameter estimation, and allows a faster learning scheme in comparison with the back-propagation techniques used for multi-layer neural network models.

In the present study we restrict ourselves to the single layer neural networks with one of the most commonly used radial basis functions, i.e. Gaussian Radius Basis Functions (RBF) of the following form,

$$w(X) = \exp\left\{ - \lambda_k \left\| X - Z_k \right\|^2 \right\}.$$

Here $X = [x_1, ..., x_d]'$ is a state variable vector and $Z_k = [z_{k,1}, ..., z_{k,d}]'$ are center vectors. Then an RBF-neural network model for prediction of x(t) based on the state values $X(t) = [x(t-1), x(t-2), ..., x(t-d)]'$ is given by

$$(3.3) \qquad \begin{aligned} x(t) &= \theta_0 + \sum_{k=1}^{m} \theta_k \exp\{-\lambda_k \|X(t-1) - Z_k\|^2\} + \varepsilon(t) \\ X(t-1) &= [x(t-1), ..., x(t-d)]' . \end{aligned}$$

On the other hand the prediction of x(t) by the RBF-AR(p, m, d) model could be given by,

$$(3.4) \qquad \begin{aligned} x(t) &= \phi_0(X(t-1)) + \sum_{i=1}^{p} \phi_i(X(t-1))x(t-i) + \varepsilon(t), \\ \phi_i(X(t-1)) &= c_{i,0} + \sum_{k=1}^{m} c_{i,k} \exp\{-\lambda_k \|X(t-1) - Z_k\|^2\}, \end{aligned}$$

which looks rather similar to the RBF neural network model. Actually they share almost the same flexibility in characterizing complex dynamics. However the models show significantly different performance at the stage of model identification. The difficulty in the iterative method for the parameter estimation of the neural network models is well known even for single-layer network models. The topic of accelerating the slow convergence rate of the parameter estimation algorithm of neural network model receives special attention by neural network researchers and is called "learning theory". An enormous effort has been devoted to accelerating the speed of convergence of the estimation algorithms without much success. We will see that RBF-AR models are free from this computational burden in the next section.

3.3. An efficient computational method. Peng et al. (2001) presented an efficient estimation algorithm for the parameter estimationof the general RBF-AR(p, m, d) model (3.4) using a so-called structured parameter optimization method. Here parameters are classified into two sets; a nonlinear parameter set, $(z_{1,1}, ..., z_{1,d}, z_{2,1}, ..., z_{2,d}, ... z_{m,1}, ..., z_{m,d}, \lambda_1, \lambda_2, ..., \lambda_m)'$ and a linear parameter set, $(c_{1,0}, c_{2,0}, ..., c_{p,0}, c_{1,1}, c_{2,1}, ..., c_{p,1}, ..., c_{1,m}, c_{2,m}, ..., c_{p,m})'$. Since the linear parameters are easily obtained, by solving a linear equation, when the nonlinear parameters are given, they suggest that we should optimize in the nonlinear parameter space instead of in the high dimensional space of whole parameters. This method drastically reduces the ill-conditioning of the optimization algorithm.

To discuss and check the performance of our new computational method, and to compare the performance of the RBF neural network model and the RBF-AR model, we use the well known difficult nonlinear time series data (Fig. 10) of the EEG trace of a subject with epilepsy (Valdes Sosa

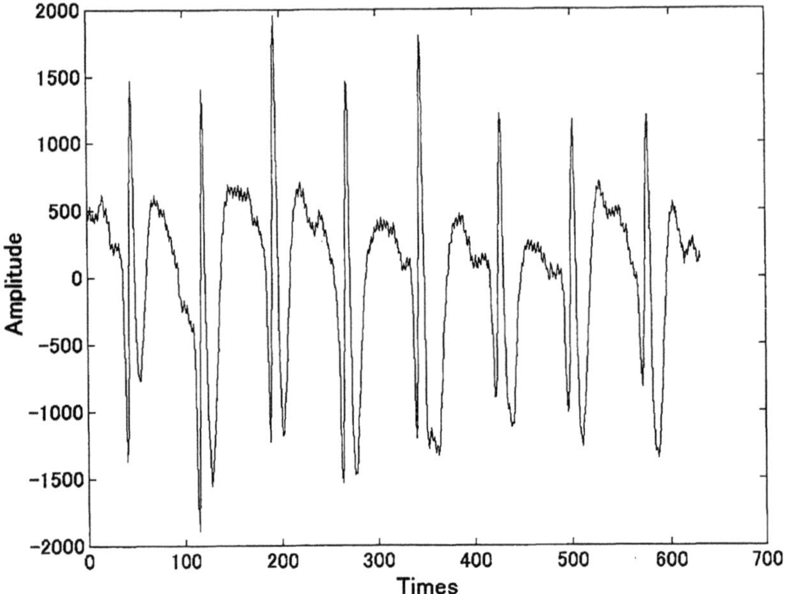

FIG. 10. *Epilepsy EEG data.*

et al., 1999b). Fig. 11 shows how the nonlinear parameters (centers and scaling parameters) and residual variance of the RBF-AR(75,3,2) model converge in the iterative algorithm. The iterative algorithm seems to have converged after about 20 iterations. The time series plot and histogram of residuals are shown in Fig. 12. Although the original data of Fig. 10 are very non-stationary and non-Gaussian looking, the resulting prediction errors are more or less homogeneous and Gaussian distributed.

The new algorithm is valid for the RBF neural net model as well as for the RBF-AR model, and computes the model coefficients very efficiently. It is interesting to compare the convergence speed of prediction error variance of the two models chosen from the two model sets. The solid lines in Fig. 13 show the convergence of the prediction error variance of the two models, RBF(5,8) and RBF-AR(8,3,4). The dotted lines show the convergence of the ordinary optimization method applied to the whole parameter space without structuring. Table 2 shows that the attained RBF-AR model gives not only smaller prediction error variance than the RBF model but also its convergence speed is much faster than the RBF model. Unfortunately neither the RBF-AR model nor the RBF neural network model identified from the data generates completely satisfactory simulation data with random white noise inputs using the estimated noise variance. The modeling effort needs to be continued further.

3.4. The application of RBF-ARX models to the modeling of nonlinear power plant control systems. The overwhelming advantage

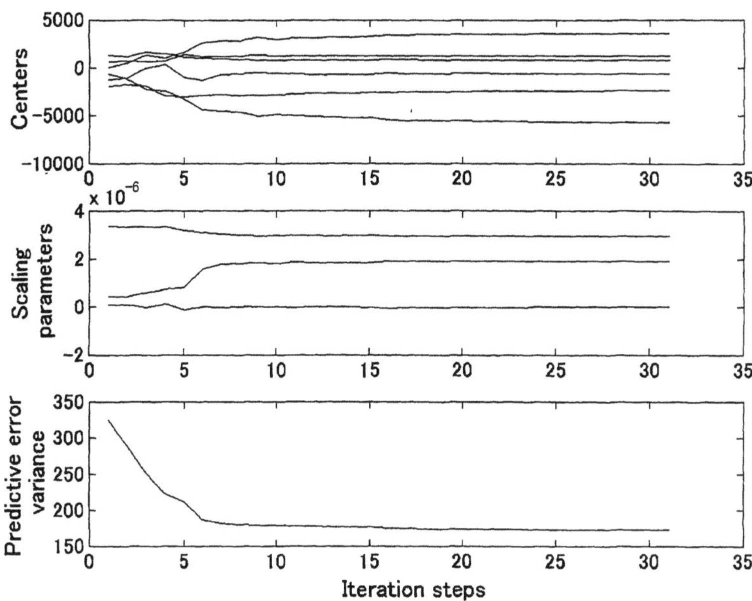

FIG. 11. *Convergence of the centers and scaling parameters for the RBF-AR(75,3,2).*

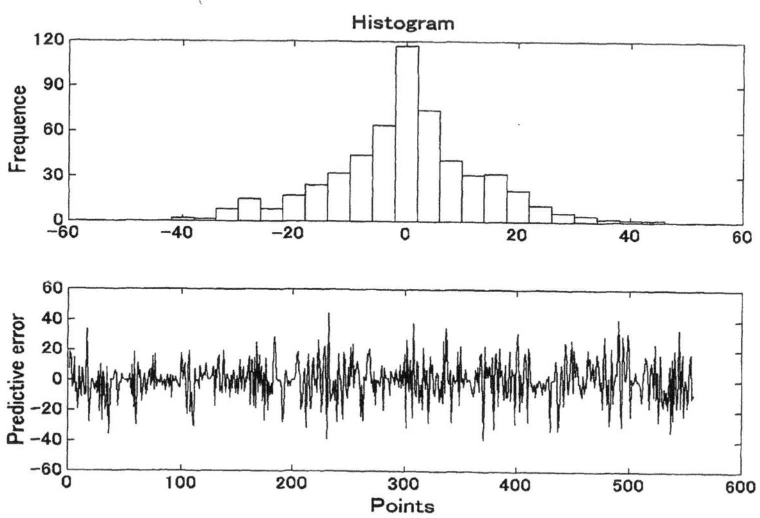

FIG. 12. *Residuals and their histogram for the RBF-AR(75,3,2).*

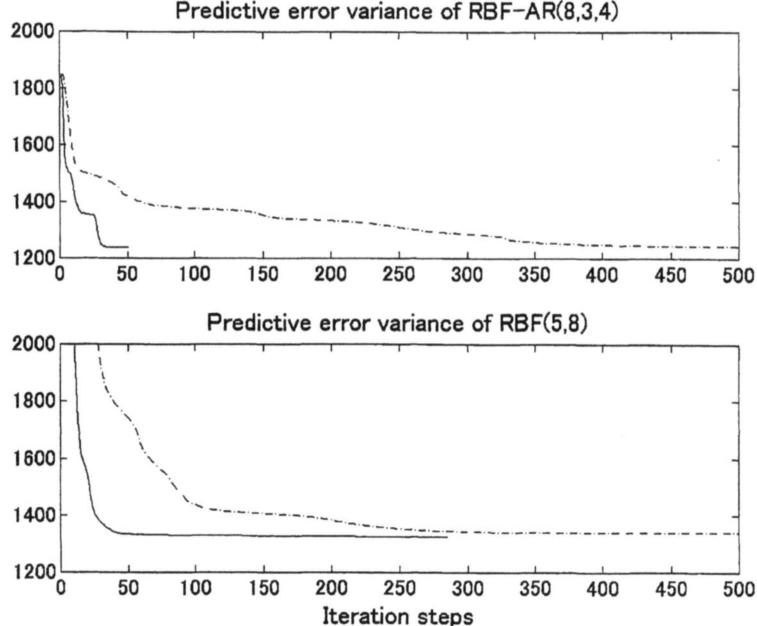

FIG. 13. *Comparison of the convergence speed of the structure parameter method and the conventional method for RBF-AR(8,3,4) and RBF(5,8).*

TABLE 2
Comparison of the estimation results of RBF(5,8) and RBF-AR(8,3,4).

Models	Initial variance	Converged variance	Total iterations	AIC
RBF(5,8)	2.91×10^5	1.32×10^3	285	4630
RBF-AR(8,3,4)	1.85×10^3	1.23×10^3	50	4595

of AR model based nonlinear dynamic models over conventional nonlinear models is more clearly seen in the application of the models in modern predictive control problems. Here control strategy needs to be designed on the basis of a nonlinear state space model, and the state dimension is usually high. A multivariate version of the ExpAR model with exogenous variables (a so-called multivariate ExpARX model) was successfully used for modeling the boiler system of thermal power plant (Toyoda et al. (1997), Peng et al. (2001)), and multivariate RBF-AR models with exogenous variables (RBF-ARX models) were used for modeling the nitrogen oxide (NOx) deconvolution process in thermal power plants in Peng et al. (2001). Here the purpose of the NOx decomposition process control in thermal power plants is to reduce the NOx concentration in fuel gas from the boiler of the plants in order to protect the environment. The process has nonlinear dynamics dependent on the power load demand of the plants, and the nonlinear characteristics are mainly caused by variation in gains with load.

The Hammerstein model (Haber and Keviczky, 1999) is usually used by control engineers to describe the dynamics with nonlinear static gain. Here we can see that a RBF-AR(p, m, n) model (3.5) is useful in characterizing the dynamics of the NOx decomposition process as follows

$$y(t) = \phi_0(\mathbf{X}(t-1)) + \sum_{i=1}^{p} \phi_{y,i}(\mathbf{X}(t-1))y(t-i)$$

$$(3.5) \qquad + \sum_{i=0}^{p-1} \phi_{u,i}(\mathbf{X}(t-1))u(t-d-i)$$

$$+ \sum_{i=1}^{p} \phi_{v,i}(\mathbf{X}(t-1))v(t-i) + e(t)$$

where $y(t), u(t),$ and$v(t)$ are the output, input and disturbance of the process respectively, $\mathbf{X}(t-1) = [x(t-1), x(t-2), \cdots, x(t-n)]^{\mathrm{T}}$ is the load demand series, p, m and n are the orders, d is the pure time-delay of the process, and the RBF-coefficients are similar to those in model (3.4). In order to compare model performance, the Hammerstein (p) model (3.6) below is also used to identify the nonlinear process

$$y(t) = c_0 + \sum_{i=1}^{p} a_i y(t-i) + \sum_{i=0}^{p-1} b_i^1 u(t-d-i)$$

$$(3.6) \qquad + \sum_{i=0}^{p-1} b_i^2 u^2(t-d-i) + \sum_{i=1}^{p} b_i^3 v(t-i) + \sum_{i=1}^{p} b_i^4 v^2(t-i) + e(t).$$

A measured data set from an actual NOx decomposition process is shown in Fig. 14. For the estimation of the model (3.5) and the model (3.6) we use the computational method proposed by Peng et al. (2001). The estimated results for the NOx decomposition process are shown in Figs. 15–18 and Table 3.

TABLE 3

The performance comparison between the RBF-AR(6,4,3) model (3.5) and the Hammerstein(6) model (3.6) for the NOx decomposition process.

Models	Predictive error Variance	AIC
Hammerstein(6)	0.0376	-9775
RBF-AR(6,4,3)	0.0303	-10256

Fig. 15 shows the centers, scaling parameters and the predictive error variance of the RBF-AR(6,4,3) model (3.5) at each iteration during

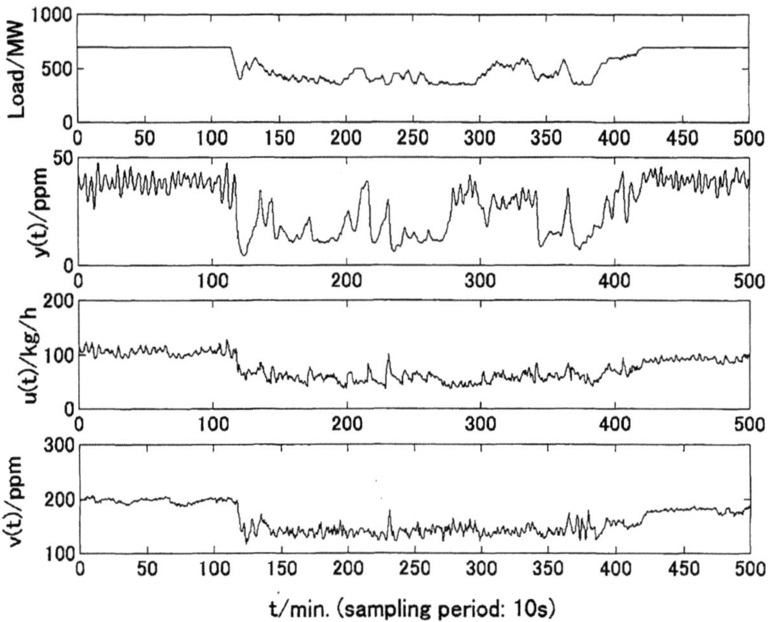

FIG. 14. *The observed data of the NOx decomposition process (d = 10).*

the search of optimal parameters using Peng-Ozaki's structured optimization method and the ordinary flat optimization method respectively. From
Fig. 15, we can see that the two types of the parameter optimization methods are all convergent, but in both the convergent speed and precision
the structured optimization method is much better than the ordinary flat
optimization method which optimizes all the parameters simultaneously.

If we compare the RBF-AR model with the Hammerstein model, the
performance of the RBF-AR model is far better than that of the Hammerstein model as is shown in Table 3. The prediction error variance and the
AIC of RBF-AR(6,4,3) are much smaller than those of the Hammerstein
model(6). The prediction errors and the histograms of both models are
shown in Fig. 16 and Fig. 17 respectively. Fig. 18 shows the eigenvalues of
the RBF-AR(6,4,3) model described in (3.5) varying with load demand of
the power plants, which show that the dynamics of the NOx decomposition
process depends in a nonlinear way on the load. However, the Hammerstein model is still nonlinear with respect to the input at any operating
point, but that of the RBF-AR model is linear at certain fixed load levels or operating points. Therefore some linear controller design methods
may be applied to control the nonlinear process generated by the RBF-AR

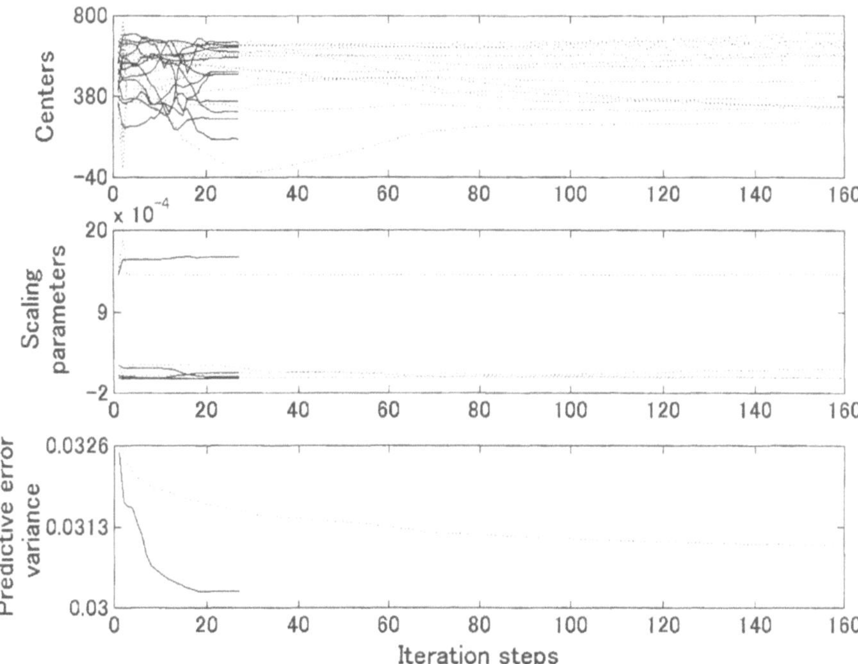

FIG. 15. *The convergence of the nonlinear parameters and prediction error variance of the RBF-AR(6,4,3) model of (3.5) during parameter optimization of the NOx decomposition process. The solid line gives the results using Peng et al. (2001)'s algorithm, and the dotted line gives the results using an ordinary optimization method. Both methods used the same initial parameters.*

model. A successful application of multivariate RBF-ARX models to the nonlinear control of power plant boiler systems is presented in Peng et al. (2001).

4. Conclusion. The innovation approach takes advantage of the Markovian nature of the process considered for the time series. Markov process theory tells us that the Gaussian assumption for the prediction error is reasonable for time series data sampled from continuous time phenomena with sufficiently small sampling interval, even though the time series is non-Gaussian distributed. At the same time a Markov model gives us the causal relations between variables through its stochastic differential equation representation. The class of potential causal models for time series analysis is very wide indeed, including stochastic or deterministic differential equation models and neural network models as well as nonlinear time series models such as ExpAR models and RBF-AR models. For time series analysts there is no reason why we should restrict the candidate models to conventional linear and nonlinear time series models in real data analysis.

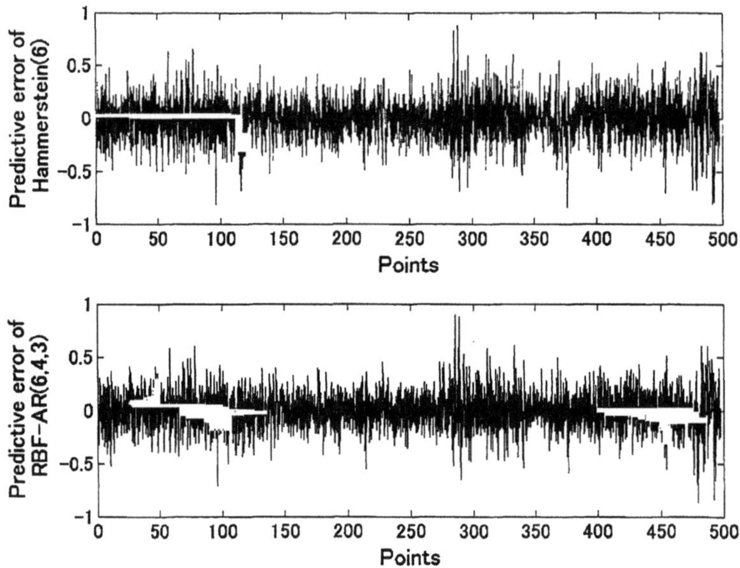

FIG. 16. *The prediction errors of the Hammerstein(6) model described in (3.6) and the RBF-AR(6,4,3) model (3.5) for the NOx decomposition process.*

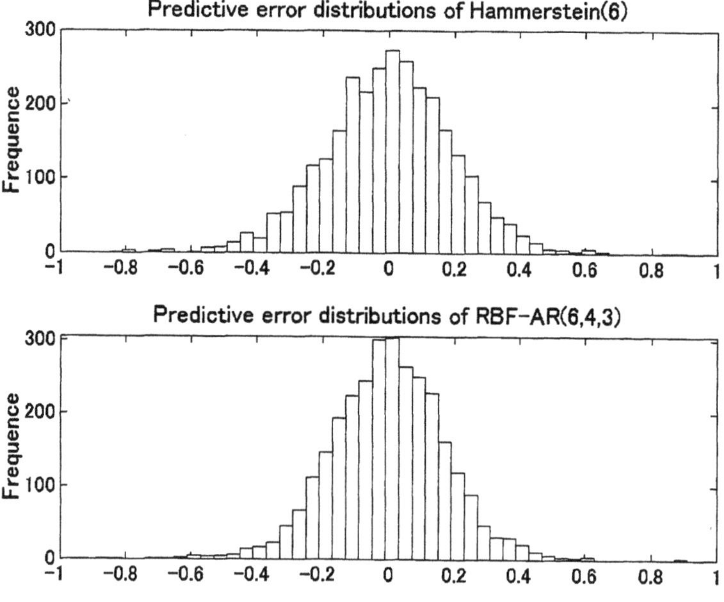

FIG. 17. *The histograms of the prediction errors of the Hammerstein(6) model (3.6) and the RBF-AR(6,4,3) model (3.5) for the NOx decomposition process.*

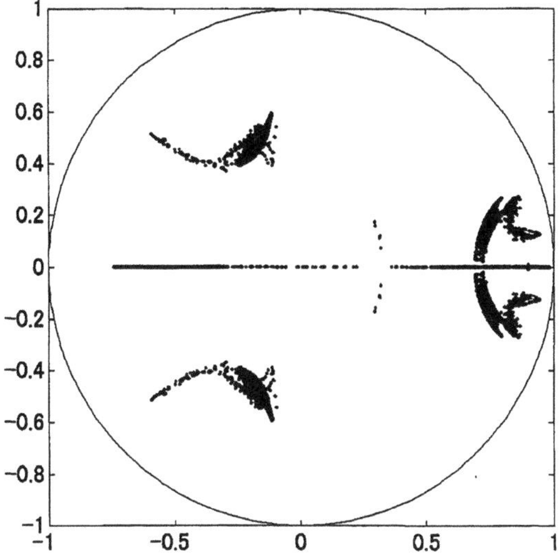

FIG. 18. *The eigenvalues of the RBF-AR(6,4,3) model (3.5) changing with load demand for the NOx decomposition process.*

The neural network models and the continuous deterministic or stochastic dynamical system models are as useful as conventional time series models. The performance of these models can be assessed on the same ground using the innovation approach and the likelihood function. It is time for time series analysts to look again at the innovation approach, and solve, together with scientists in various fields, problems of prediction, smoothing and control in modern science using recently developed nonlinear models.

Acknowledgements. The authors are grateful to the following organizations for supporting the current research; i) Ministry of Education, Culture, Sports, Science and Technology, Japan, ii) Mizuho Trust and Banking Co. Ltd. and iii) The Japanese Society for Promotion of Sciences.

REFERENCES

[1] AKAIKE H. (1969), "Fitting autoregressive models for prediction," Ann. Inst. Stat. Math., **21**, 243–247.

[2] BISCAY R., J.C. JIMENEZ, J.J. RIERA, AND P.A. VALDES SOSA (1996), "Local linearization method for the numerical solution of stochastic differential equations," Ann. Inst. Stat. Math., **48**, 631–644.

[3] BOUCHAUD J.P. AND R. CONT (1998), "A Langevin approach to stock market fluctuations and crashes," Eur. Phys. J., **B6**, pp. 543–550.

[4] BOX G.E.P. AND G.M. JENKINS (1970), *Time Series Analysis: Forecasting and Control*, Holden-Day, San Francisco.

[5] CAMPBELL M.J. AND A.M. WALKER (1977), "A survey of statistical work on the McKenzie river series of annual Canadian lynx trappings for the year 1821–1934, and a new analysis," J. Roy. Statist. Soc., A **140**, 411–431.

[6] CICHOCKI A. AND R. UNBEHAUEN (1993), "Neural Networks for Optimization and Signal Processing," J. Wiley and Sons, Chichester/New York/Brisbane/Toronto/Singapore.

[7] DOOB J.L. (1953), *Stochastic Processes*, John Wiley and Sons Inc.

[8] DYN N. (1989a), "Interpolation and approximation by radial and related functions," in Approximation theory, VI, Academic Press, London, pp. 211–234.

[9] DYN N. (1989b), "Interpolation by piecewise-linear radial basis functions," J. Approximation Theory, **59**, 202–223.

[10] FROST P.A. AND T. KAILATH (1971), "An innovation approach to least squares estimation Part III: nonlinear estimation in white noise," IEEE Trans on Aut. Control, AC-13, pp. 646–655.

[11] HABER R. AND L. KEVICZKY (1999), *Nonlinear system identification - input-output modeling approach, Volume 1: Nonlinear system structure identification.* Kluwer Academic Publishers, The Netherlands.

[12] IINO M. AND T. OZAKI (2000), "A Nonlinear Model for Financial Dynamics," in Proceeding of the International Symposium on Frontiers of Time Series Modeling, ISM Tokyo Japan.

[13] ITO K. (1980), "Chaos in the Rikitake two-disc dynamo system," Earth Planet. Sci. Lett., 51, 451–456.

[14] JIMENEZ J.C., I. SHOJI, AND T. OZAKI (1999), "Simulation of stochastic differential equations through the local linearization method: a comparative study," J. Stat. Phys., **94**, 587–602.

[15] JIMENEZ J.C. AND T. OZAKI (2000a), "Linear estimation of continuous-discrete linear state space models with multiplicative noise," Res. Memo., **772**, The Institute of Statistical Mathematics, Tokyo.

[16] JIMENEZ J.C. AND T. OZAKI (2000b), "Local Linearization filters for nonlinear continuous-discrete state space models with multiplicative noise," Res. Memo., **773**, The Institute of Statistical Mathematics, Tokyo.

[17] KAILATH T. (1968), "An innovation approach to least squares estimation Part I: linear filtering in additive white noise," IEEE Trans on Aut. Control, AC-13, pp. 646–655.

[18] KALMAN R.E. (1960), A New Approach to Linear Filtering and Prediction Problems, Trans. ASME, Series D, J. of Basic Eng., **82**, 35–45.

[19] KLOEDEN P.E. AND E. PLATEN (1995), *Numerical Solution of Stochastic Differential Equations*, New York, Springer.

[20] KOLMOGOROV A.N. (1931), "On analytical methods in probability theory," Math. Ann., **104**, 415–458.

[21] KOLMOGOROV A.N. (1941), "Interpolation and extrapolation of stationary random sequences," Bull. Acad. Sci. USSR, Math. Ser., **5**, 3–14.

[22] LAPEDES A. AND R. FARBER, (1987), "Neural Networks work," *Neural Information Systems*, Academic Institute of Physics, pp. 442–456.

[23] LOPES DA SILVA F.H., A. VAN ROTTERDAM, P. BARTS, E. VAN HEUSDEN, AND W. BURR (1976), "Models of neuronal populations: the basic mechanisms." In: M.A. Corner and D.F. Swaab (Eds.), Perspective in Brain Research, Prog. Brain Res., **45**, 281–308.

[24] MASANI P. AND N. WIENER (1959), "Nonlinear prediction," in Probability and Statistics, John Wiley and Sons, pp. 190–212.

[25] MAY R. (1976), "Simple mathematical models with very complicated dynamics," Nature, **261**, 459–467.

[26] MIL'STEIN G.N. (1995), *Numerical Integration of Stochastic Differential Equations*, Dordrecht: Kluwer Academic Publishers.

[27] OZAKI T. (1980), "Nonlinear time series models for nonlinear random vibrations," *J. of Appl. Prob.*, **17**, 84–93.

[28] OZAKI T. (1985a), "Nonlinear time series models and dynamical systems," in *Handbook of Statistics*, 5, E.J. Hannan et al. (Eds.), North-Holland.

[29] OZAKI T. (1985b), "Statistical identification of storage models with application to stochastic hydrology," *Water Resources Bulletin*, **21**.

[30] OZAKI T. (1986), "Local Gaussian modelling of stochastic dynamical systems in the analysis of nonlinear random vibrations," in *Essays in Time Series and Allied Processes*, Festschrift in honour of Prof. E.J. Hannan, Probability Trust.

[31] OZAKI T. (1992), "Identification of nonlinearities and non-Gaussianities in time series," in *New Directions in Time Series Analysis, Part I*, D.R. Brillinger et al. (Eds.), IMA Volumes in Mathematics and Its Application, Springer Verlag, **45**, 227–264.

[32] OZAKI T. (1993a), " A local linearization approach to nonlinear filtering," Int. J. Control, **57**, 75–96.

[33] OZAKI T. (1993b), "Non-Gaussian characteristics of Exponential Autoregressive processes," in *Developments in Time Series Analysis, Festschrift in honour of Prof. Maurice Priestley*, T. Subba Rao (Ed.), Chapman and Hall, pp. 257–273.

[34] OZAKI T. AND H. ODA (1978), "Nonlinear time series model identification by Akaike's information criterion," Information and Systems, B. Dubuisson (Ed.), Pergamon Press, pp. 83–91.

[35] OZAKI T. AND V.H. OZAKI (1989), "Statistical identification of nonlinear dynamics in macroeconomics using nonlinear time series models," in *Statistical Analysis and Forecasting of Economic Structural Change*, P. Hackl (Ed.), Springer-Verlag, pp. 345–365.

[36] OZAKI T., P.A. VALDES SOSA, AND V. HAGGAN-OZAKI (1997), "Reconstructing Nonlinear Dynamics from Time Series: with Application to Epilepsy Data Analysis," J. Signal Processing, **3**, 153–162.

[37] OZAKI T., J.C. JIMENEZ, AND V. HAGGAN-OZAKI (2000), "Role of the likelihood function in the estimation of chaos models," J. Time Series Analysis, **21**, 363–387.

[38] OZAKI T., J.C. JIMENEZ, M. IINO, Z. SHI, AND S. SUGAWARA (2001), "Use of stochastic differential equation models in financial time series analysis : Monitoring and control of currencies in exchange market," Proceeding of 3rd Japan-US joint seminar on statistical time series analysis, Kyoto, Japan, June 18–22, Sponsored by JSPS–NSF, June 2001, pp. 17–24.

[39] PARK J. AND I.W. SANDBERG (1991), "Universal approximation using radial basis function networks," *Neural Comput.*, **3**, 247–257.

[40] PARK J. AND I.W. SANDBERG (1993), "Approximation and radial basis networks," Neural Comput., **5**, 305–306.

[41] PENG. H., T. OZAKI, AND Y. TOYODA (2001), " Nonlinear system identification using radial basis function-based signal-dependent ARX model," Proc. of 5th IFAC Symposium on Nonlinear Control Systems, Saint-Petersburg, Russia, July 4–6, 2001.

[42] POGIO T. AND F. GIROSI (1990), "Networks for Approximation and Learning," Proceeding of the IEEE, **78**, 1461–1497.

[43] ROBERTS S. AND L. TARASSENKO (1995), "Automated sleep EEG analysis using an RBF network," in *Applications of Neural Networks*, A. Murray (Ed.), Kluwer Academic Publishers, pp. 305–322.

[44] RIKITAKE T (1958), "Oscillations of a system of disc dynamos," Proc. Camb. Philos. Soc., **54**, 89–105.

[45] SHI Z., Y. TAMURA, AND T. OZAKI (1999), "Nonlinear time series modelling with radial basis function-based state-dependent autoregressive model," Intern. J. of System science, **30**, 717–727.

[46] SHOJI I. AND T. OZAKI (1997), "Comparative study of estimation methods for continuous time stochastic processes," Journal of Time Series Analysis, **18**, 485–506.

[47] SHOJI I. AND T. OZAKI (1998), " A statistical method of estimation and simulation for systems of stochastic differential equations," Biometrika, **85**, 240–243.

[48] SHOJI I. (1998), "A comparative study of maximum likelihood estimators for nonlinear dynamical system models," Int. J. Control, **71**, 391–404.

[49] TONG H. (1977), "Some comments on the Canadian lynx data," J. Roy. Soc., A **140**, 432–436.

[50] TOYODA Y., K. ODA, AND T. OZAKI (1997), "The nonlinear system identification method for advanced control of the fossil power plants". *Proceedings of 11 th IFAC Symposium on System Identification,* Fukuoka, Japan, pp. 8–11.

[51] VALDES SOSA P.A., J.C. JIMENEZ, J. RIERA, R. BISCAY, AND T. OZAKI (1999a), "Nonlinear EEG analysis based on a neural mass model," Biol. Cybern., **81**, 415–424.

[52] VALDES SOSA P., J. BOSCH, J.C. JIMENEZ, N. TRUJILLO, R. BISCAY, F. MORALES, J.L. HERNANDEZ, AND T. OZAKI (1999b), "The Statistical Identification of Nonlinear Brain Dynamics: A Progress Report," in "Nonlinear Dynamics and Brain Functioning," Nova Science Publishers, edited by N. Pradhan, P.E. Rapp, and R. Sreenivasan.

[53] VESIN J. (1993), "An amplitude-dependent autoregressive signal model based on a radial basis function expansion," In Proc. Intern. Conf. Acoust. Speech Signal Process., Minnesota, **3**, 129–132.

[54] WOLD H. (1938), *A Study in the Analysis of Stationary Time Series,* Uppsala, Sweden: Almqvist and Wicksell.

[55] WIENER N. (1949), *Extrapolation, Interpolation, and Smoothing of Stationary Time Series with Engineering Applications,* New York: Wiley. Originally issued in MIT Radiation Lab. Report, Cambridge, Mass., February 1942.

[56] WIENER N. (1961), *Cybernetics, 2nd edition,* MIT Press, Cambridge, Mass.

[57] WONG E. (1963), "The construction of a class of stationary Markoff process," Proc. Amer. Math. Soc. Symp. Appl. Math., **16**, 264–276.

[58] ZETTERBERG L.H., L. KRISTIANSSON, AND K. MOSSBERG (1978), "Performance of a model for a local neural populations," Biol. Cybern., **31**, 15–26.

NON-GAUSSIAN TIME SERIES MODELS

MURRAY ROSENBLATT*

Abstract. Non-Gaussian linear time series models are discussed. The ways in which they differ from Gaussian models are noted. This is particularly the case for prediction and parameter or transfer function estimation.

Key words. Non-Gaussian, minimum and nonminimum phase, prediction, estimation, time series.

AMS(MOS) subject classifications. 11K70, 62M10, 62M15, 62M20, 60GXX, 62G05.

1. Introduction. Linear time series models have been studied extensively and used in many applied areas (see Yaglom 1986). However, much of the investigation has been focused on Gaussian models. Applications in a geophysical context are discussed in Robinson 1981 and Aki 1980. More recently non-Gaussian models have been considered (see Wiggins 1978 and Donoho 1981). The non-Gaussian linear models have a more complicated structure than the Gaussian models. We mention some results and some open questions. These relate to prediction and parameter estimation.

2. Initial remarks. A result of Q. Cheng 1992 indicates the difference between Gaussian and non-Gaussian linear models. Let $\{\xi_t\}$ be a sequence of independent, identically distributed (i.i.d.) random variables with $E\xi_t \equiv 0$, $0 < E\xi_t^2 = \sigma^2 < \infty$. A linear process $\{x_t\}$ has the form

$$(2.1) \qquad x_t = \sum_{k=-\infty}^{\infty} a_k \xi_{t-k}$$

with the a_k's real and

$$\sum a_k^2 < \infty.$$

Let

$$a(e^{-i\lambda}) = \sum a_k e^{-ik\lambda}.$$

If the process $\{x_t\}$ is Gaussian, the full probability structure is determined by

$$|a(e^{-i\lambda})|$$

---

*University of California, San Diego, La Jolla, CA 92093-0112 (mrosenblatt@ucsd.edu).

since the spectral density of the process

$$f(\lambda) = \frac{1}{2\pi}|a(e^{-i\lambda})|^2.$$

Now assume $a(e^{-i\lambda}) \neq 0$ almost everywhere. Cheng's result then says that if $\{x_t\}$ is non-Gaussian, if x_t has two representations

$$x_t = \sum a_j \xi_{t-j} = \sum a'_j \xi'_{t-j}$$

with $\{\xi_t\}$, $\{\xi'_t\}$ i.i.d. sequences,

$$\sum a_j^2, \sum a_j'^2 < \infty,$$

there is a value $c \neq 0$ and an integer t_0 such that

$$\xi'_t = c\xi_{t-t_0}, \qquad a'_t = \frac{1}{c}a_{t+t_0}.$$

So except for scalar multiplication and time shift non-Gaussian processes with different $a(\lambda) \neq 0$ a.e. represent different probability structures. This is quite different from the Gaussian case. This result holds with time d-dimensional, $d \geq 1$.

It is of some interest to look at the prediction and estimation problems for non-Gaussian autoregressive moving average models (ARMA)

$$x_t - \phi_1 x_{t-1} - \ldots - \phi_p x_{t-p} = \phi(B)x_t$$
$$= \xi_t + \theta_1 \xi_{t-1} + \ldots + \phi_q \xi_{t-q} = \theta(B)\xi_t$$

where the polynomials

$$\phi(z) = 1 - \phi_1 z - \ldots - \phi_p z^p,$$
$$\theta(z) = 1 + \theta_1 z + \ldots + \theta_q z^q.$$

have no zeros in common. Here B is the one step backshift operator. If $\phi(z)$ has all its zeros outside the unit disc $|z| \leq 1$ in the complex plane the process is called causal since it can be written

$$x_t = \sum_{j=0}^{\infty} \alpha_j \xi_{t-j} \qquad \text{with } \sum_{j=0}^{\infty} |\alpha_j| < \infty.$$

If $\theta(z)$ has all its zeros outside the unit disc the process $\{x_t\}$ is called invertible because one can write

$$\xi_t = \sum_{j=0}^{\infty} \beta_j x_{t-j} \qquad \text{with } \sum_{j=0}^{\infty} |\beta_j| < \infty.$$

If $\{x_t\}$ is both causal and invertible we shall call the process minimum phase.

3. Prediction. In the minimum phase case ARMA non-Gaussian stationary processes still have the best predictor in mean square

$$E(x_t \mid x_\tau, \tau < t)$$

linear. It is also plausible that if

$$x_t = \sum_{k=0}^{\infty} a_k \xi_{t-k}$$

with $\sum_{k=0}^{\infty} a_k^2 < \infty$, the best predictor of x_1 in terms of $\{x_s, s \leq 0\}$ is linear if $a(z) = \sum_{k=0}^{\infty} a_k z^k$ is an outer analytic function on the open unit disc in the complex plane. $a(z)$ is an outer function if $|a(z)| \geq |b(z)|$ on $|z| < 1$ for any function $b(z) = \sum_{k=0}^{\infty} b_k z^k$, $\sum b_k^2 < \infty$ with $b(e^{-i\lambda}) = a(e^{-i\lambda})$ almost everywhere.

In the converse direction there are partial results. If $\{x_t\}$ is stationary autoregressive nonminimum phase with the zeros of $\phi(z)$ inside the unit disc simple and for some integer $a \geq 2$ the $(a+1)$st cumulant μ_{a+1} of ξ_t is nonzero, then the best one-step predictor in mean square must be nonlinear. Using essentially the same argument one can show that for a stationary nonminimum phase ARMA scheme x_t with zeros of $\phi(z)$ inside $|z| < 1$ simple and all zeros of $\theta(z)$ outside $|z| \leq 1$ and for some $a \geq 2$ the $(a+1)$st cumulant μ_{a+1} of ξ_t nonzero, the best one-step predictor in mean square must be nonlinear (see Rosenblatt 2000). Also if x_t is a moving average with some zeros $\theta(z)$ inside the unit disc and for some $a \geq 2$ the $(a+1)$st cumulant μ_{a+1} of ξ_t nonzero, the best one step predictor in mean square must be nonlinear.

A result of M. Kanter 1979 gives a lower bound for the mean square error under appropriate conditions. Assume that $\{x_t\}$ is an infinite moving average

$$x_t = \sum_{j=-\infty}^{\infty} a_j \xi_{t-j}$$

in terms of the i.i.d. random variables ξ_t with $E\xi_t^2 = \sigma^2 < \infty$ and the differential entropy

$$H(\xi) = \int p_\xi(x) \log p_\xi(x)\, dx$$

where $p_\xi(x)$ is the density function of the ξ_t's. Let ε^2 be the prediction error variance of the best linear predictor. If $\hat{x}_n$ is the best predictor of x_n in mean square then

$$E(x_n - \hat{x}_n)^2 \geq \varepsilon^2 C e^{2H(\xi)}$$

with C a positive constant depending only on σ^2. This implies that the nonlinear prediction error variance is positive if the linear prediction error variance is positive, $H(\xi)$ is finite and $\sigma^2 > 0$.

Simple examples can be given with linear prediction error positive and nonlinear prediction perfect—examples with the ξ distribution discrete. Consider

$$x_n = \frac{1}{2} x_{n-1} + \xi_n$$

where the ξ_n's are i.i.d. with

$$\xi_n = \begin{cases} 0 & \text{with probability } \dfrac{1}{2} \\[2mm] 1 & \text{with probability } \dfrac{1}{2} \, . \end{cases}$$

The best linear predictor is

$$\frac{1}{2} x_{n-1} + E\xi_n.$$

Predicting forwards this is the best nonlinear predictor in mean square. The stationary x distribution is uniform. x_n has the binary representation

$$x_n = \xi_n \cdot \xi_{n-1} \xi_{n-2} \cdots .$$

From this one can see that one can predict backwards perfectly but not with the best backwards linear predictor.

4. Parameter estimation. The spectral density of the stationary ARMA $\{x_t\}$ satisfying (2.1) is

$$f(\lambda; \beta) = \frac{\sigma^2}{2\pi} g(\lambda; \beta)$$

with

$$g(\lambda, \beta) = \left| \frac{\theta(e^{-i\lambda})}{\phi(e^{-i\lambda})} \right|^2$$

where

$$\beta = (\phi_1, \ldots, \phi_p, \theta_1, \ldots, \theta_q)'.$$

The object is to estimate σ^2 and β. If $\{x_t\}$ is Gaussian the process can be assumed to be minimum phase since one cannot distinguish between zeros of $\theta(z)$, $\phi(z)$ inside the unit disc in the complex plane and those outside. The maximum likelihood estimates of β and σ^2 are asymptotically normal and independent. Let

$$W(\beta_0) = \frac{1}{4\pi} \int_{-\pi}^{\pi} \left\{ \frac{\partial \ln g(\lambda; \beta_0)}{\partial \beta} \right\} \left\{ \frac{\partial \ln g(\lambda; \beta_0)}{\partial \beta} \right\}' d\lambda.$$

The maximum likelihood estimate $\hat{\beta}_n$ of β is asymptotically normal with mean β_0 and covariance matrix

$$n^{-1}W^{-1}(\beta_0)$$

(see Brockwell and Davis 1991). The maximum likelihood estimate of the variance $\hat{\sigma}_n^2$ is asymptotically independent of $\hat{\beta}_n$ and asymptotically normal with mean σ^2 and variance $2\sigma^4 n^{-1}$. In the non-Gaussian minimum phase case one can still consider $\hat{\beta}_n$ as an estimate of the true parameter vector β_0 and its asymptotic properties will be the same as in the Gaussian case. Then $\hat{\sigma}_n^2$ is still asymptotically independent of $\hat{\beta}_n$ and normal as before but the asymptotic variance is now $(m_4 - \sigma^4)n^{-1}$ with m_4 the 4th moment of the ξ random variables (see Rosenblatt 2000). However, in the non-Gaussian minimum phase case the actual maximum likelihood estimate will be more efficient asymptotically. In the minimum phase non-Gaussian case consider an autoregressive scheme

$$x_t - \phi_1 x_{t-1} - \ldots - \phi_p x_{t-p} = \xi_t$$

as a simple example. The polynomial $\phi(z)$ is assumed to have all its zeros with absolute value greater than one. Assume that the ξ's are i.i.d. with a density h that is absolutely continuous and positive with finite Fisher information

$$I(h) = \int \left(\frac{h'}{h}\right)^2 h \, d\lambda.$$

Assuming h is known the maximum likelihood estimate (approximate) of

$$\theta = (\phi_1, \ldots, \phi_p)$$

is asymptotically normal and unbiased with covariance matrix

$$n^{-1}\Gamma^{-1}I(h)^{-1}$$

where Γ is the $p \times p$ covariance matrix of the stationary $AR(p)$ sequence with parameter θ. Generally we do not know h so we have a semiparametric problem. We briefly sketch aspects of an adaptive procedure suggested by Kreiss that assumes one already has a $\sqrt{n}$ consistent procedure available. The estimate already discussed based on the Gaussian likelihood is such a $\sqrt{n}$ consistent procedure. The object is to improve this $\sqrt{n}$ consistent procedure adaptively so as to get one that is asymptotically as good as the efficient procedure that used knowledge of h. Let

$$\Delta_n = \frac{1}{\sqrt{n}} \sum_{j=1}^{n} \dot{\phi}(\xi_j) x(j-1)$$

with

$$\dot{\phi} = -\frac{h'}{h}, \qquad x(j-1) = (x_{j-1}, \ldots, x_{j-p}).$$

Set

$$\theta_n = \theta + n^{-1/2}r,$$

$$h_n = \left[\left(1 - \frac{\int \beta^2 d\lambda}{n}\right)^{1/2} \sqrt{h} + \frac{\beta}{\sqrt{n}}\right]^2, \qquad \beta \in L^2, \quad \beta \perp \sqrt{h}.$$

The distribution $p_{n,(r,\beta)}$ of $(x_{1-p}, \ldots, x_n)$ has density

$$g_n(x_{1-p}, \ldots, x_0; h, \beta) \prod_{j=1}^{n} h_n(x_j - \theta'_n x(j-1)).$$

The claim is that if the distribution is $p_{n,(r,\beta)}$

$$\Gamma^{-1/2}I(h)^{-1/2}\Delta_n - \Gamma^{1/2}I(h)^{1/2}r$$

is asymptotically normal with mean zero and covariance matrix I_p. Also that if

$$\theta_n - \theta = 0(n^{-1/2})$$

that

$$\Delta_n(\theta_n) - \Delta_n(\theta) + \Gamma I(h)\sqrt{n}(\theta_n - \theta) = o(1)$$

with respect to distribution $p_{n,0}$. If $\bar{\theta}_n$ is an initially given $\sqrt{n}$ consistent estimate and

$$\hat{\theta}_n = \bar{\theta}_n + n^{-1/2}\frac{\Gamma^{-1}}{I(h)}\Delta_n(\bar{\theta}_n)$$

then it follows that

$$\sqrt{n}(\hat{\theta}_n - \theta)$$

is asymptotically normal with mean 0 and covariance matrix $\Gamma^{-1}I(h)^{-1}$. To get an adaptive estimate one needs consistent estimates $\hat{\Gamma}_n$ and $\hat{I}_n$ of Γ and I as well as a $\hat{\Delta}_n(\bar{\theta}_n)$ not using knowledge of h such that

$$\hat{\Delta}_n(\bar{\theta}_n) - \Delta_n(\bar{\theta}_n) = o(1).$$

An appropriate estimate of h by simply deconvolving is used appropriately to effect this. If

$$\xi_j(\theta) = x_j - \theta' x(j-1)$$

a density estimate of h is given by

$$\hat{f}_{\sigma_n}(x) = \frac{1}{n} \sum_{j=1}^{n} \phi(x - \xi_j(\theta_n); \sigma_n)$$

where it is understood that $\phi(x; \sigma)$ is the Gaussian density with mean 0 and variance σ^2 and $\sigma_n \downarrow 0$ as $n \to \infty$. Let g be a continuous density with support $[-1, 1]$ and $g(x) \le g(0) = 1$. Then under proper conditions

$$\hat{\Delta}_n(\theta_n)$$
$$= \sqrt{n} \int \frac{-\hat{f}'_{\sigma_n}(x)}{\hat{f}_{\sigma_n}(x)} \frac{1}{n} \sum_{j=1}^{n} \{[\phi(x - \xi_j(\theta_n); \sigma_n) - \hat{f}_{\sigma_n}(x)]x(j-1)\}g\left(\frac{x}{c_n}\right) dx$$

with $c_n \to \infty$ as $n \to \infty$ will do. An extended discussion is given in J.P. Kreiss 1987. It is not clear how well such procedures work for moderate size samples.

In Breidt et al. 1991 maximum likelihood estimation for noncausal autoregressive schemes is taken up assuming knowledge of the density function f of the noise. Related questions are discussed in Gassiat 1993. The corresponding question of maximum likelihood estimation for nonminimum phase ARMA schemes assuming knowledge of f is considered in Lii and Rosenblatt 1996. An important question is resolving efficient and possibly adaptive estimation when f is not known. In the nonminimum phase case the estimates based on the Gaussian likelihood are not consistent.

Let us now consider the autoregressive scheme

$$(4.1) \qquad \sum_{k} \phi_k x_{t-k} = \xi_t, \qquad \phi_0 = 1$$

with the indices t, k d-dimensional, $d \ge 1$. There is a stationary solution x_t of (4.1) if

$$\phi(\lambda) = \sum \phi_k e^{-ik \cdot \lambda}$$

is bounded away from 0. Then

$$\phi(\lambda)^{-1} = \sum_{k} \alpha_k e^{-ik \cdot \lambda}, \qquad \sum |\alpha_k| < \infty$$

and

$$x_t = \sum_{k} \alpha_k \xi_{t-k}.$$

If $d \ge 2$ the minimum phase condition does not necessarily make sense.

Wiggins 1978 (a geophysicist) on intuitive grounds suggested a way of estimating the parameters ϕ_k in the non-Gaussian case. Let

$$c_\nu(\xi) = cum \underbrace{(\xi, \ldots, \xi)}_{\nu}$$

with $E|\xi|^\nu < \infty$ for some integer $\nu > 2$. Consider

$$(4.2) \qquad M(F_\xi) = \left| \frac{c_\nu(\xi)}{c_2(\xi)^{\nu/2}} \right|$$

where it is assumed $c_\nu(\xi) \neq 0$ and F_ξ is the ξ distribution. Let

$$\xi_t(a) = \sum_k a_k x_{t-k}.$$

Given the a's compute

$$\xi_t(a), t = (t_1, \ldots, t_d),$$

$1 \le t_i \le n$, $i = 1, \ldots, d$. Estimate the νth and 2nd cumulants of $\xi_t(a)$'s. The absolute value of the normalized νth cumulant

$$(4.3) \qquad M_n(a) = M({}_nF_{\xi(a)})$$

with ${}_nF_{\xi(a)}$ the sample distribution of the $\xi_t(a)$'s is maximized as a function of the a's. Let

$$R_1 = \{a : a_0 = 1\}.$$

If $a_n = \{a : M_n(a) \text{ maximized on } R_1\}$ then a_n tends to the true parameter $\phi \in R_1$ in probability as $n \to \infty$. Instead of $M(F_\xi)$ as defined earlier consider

$$(4.4) \qquad M(F_\xi) = \frac{c_\nu(\xi)^2}{c_2(\xi)^\nu}.$$

Let

$$\hat{a}_n = \{a : M_n(a) \text{ maximized on } R_1\}$$

with $M_n(a)$ given by (4.3) but with M given by (4.4) instead of (4.2). Notice that

$$M(a) = H(Eh_j(\xi(a)_0), j = 1, \ldots, q)$$

with H the proper rational function and h_j's powers of the $\xi(a)$'s, $h = (h_1, \ldots, h_q)$. Given the function g let $E_n(g(\xi(a))$ be the sample mean

$$E_n g(\xi(a)) = n^{-d} \sum_{t \in S_n} h(\xi(a)_t)$$

with the sum over t taken over the sampled range $S_n = \{1 \le t_i \le n, i = 1, \ldots, d\}$. Then $M_n(a)$ can also be given by

$$M_n(a) = H(E_n h_j(\xi(a)), j = 1, \ldots, q).$$

Set

$$\psi(u) = \sum_{k=1}^{q} h_k'(u) D_k H(Eh(\xi(a)_0))$$

where it is understood that D_k is the partial derivative with respect to the kth entry in H. The matrices A and B are given by

$$A_{i,j} = \sum_{k \ne 0} \alpha_{k-i}\alpha_{k-j} E(\psi'(\xi_0)) var(\xi_0),$$

$$B_{i,j} = \sum_{k \ne 0} \alpha_{k-i}\alpha_{k-j} E(\psi(\xi_0)^2) var(\xi_0).$$

One can then show that if $E(\xi^{2\nu+2}) < \infty$, $\nu > 2$, the estimate $\hat{a}_n$ of the true coefficient vector ϕ is consistent and that $n^{d/2}(\hat{a}_n - \phi)$ is asymptotically normal with mean zero and covariance matrix $A^{-1}BA^{-1}$.

The asymptotic distribution of these estimates was initially given in Gassiat 1990. Our version of the derivation is given in Rosenblatt 2000.

Consider an ARMA model

$$\phi(B)x_t = \phi(B^{-1})\xi_t$$

with $\phi_p \ne 0$ and the zeros of $\phi(z)$ all outside $\{z : |z| \le 1\}$. Such a scheme is causal and is called all-pass because its spectral density is constant. The process is white noise but not a sequence of independent, identically distributed random variables if $p > 0$. In Breidt et al. 2001 a modified log likelihood with the noise distribution formally assumed two-sided exponential is used leading to a least absolute deviation estimate of the parameters of such causal all-pass ARMA schemes. They then show how to estimate parameters of noncausal autoregressive schemes by simply applying these methods. In a number of specific computational examples the methods are shown to give greater resolution than the cumulant estimates.

5. Estimation of the transfer function $a(e^{-i\lambda})$. More generally one can consider estimation of the transfer function $a(e^{-i\lambda})$ of a non-Gaussian linear process (2.1). Under appropriate conditions this is discussed in some detail in Lii and Rosenblatt 1982. We briefly sketch some of the details here. It should be noted that the rate of convergence is appreciably slower than in the parametric case and this is to be expected since the problem can be thought of as a nonparametric problem. First assume that

$$\sum |j||a_j| < \infty.$$

If a cumulant γ_k of the random variables ξ_t is finite for some $k > 2$ the corresponding cumulant spectral density of the linear process $\{x_t\}$ exists and is given by

$$f_k(\lambda_1, \ldots, \lambda_{k-1}) = (2\pi)^{-k+1}\gamma_k a(e^{-i\lambda_1}) \ldots a(e^{-i\lambda_{k-1}}) a(e^{i(\lambda_1 + \ldots + \lambda_{k-1})}).$$

Assume that $a(e^{-i\lambda}) \neq 0$ for all λ and introduce

$$h(\lambda) = arg\left\{ a(e^{-i\lambda}) \frac{a(1)}{|a(1)|} \right\}.$$

Notice that

$$\{a(1)/|a(1)|\}^k \gamma_k = (2\pi)^{(k/2)-1} f_k(0, \ldots, 0)\{f(0)\}^{-k/2}$$

where it is understood that $f(\lambda) = f_2(\lambda)$, the usual second order spectral density. But the more important fact is that

$$h(\lambda_1) + \ldots + h(\lambda_{k-1}) - h(\lambda_1 + \ldots + \lambda_{k-1})$$
$$= arg\left[\left\{ \frac{a(1)}{|a(1)|} \right\}^k \gamma_k^{-1} f_k(\lambda_1, \ldots, \lambda_{k-1}) \right]$$

with $h(-\lambda) = -h(\lambda)$. Also

$$h'(0) - h'(\lambda) = \lim_{\Delta \to 0} \frac{1}{(k-2)\Delta}\{h(\lambda) + (k-2)h(\Delta) - h(\lambda + (k-2)\Delta)\}.$$

Set

$$h_1(\lambda) = \int_0^\lambda \{h'(u) - h'(0)\}\, du$$

so that

$$h(\lambda) = h_1(\lambda) + c\lambda$$

with $c = h'(0)$. Since the coefficients a_j are real it follows that $h(\pi) = a\pi$ for some integer a. The integer a cannot be determined without more information because it corresponds to the subscripting of the ξ_t's. $a(1)$'s sign is also not determined since the a_j's and ξ_t's can be multiplied by (-1) without changing the process x_t observed. Therefore up to sign and a factor $\exp(i\alpha\lambda)$ with α an integer

$$a(e^{-i\lambda}) = |2\pi f(\lambda)|^{1/2} \exp\{ih(\lambda)\}.$$

If $\gamma_k \neq 0$ for some $k > 2$ and $Ex_t^{2k} < \infty$ $a(e^{-i\lambda})$ can be estimated up to sign. Remarks are made in the case $k = 3$ but they are analogous in the case $k > 3$. On the basis of observations $x_1, \ldots, x_n$ compute a spectral estimate $f_n(\lambda)$ of $f(\lambda)$. Let $_nf(\lambda, \mu)$ be a bispectral estimate of $f_3(\lambda, \mu)$. Consider

$$H_n(\lambda) = -\sum_{j=1}^{k-1} arg\, _nf(j\Delta, \Delta)$$

with $k\Delta = \lambda$. This is a consistent estimate of $h_1(\lambda)$ if $\Delta \to 0$ and $\Delta^3 n \to \infty$ as $n \to \infty$ and the weight function of the estimate $_nf$ is properly chosen.

REFERENCES

[1] AKI K., Quantitative Seismology Theory and Methods, Vol. 2, W. Freeman and Co., 1980.

[2] BREIDT F.J., DAVIS R., AND TRINDADE A., *Least absolute deviation estimation for all-pass time series*, Ann. Stat. **29** (2001), 919–946.

[3] BREIDT F.J., DAVIS R.A., LII K.S., AND ROSENBLATT M., *Maximum likelihood estimation for noncausal autoregressive processes*, J. Multivar. Anal. **36** (1991), 175–198.

[4] BROCKWELL P. AND DAVIS R., Time Series: Theory and Methods, 2nd edition, Springer, 1991.

[5] CHENG Q., *On the unique representation of non-Gaussian linear processes*, Ann. Stat. **20** (1992), 1143–1145.

[6] DONOHO D., *On minimum entropy deconvolution* in Applied Time Series Analysis (D. Findley, ed.), 1981, pp. 565–608.

[7] GASSIAT E., *Estimation semi-paramétrique d'un modèle autorégressif stationnaire multi-indice non nécessairement causal*, Ann. Inst. H. Poincaré Probab. Statist. **26** (1990), 181–205.

[8] GASSIAT E., *Adaptive estimation in noncausal stationary AR processes*, Ann. Stat. **21** (1993), 2022–2042.

[9] KANTER M., *Lower bounds for nonlinear prediction error in moving average processes*, Ann. Prob. **7** (1979), 128–138.

[10] KREISS J., *On adaptive estimation in autoregressive models when there are nuisance functions*, Stat. and Decisions **5** (1987), 59–76.

[11] LII K.S. AND ROSENBLATT M., *Deconvolution and estimation of transfer function phase and coefficients for non-Gaussian linear processes*, Ann. Stat. **10** (1982), 1195–1208.

[12] LII K.S. AND ROSENBLATT M., *Maximum likelihood estimation of non-Gaussian nonminimum phase ARMA sequences*, Stat. Sin. **6** (1996), 1–22.

[13] ROBINSON E.A., Time Series Analysis and Applications, Goose Pond Press, 1981.

[14] ROSENBLATT M., Gaussian and Non-Gaussian Linear Time Series and Random Fields, Springer, 2000.

[15] WIGGINS R.A., *Minimum entropy deconvolution*, Geoexploration16 (1978), 21–35.

[16] YAGLOM A.M., Correlation Theory of Stationary and Related Random Functions, Vols. 1, 2, Springer, 1986.

MODELING CONTINUOUS TIME SERIES DRIVEN BY FRACTIONAL GAUSSIAN NOISE

WINSTON C. CHOW* AND EDWARD J. WEGMAN†

Abstract. We consider the stochastic differential equations, $dX(t) = \theta X(t)dt + dB_H(t);\ t > 0$, and $dX(t) = \theta(t)X(t)dt + dB_H(t);\ t > 0$ where $B_H(t)$ is fractional Brownian motion. We find solutions for these differential equations and show the existence of the integrals related to these solutions. We then show that $B_H(t)$ is not a martingale. This implies that several conventional methods for defining integrals on fractional Brownian motion are inadequate. We demonstrate the existence of an estimator for θ which depends on the existence of integrals of certain integrals with respect to fractional Brownian motion. We conclude by showing the existence and Riemann sum approximations for these integrals.

1. Introduction. In this paper, we demonstrate the existence of optimal statistical estimators for parameters of certain forms of stochastic differential equations driven by fractional Gaussian noise. Dobrushin (1979) and Major (1981) both consider linear and nonlinear functionals of self-similar Gaussian fields with stationary increments. Fractional Brownian motion is such a process. This type of random noise appears in certain physical processes that exhibit correlations that decrease slowly with time and low frequency power. Some physical processes possess the fractal property of self-similarity, which is a basic property of fractional Brownian motion. Previously established parametric estimators mainly deal with random noise in the form of Gaussian white noise and its standard Brownian motion, although algorithms have also been derived to handle random processes in the form of square-integrable martingales, which generalize the Brownian motion noise process. Both man-made and natural processes appear to exhibit randomness in the form of fractional Brownian motion or fractional Gaussian noise.

The fractal property of statistical self-similarity often appears in geophysical processes. In geology and hydrology, models with fractional random processes prove useful. River discharges tend to exhibit clusters of high periods and low periods and thus exhibit long-term dependencies (Mandelbrot, 1983). Gregotski, Jensen, and Arkani-Hamed (1991) demonstrate experimental data indicating that spatial magnetic patterns of certain geographical locations behave in a statistical self-similar way where the independent variables are spatial processes. Self-similarity also is modeled for

---

*Winston C. Chow is with the Naval Network and Space Operations Command based in Dahlgren, VA. This work is derived in part from his Ph.D. dissertation completed at George Mason University. Dr. Chow's work was supported by the NSWC Training Program.

†Center for Computational Statistics George Mason University, MS 4A7, 4400 University Drive, Fairfax, VA 22030-4444. Dr. Wegman's work was supported by the Army Research Office. This work was completed while Dr. Wegman was a Navy-ASEE Distinguished Faculty Fellow at the Naval Surface Warfare Center/Dahlgren Division.

communication channels and internet communication. Random errors in communication channels may occur in groups of bursts, where this groups of bursts are themselves grouped in bursts (Barton and Poor, 1988). Stewart et al. (1993) show that radar images from natural "clutter sources" have a texture that looks like fractional Brownian motion in two dimensions with the independent variables being distances. Finally, we note that Wegman and Habib (1992) apply the class of stochastic differential equation models we describe here to describe sub-threshold neuron-firing processes.

2. Solution of the stochastic differential equations. We consider first the parametric model as the stochastic differential equation

$$(2.1) \qquad dX(t) = \theta X(t)dt + dB_H(t); \qquad t > 0.$$

$B_H(t)$ is fractional Brownian motion. Let $\{B(t) : t \in R\}$ be a standard Brownian motion process, then fractional Brownian motion, B_H for given $H \in (1/2, 1)$ is defined as follows:

$$(2.2) \qquad B_H(t) = \frac{1}{\Gamma(H + 1/2)} \left\{ \int_{-\infty}^{0} (|t - \tau|^{H-1/2} - |\tau|^{H-1/2})dB(\tau) + \int_{0}^{t} |t - \tau|^{H-1/2} dB(\tau) \right\}.$$

Notice for $H = 1/2$, fractional Brownian motion coincides with ordinary Brownian motion.

To develop the solution to (2.1), first of all, consider the homogeneous form of this differential equation $dX(t) = \theta X(t)dt$. It is straightforward to see that $X(t) = e^{\theta t}X(0)$ is the homogeneous solution. Assume, then, that the particular solution has the form $X(t) = e^{\theta t}Y(t)$. Under this assumption we have the following differential equation

$$dY(t) = e^{-\theta t}dB_H(t).$$

This equation is formally equivalent to the integral equation

$$Y(t) = \int_{0}^{t} e^{-\theta \tau}dB_H(\tau).$$

Substituting this solution for $Y(t)$ back into the original yields the particular solution

$$X(t) = e^{\theta t} \int_{0}^{t} e^{-\theta \tau}dB_H(\tau).$$

Thus formally the general solution is

$$(2.3) \qquad X(t) = e^{\theta t}X(0) + e^{\theta t} \int_{0}^{t} e^{-\theta \tau}dB_H(\tau).$$

For the case where $X(0)$ is zero the solution is

$$(2.4) \qquad X(t) = e^{\theta t} \int_0^t e^{-\theta \tau} dB_H(\tau).$$

These are formal solutions to the stochastic differential equation (2.1) since existence of the stochastic integrals in Equations (2.3) and (2.4) have not been established. In the case of an equation driven by a martingale, the existence of these integrals has been established. However, as we shall shortly see, fractional Brownian motion is not a martingale, hence, we need to establish the existence of these integrals separately.

These solutions can easily be generalized to a nonparametric form, where the θ term is an unknown function rather than an unknown constant

$$(2.5) \qquad dX(t) = \theta(t)X(t)dt + dB_H(t).$$

Now consider a solution to the homogeneous differential equation of the form $X(t) = A(t)X(0)$. It is straightforward to show that

$$A(t) = e^{\int_0^t \theta(\tau)d\tau}.$$

Hence, substituting for $A(t)$, the homogeneous solution is as follows:

$$X(t) = e^{\int_0^t \theta(\tau)d\tau} X(0).$$

Now, assuming a particular solution to be of the form

$$X(t) = A(t)Y(t)$$

where $A(t)$ is as before and $Y(t)$ is an unknown process, we find

$$X(t) = A(t) \int_0^t (1/A(\tau))dB_H(\tau).$$

So the general solution is

$$X(t) = e^{\int_0^t \theta(\alpha)d\alpha} X(0) + e^{\int_0^t \theta(\alpha)d\alpha} \int_0^t e^{-\int_0^\tau \theta(\alpha)d\alpha} dB_H(\tau),$$

or assuming $X(0) = 0$,

$$X(t) = e^{\int_0^t \theta(\alpha)d\alpha} \int_0^t e^{-\int_0^\tau \theta(\alpha)d\alpha} dB_H(\tau)$$

or equivalently

$$(2.6) \qquad X(t) = \int_0^t e^{\int_\tau^t \theta(\alpha)d\alpha} dB_H(\tau).$$

As before, these are formal manipulations since we have not yet proved the existence of the integrals involved. As mentioned above if B_H were a martingale, the existence of integrals in expressions (2.3), (2.4), and (2.6) would be demonstrated. However, B_H is not a martingale, and hence we need to appeal to first principles in order to demonstrate the existence of these integrals. We base the result on the following theorem.

THEOREM 2.1 (Cramer and Leadbetter, 1967, p. 90). *If the covariance function $R(s,r)$ of X is of bounded variation in $[0,t] \times [0,t]$ and f is a deterministic function f is such that $\int_0^t \int_0^t f(s)f(r)d_{s,r}R(s,r)$ exists as a Riemann-Stieltjes integral, then $\int_0^t f(s)dX(s)$ is well defined.*

The covariance of fractional Brownian motion is given by

$$(2.7) \qquad R_{B_H}(s,t) = \frac{V_H}{2}\left(\,|\,s\,|^{2H} + |\,t\,|^{2H} - |\,t-s\,|^{2H}\,\right)$$

where $V_H = \mathrm{var}\,[B_H(1)] = \frac{-\Gamma(2-2H)\cos(\pi H)}{\pi H(2H-1)}$ such that $H \in (1/2, 1)$ (Barton and Poor, 1988). For $H > 1/2$, this $R_{B_H}(s,t)$ is clearly of bounded variation so that by Theorem 2.1, the integrals in (2.3), (2.4), and (2.6) exist and are well-defined.

3. B_H is not a martingale. As we have just indicated, integrals of a continuous process with respect to B_H are well defined under mild conditions if B_H is a square-integrable martingale or a local square-integrable martingale. Unfortunately, this not the case will be seen in the theorems to follow. Although a martingale is a local martingale, what follows first is a proof that fractional Brownian motion is not a martingale, which can be easily generalized to show that B_H is also not a local martingale.

THEOREM 3.1. *Let $\{B_H(t): -\infty < t < \infty\}$ be a fractional Brownian motion. Let the σ-algebra filtration $\{\mathcal{A}_t: -\infty < t < \infty\}$ be the filtration to which a Brownian motion B is adapted and, let B_H be derived from B. $\{B_H(t), \mathcal{A}_t : t \geq 0\}$ is not a martingale.*

Proof. Let $t > s \geq 0$.

$$E[B_H(t) \mid \mathcal{A}(s)]$$
$$= \frac{1}{\Gamma(H+1/2)}\left\{E\left(\int_{-\infty}^0 (|\,t-\tau\,|^{H-1/2} - |\,\tau\,|^{H-1/2})dB(\tau) \mid \mathcal{A}(s)\right)\right.$$
$$\left. + E\left(\int_0^t |\,t-\tau\,|^{H-1/2}\,dB(\tau) \mid \mathcal{A}(s)\right)\right\}$$
$$(3.1) \quad = \frac{1}{\Gamma(H+1/2)}\int_{-\infty}^0 (|\,t-\tau\,|^{H-1/2} - |\,\tau\,|^{H-1/2})dB(\tau)$$
$$+ \int_0^s |\,t-\tau\,|^{H-1/2}\,E[dB(\tau) \mid \mathcal{A}(s)]$$
$$+ \int_s^t |\,t-\tau\,|^{H-1/2}\,E[dB(\tau) \mid \mathcal{A}(s)].$$

Since $B(\tau)$ has independent increments, $E[dB(\tau) \mid \mathcal{A}(s)] = 0$ for all $\tau \geq s$. Hence, the last term on the right-side of the equation (3.1) equals 0, and therefore we have as follows:

$$\frac{1}{\Gamma(H+1/2)} \, E\left(\int_{-\infty}^{0} (|\, t - \tau \,|^{H-1/2} - |\, \tau \,|^{H-1/2}) dB(\tau) \right.$$

$$\left. + \int_{0}^{t} |\, t - \tau \,|^{H-1/2} \, dB(\tau) \mid \mathcal{A}(s) \right)$$

$$= \frac{1}{\Gamma(H+1/2)} \int_{-\infty}^{0} (|\, t - \tau \,|^{H-1/2} - |\, \tau \,|^{H-1/2}) dB(\tau)$$

$$+ \int_{0}^{s} |\, t - \tau \,|^{H-1/2} \, dB(\tau).$$

Notice that the right-hand side of the above expression depends explicitly on t; this is not equal to $B_H(s)$ since $B_H(s)$ is

$$B_H(s) = \frac{1}{\Gamma(H+1/2)} \int_{-\infty}^{0} (|\, s - \tau \,|^{H-1/2} - |\, \tau \,|^{H-1/2}) dB(\tau)$$

$$+ \int_{0}^{s} |\, s - \tau \,|^{H-1/2} \, dB(\tau).$$

Hence, $\{B_H(t), \mathcal{A}(t) : t \geq 0\}$ is not a martingale, and the theorem is proved. $\qquad\qquad\square$

COROLLARY 3.2. *Let* $\{B_H(t) : -\infty < t < \infty\}$ *be a fractional Brownian motion. Let the σ-algebra filtration $\{\mathcal{A}_t : -\infty < t < \infty\}$ be the filtration to which a Brownian motion B is adapted and, let B_H be derived from B. $\{B_H(t), \mathcal{A}_t : t \geq 0\}$ is not a local martingale.*

Proof. Suppose $\{B_H(t), \mathcal{A}(t) : t \geq 0\}$ is a local martingale. There exists $\{T_n\}$ a sequence of stopping times such that $T_n \to \infty$ and $T_n \leq T_{n+1}$. Then $B_H(T_n \wedge t) I_{(T_n > 0)}$, where the I function is an indicator function, is a martingale. By the definition of a martingale the following must then hold:

$$E[B_H(T_n \wedge t) I_{(T_n > 0)} \mid \mathcal{A}(s)]$$

$$= B_H(T_n \wedge s) I_{(T_n > 0)} B_H(T_n \wedge s) I_{(T_n > 0)}$$

$$= \frac{1}{\Gamma(H+1/2)} I_{(T_n > 0)} \left\{ \int_{-\infty}^{0} |\, T_n \wedge s - \tau \,|^{H-1/2} - |\, \tau \,|^{H-1/2}) dB(\tau) \right.$$

$$\left. + \int_{-\infty}^{T_n \wedge s} |\, T_n \wedge s - \tau \,|^{H-1/2} \, dB(\tau) \right\}.$$

Let f be defined such that

$$f(r, \alpha) = I_{(-\infty, 0)}\left(|\, r - \alpha \,|^{H-1/2} - |\, \alpha \,|^{H-1/2} \right) + I_{[0, r)} |\, r - \alpha \,|^{H-1/2}.$$

Using this definition of f to simplify formulas,

$$B_H(T_n \wedge s) I_{(T_n > 0)} = I_{(T_n > 0)} \int_{-\infty}^{T_n \wedge s} f(T_n \wedge s, \tau) dB(\tau)$$

or

$$B_H(T_n \wedge s) = \int_0^{T_n \wedge s} f(T_n \wedge s, \tau) dB(\tau).$$

Hence, if $B_H(T_n \wedge t)I_{(T_n>0)}$ is assumed to be a martingale, then the following relationship has been shown to be true:

$$E[B_H(T_n \wedge t)I_{(T_n>0)} \mid \mathcal{A}(s)] = I_{(T_n>0)} \int_{-\infty}^{T_n \wedge s} f(T_n \wedge s, \tau) dB(\tau), \quad \text{for } t \geq s$$

$$= \int_0^{T_n \wedge s} f(T_n \wedge s, \tau) dB(\tau).$$

Letting Ω be the sample space, by the definitions of the expected value and the indicator function,

$$E[B_H(T_n \wedge t)I_{(T_n>0)}] = \int_\Omega \int_0^{T_n \wedge t} f(t \wedge T_n, \tau) dB(\tau, \omega) dP(\omega)$$

where P is the probability measure and $\omega \in \Omega$. By the measure theoretic definition of conditional expected value, given $A \in \mathcal{A}(s)$,

$$\int_A E[B_H(T_n \wedge t)I_{(T_n>0)} \mid \mathcal{A}(s)] dP(\omega) = \int_A \int_0^{T_n \wedge t} f(T_n \wedge t, \tau) dB(\tau, \omega) dP(\omega)$$

so that

$$\int_A \int_0^{T_n \wedge s} f(T_n \wedge s, \tau) dB(\tau, \omega) dP(\omega)$$

$$= \int_A \int_0^{T_n \wedge t} f(T_n \wedge t, \tau) dB(\tau, \omega) dP(\omega), \quad \text{for } t \geq s.$$

However, this cannot be true since t is not included in the deterministic function f of the integral on the left-hand side of the last equation. Therefore, we have a contradiction and $\{B_H(t), \mathcal{A}(t): t \geq 0\}$ must not be a local martingale, and the theorem is proved. $\square$

Using the equation for $E[B_H(T_n \wedge t) \mid \mathcal{A}(s)]$ in the proof that fractional Brownian motion is not a local martingale, we can generalize one step further and claim that B_H is not a semimartingale. In proving that B_H is not a semimartingale, the following result is needed:

THEOREM 3.3 (Shiryayev, 1984, p. 213). *If W and Y are to random variables such that $W \leq Y$ a.s., then*

$$E[W \mid \mathcal{A}] \leq E[Y \mid \mathcal{A}] \quad a.s.$$

Now the theorem claiming that fractional Brownian motion is not a semimartingale along with its proof will be given.

COROLLARY 3.4. $B_H = \{B_H(t) : t \in (-\infty, \infty)\}$ *is not a semi-martingale.*

Proof. Suppose B_H is a semimartingale. Then

$$B_H(t) = B_H(0) + M(t) + A(t) \qquad t \geq 0 \quad \text{a.s.}$$

or

$$B_H(t) = M(t) + A(t) \qquad \text{since} \quad B_H(0) = 0$$

where M is a local martingale and A is a right-continuous adapted process with locally bounded variation sample paths. Thus,

$$B_H(t) - A(t) = M(t), \quad t \geq 0$$

is a local martingale. So there exists an increasing stopping time sequence $\{T_n\}$ such that $T_n \to \infty$ as $n \to \infty$ and $B_H(T_n \wedge t) - A(T_n \wedge t)$ is a martingale. Given the adapting σ-algebra $\mathcal{A} = \{\mathcal{A}(t) : t \geq 0\}$ and using the definition of a martingale,

$$E[B_H(T_n \wedge t) \mid \mathcal{A}(s)] - E[A(T_n \wedge t) \mid \mathcal{A}(s)]$$
$$= B_H(T_n \wedge s) - A(T_n \wedge s) \qquad \text{for all} \quad s < t.$$

But $B_H(r)$ is $\int_{-\infty}^{r} f(r, \alpha) dB(\alpha)$ where

$$f(r, \alpha) = I_{(-\infty, 0)} \left(\mid r - \alpha \mid^{H-1/2} - \mid \alpha \mid^{H-1/2} \right) + I_{[0, r)} \mid r - \alpha \mid^{H-1/2}.$$

Substituting the definition of B_H, using f for the needed integrand, and substituting the expression for $E\{B_H(T_n \wedge t) \mid \mathcal{A}(s)\}$ as given in the proof that B_H is not a local martingale, we have:

$$E[B_H(T_n \wedge t) \mid \mathcal{A}(s)] - E[A(T_n \wedge t) \mid \mathcal{A}(s)]$$
$$= B_H(T_n \wedge s) - A(T_n \wedge s) \qquad \text{for all} \quad s < t \quad \Longrightarrow$$
$$\int_{-\infty}^{T_n \wedge s} f(T_n \wedge t, \alpha) dB(\alpha) - E[A(T_n \wedge t) \mid \mathcal{A}(s)]$$
$$= \int_{-\infty}^{T_n \wedge s} f(T_n \wedge s, \alpha) dB(\alpha) - A(T_n \wedge s) \quad \Longrightarrow$$
$$E[A(T_n \wedge t) \mid \mathcal{A}(s)] - A(T_n \wedge s)$$
$$= \int_{-\infty}^{T_n \wedge s} f(T_n \wedge t, \alpha) dB(\alpha) - \int_{-\infty}^{T_n \wedge s} f(T_n \wedge s, \alpha) dB(\alpha).$$

Since A is of locally bounded variation, on every finite interval, it must be the difference of two monotonic functions. This implies that $E[A(T_n \wedge t) \mid \mathcal{A}(s)]$ must also be the difference of two monotone functions for $s \in [0, t]$ by the theorem that immediately preceded this present result. This means that $E[A(T_n \wedge t) \mid \mathcal{A}(s)]$ must also be of locally bounded variation, and so $E[A(T_n \wedge t) \mid \mathcal{A}(s)] - A(T_n \wedge s)$ must be of locally bounded variation.

Since B is almost surely not differentiable for all $t \in (-\infty, \infty)$, it is not of bounded variation for all intervals. This implies by definition that for all $r \in (-\infty, \infty)$,

$$\int_0^{T_n \wedge r} \mid dB(\alpha) \mid = \infty \qquad \text{(Shiryayev, 1981, p. 201).}$$

But

$$\min_{\alpha \in (0, T_n \wedge s)} [(T_n \wedge t - \alpha)^{H-1/2} - (T_n \wedge s - \alpha)^{H-1/2}]$$
$$= (T_n \wedge t)^{H-1/2} - (T_n \wedge s)^{H-1/2}$$
$$= (T_n \wedge t)^{H-1/2} - s^{H-1/2} > 0$$

for the case where the random process $T_n > s$. There is no loss of generality in the arguments to follow by assuming the special case, for which $T_n > s$, since in order for B_H to be a semimartingale, the arguments must not lead to a contradiction under any circumstance. Now

$$\int_0^{T_n \wedge s} \mid (T_n \wedge t)^{H-1/2} - s^{H-1/2} dB(r) \mid$$
$$= [(T_n \wedge t)^{H-1/2} - s^{H-1/2}] \int_0^{T_n \wedge s} \mid dB(r) \mid = \infty.$$

This implies

$$\int_0^{T_n \wedge s} \left| d_r \int_0^{T_n \wedge r} \left\{ (T_n \wedge t)^{H-1/2} - s^{H-1/2} \right\} dB(\alpha) \right| = \infty,$$

where d_r is the differential with respect to r symbol. In other words, this last equation states that the limiting sum of the variations of the random process, $\int_0^{T_n \wedge r} \{ (T_n \wedge t)^{H-1/2} - s^{H-1/2} \} dB(\alpha)$, is unbounded. Since

$$[(T_n \wedge t - \alpha)^{H-1/2} - (T_n \wedge s - \alpha)^{H-1/2}] \geq (T_n \wedge t)^{H-1/2} - s^{H-1/2},$$

the limiting sum of the variations of the stochastic process represented by $\int_0^{T_n \wedge r} \{ (T_n \wedge t - \alpha)^{H-1/2} - (T_n \wedge s - \alpha)^{H-1/2} \} dB(\alpha)$ must also be unbounded. Moreover, this means that the random process

$$\int_{-\infty}^{T_n \wedge r} [f(T_n \wedge t, \alpha) - f(T_n \wedge s, \alpha)] dB(\alpha)$$
$$= \int_{-\infty}^0 [f(T_n \wedge t, \alpha) - f(T_n \wedge s, \alpha)] dB(\alpha)$$
$$+ \int_0^{T_n \wedge r} \left\{ (T_n \wedge t - \alpha)^{H-1/2} - (T_n \wedge s - \alpha)^{H-1/2} \right\} dB(\alpha)$$

must also be of unbounded variation in the interval $[0, T_n \wedge s]$, that is,

$$\int_0^{T_n \wedge s} \left| d_r \int_{-\infty}^{T_n \wedge r} [f(T_n \wedge t, \alpha) - f(T_n \wedge s, \alpha)] dB(\alpha) \right| = \infty.$$

In other words, $\int_{-\infty}^{T_n \wedge s} f(T_n \wedge t, \alpha) dB(\alpha) - \int_{-\infty}^{T_n \wedge s} f(T_n \wedge s, \alpha) dB(\alpha)$ is not of locally bounded variation. This is a contradiction to the fact that this process was set equal to $E[A(T_n \wedge t) \mid \mathcal{A}(s)] - A(T_n \wedge s)$, which was shown to be of locally bounded variation. Hence, B_H must not be a semimartingale, and the theorem is proved. $\qquad\square$

4. Christopeit's quasi-least-squares methods and its implications. Given the fractional Brownian motion process B_H for $H \in (1/2, 1)$, we now consider the estimation problem for parametric model given by

$$dX(t) = \theta X(t) dt + dB_H(t)$$

by first considering a continuous extension of a least squares method. The integral form of the model fits the stochastic process regression model as given in Christopeit (1986), except for the fact that the noise, which is fractional Brownian motion here, is not a martingale.

Christopeit's model is represented by

$$Y(t) = Y(0) + \theta \int_0^t X(s) dF(s) + M(t)$$

where F is an increasing process and M is a martingale. The quasi-least-squares estimate of θ as given in Christopeit is as follows:

$$\widehat{\theta}(t) = \left[\int_0^t X^2(s) dF(s) \right]^{-1} \int_0^t X(s) dY(s)$$

for the sample path in $[0, t]$. This method is called quasi-least squares because given a discrete partition of the time interval involved a least-squares estimate converges to the above estimate.

Although B_H is not a martingale, the quasi-least-squares estimator as given for the model that we are considering is given by

$$\widehat{\theta} = \int_0^t X(s) dX(s) \Big/ \int_0^t X^2(s) ds.$$

The integral in the numerator will be shown to be well defined in what follows.

The fact that the noise, being fractional Brownian motion, is not a martingale only affects the asymptotic properties and not the fact that the estimator is a quasi-least-squares estimate as long as the integrals in the estimator are well defined. Thus, the above estimator may still be a legitimate quasi-least-squares estimator although its asymptotic properties may not be as desirable. But the existence of integral in the numerator, $\int_0^t X(s)dX(s)$, must be demonstrated when the noise is not a martingale.

In order to determine whether $\int_0^t X(s)dX(s)$ exists as well as to decompose the estimator into the sum of the true value of the parameter, θ, and an error term, note that the estimator derived above can be formally represented by

$$\widehat{\theta} = \int_0^t X(s)[\theta X(s)ds + dB_H(s)] \Big/ \int_0^t X^2(s)ds$$

or equivalently

$$\widehat{\theta} = \theta + \left\{ \int_0^t X(s)dB_H(s) \Big/ \int_0^t X^2(s)ds \right\}.$$

This means that $\int_0^t X(s)dX(s)$ may be defined in terms of $\int_0^t X^2(s)ds$ and $\int_0^t X(s)dB_H(s)$ where θ is the true parameter value. The first integral $\int_0^t X^2(s)ds$ can be interpreted as either a quadratic mean integral or a sample path (Lebesgue or Riemann) integral, and it is finite since $X^2(s)$ is bounded almost surely in $[0, t]$. This is also why the denominator of the estimator, which is this same integral, is not of concern. The second integral, namely, $\int_0^t X(s)dB_H(s)$, will be shown to exist in the next section.

Since B_H is not a martingale, a local martingale, nor a semimartingale, the integrals $\int_0^t X(s)dB_H(s)$, where $dX(s) = \theta(s)X(s)ds + dB_H(s)$ are not defined in the conventional sense of stochastic integrals defined with respect to martingales or their variants. Thus in order for this estimator to make sense, we must develop a rigorous definition for this type of stochastic integral.

5. Defining the integrals. First recall from the previous section that given the stochastic differential equation as stated above,

$$X(s) = \int_0^s e^{\int_\tau^s \theta(\alpha)d\alpha} dB_H(\tau) \qquad \text{for} \quad X(0) = 0, \ \tau \geq 0.$$

Thus $\int_0^t X(s)dB_H(s)$ may be defined as

$$(5.1) \qquad \int_0^t X(s)dB_H(s) = \int_0^t \int_0^s e^{\int_\tau^s \theta(\alpha)d\alpha} dB_H(\tau)dB_H(s).$$

Thus we would like to show the existence of the integral on the right-hand side of (5.1). Define a function ζ represented by

$$\zeta(s,\tau) = e^{\int_\tau^s \theta(\alpha)d\alpha}.$$

Partition $[0,t]$ such that $\pi_n = \{0 = v_0, v_1, v_2, \ldots, v_n = t \leq T\}$ for $T \in (-\infty, \infty)$. Define a step function $\zeta_n(s,\tau) = e^{\int_{v_{k-1}}^{v_j} \theta(\alpha)d\alpha}$ if $\tau \in [v_{k-1}, v_k), s \in (v_{j-1}, v_j]$, $j,k = 0, 1, \ldots, n$ and $\zeta_n(s,\tau) = 0$ if $\tau > s$ or $s > t$. For this step function and analogously for any step function, we define the stochastic integral in the following way:

$$\text{(5.2)} \quad \begin{aligned} &\int_0^t \int_0^s \zeta_n(s,w)dB_H(w)dB_H(s) \\ &\equiv \sum_{j=1}^n \sum_{k=1}^j \zeta_n(v_{j-1}, v_{k-1})[B_H(v_k)-B_H(v_{k-1})][B_H(v_j)-B_H(v_{j-1})]. \end{aligned}$$

where $v_j, v_k \in \{v_0 = 0, v_1, v_2, \ldots, v_n = t \leq T\}$. Thus $\zeta_n(s,u) \to e^{\int_u^s \theta(\alpha)d\alpha}$ if $u \leq s$ and $\zeta_n(s,u) \to 0$ if $u > s$. Since ζ_n is uniformly bounded by $\max(e^{\int_u^s \theta(\alpha)d\alpha})$, for $s, u \in [0,t]$, it converges uniformly in $s, u \in [0,t]$. We now wish to show that the right-hand side of (5.2) converges as the norm of the partition, π_n, approaches 0.

To see this, we will want to show the right-hand side of (5.2) is a Cauchy sequence in quadratic mean. Since the space on which B_H lives is a complete Hilbert space, each Cauchy sequence must converge to a limit. This limit will be by definition the integral. Let us begin by observing the following Theorem.

THEOREM 5.1 (Soong, 1973, p. 28 and p. 32). *Let $W_1, \ldots, W_4$ be 4 jointly Gaussian zero mean random variables. Then,*

$$E[W_1 \ldots W_4] = E[W_1 W_2]E[W_3 W_4] + E[W_1 W_3]E[W_2 W_4]$$
$$+ E[W_1 W_4]E[W_2 W_3].$$

Let π_n and π_m be two partitions of $[0,t]$. Without loss of generality, we may consider the union of these partitions, $\pi_n \cup \pi_m = \pi_{nm} = \{v_1 \leq \ldots \leq v_N\}$ where $N = m + n$. Let $h = \|\pi_{nm}\|$. Some of the v_j's may be redundant. However, the differences, $B_H(v_k) - B_H(v_{k-1})$, in this case will be 0. We have the following result.

LEMMA 5.2.

1) $E[B_H(v_i) - B_H(v_{i-1})][B_H(v_j) - B_H(v_{j-1})]$

$$= \frac{V_H}{2}[-|v_i - v_j|^{2H} + |v_{i-1} - v_j|^{2H} + |v_i - v_{j-1}|^{2H} - |v_{i-1} - v_{j-1}|^{2H}].$$

2) $E[B_H(v_i) - B_H(v_{i-1})][B_H(v_j) - B_H(v_{j-1})][B_H(v_k) - B_H(v_{k-1})]$

$$\times [B_H(v_l) - B_H(v_{l-1})] \leq 3[(2t+1)h]^2 = O(h^2).$$

Proof. By the Soong Theorem 5.1, since B_H is a Gaussian random variable

$$E[B_H(v_i) - B_H(v_{i-1})][B_H(v_j) - B_H(v_{j-1})][B_H(v_k) - B_H(v_{k-1})]$$

$$\times [B_H(v_l) - B_H(v_{l-1})] = E[B_H(v_i) - B_H(v_{i-1})][B_H(v_j) - B_H(v_{j-1})]$$

$$\times E[B_H(v_k) - B_H(v_{k-1})][B_H(v_l) - B_H(v_{l-1})]$$

$$+ E[B_H(v_i) - B_H(v_{i-1})][B_H(v_k) - B_H(v_{k-1})]$$

$$\times E[B_H(v_j) - B_H(v_{j-1})][B_H(v_l) - B_H(v_{l-1})]$$

$$+ E[B_H(v_i) - B_H(v_{i-1})][B_H(v_l) - B_H(v_{l-1})]$$

$$\times E[B_H(v_j) - B_H(v_{j-1})][B_H(v_k) - B_H(v_{k-1})].$$

Let us consider expressions of the form

$$E[B_H(v_i) - B_H(v_{i-1})][B_H(v_j) - B_H(v_{j-1})]$$

$$(5.3) \qquad = E\big[B_H(v_i) B_H(v_j) - B_H(v_{i-1}) B_H(v_j)$$

$$- B_H(v_i) B_H(v_{j-1}) + B_H(v_{i-1}) B_H(v_{j-1})\big].$$

Since B_H is a zero mean Gaussian process, the right-hand side of (5.3) represents four covariances. From Equation (2.7) we have

$$E[B_H(v_i) - B_H(v_{i-1})][B_H(v_j) - B_H(v_{j-1})]$$

$$= \frac{V_H}{2}\Big[|v_i|^{2H} + |v_j|^{2H} - |v_i - v_j|^{2H}$$

$$- |v_{i-1}|^{2H} - |v_j|^{2H} + |v_{i-1} - v_j|^{2H}$$

$$- |v_i|^{2H} - |v_{j-1}|^{2H} + |v_i - v_{j-1}|^{2H}$$

$$+ |v_{i-1}|^{2H} + |v_{j-1}|^{2H} - |v_{i-1} - v_{j-1}|^{2H}\Big]$$

$$= \frac{V_H}{2}\Big[-|v_i - v_j|^{2H} + |v_{i-1} - v_j|^{2H}$$

$$+ |v_i - v_{j-1}|^{2H} - |v_{i-1} - v_{j-1}|^{2H}\Big].$$

Let us consider $|v_{i-1} - v_j|^{2H} - |v_i - v_j|^{2H}$ and let us assume for the moment that $v_j > v_i$. Then

$$|v_{i-1} - v_j|^{2H} - |v_i - v_j|^{2H} = (v_j - v_{i-1})^{2H} - (v_j - v_i)^{2H}$$

$$= (v_j - v_i + v_i - v_{i-1})^{2H} - (v_j - v_i)^{2H}$$

$$\leq (v_j - v_i + h)^{2H} - (v_j - v_i)^{2H}$$

$$\leq \max\{h^2 + 2h(v_j - v_i),\ h\}$$

$$\leq (2t + 1)h.$$

If $v_{i-1} \leq v_j \leq v_i$, then either $v_j = v_{i-1}$ or $v_j = v_i$ so that

$$|v_{i-1} - v_j|^{2H} - |v_i - v_j|^{2H} \leq (v_i - v_{i-1}) \leq h.$$

If $v_j \leq v_{i-1}$, then as before

$$|v_{i-1} - v_j|^{2H} - |v_i - v_j|^{2H} \leq (2t+1)h.$$

It follows then that

$$E[B_H(v_i) - B_H(v_{i-1})][B_H(v_j) - B_H(v_{j-1})] \leq (2t+1)h.$$

Similarly for the other five combinations, so that

$$\begin{aligned}
E[B_H(v_i) &- B_H(v_{i-1})][B_H(v_j) - B_H(v_{j-1})] \\
&\times [B_H(v_k) - B_H(v_{k-1})][B_H(v_l) - B_H(v_{l-1})] \leq 3[(2t+1)h]^2 \\
&= O(h^2).
\end{aligned}$$
$\square$

We are now in a position to prove the following result.

LEMMA 5.3.

$$\sum_{j=1}^{n}\sum_{k=1}^{j} \zeta_n(v_{j-1}, v_{k-1})[B_H(v_k) - B_H(v_{k-1})][B_H(v_j) - B_H(v_{j-1})]$$

is a Cauchy sequence in quadratic mean.

Proof. First note that for $a, b \in (-\infty, \infty)$, $|a - b|^2 \leq 2|a|^2 + |b|^2$.
Thus we have

$$
\begin{aligned}
E\Bigg| \sum_{i=1}^{n}\sum_{j=1}^{i} &\zeta_n(v_{i-1}, v_{j-1})\big[B_H(v_i) - B_H(v_{i-1})\big]\big[B_H(v_j) - B_H(v_{j-1})\big] \\
&- \sum_{k=1}^{m}\sum_{l=1}^{k} \zeta_m(v_{k-1}, v_{l-1})\big[B_H(v_k) - B_H(v_{k-1})\big]\big[B_H(v_l) - B_H(v_{l-1})\big] \Bigg|^2 \\
\leq E\Bigg\{ 2\Bigg| &\sum_{i=1}^{n}\sum_{j=1}^{i} \zeta_n(v_{i-1}, v_{j-1})\big[B_H(v_i) - B_H(v_{i-1})\big]\big[B_H(v_j) - B_H(v_{j-1})\big]\Bigg|^2 \Bigg\} \\
+ E\Bigg\{ 2\Bigg| &\sum_{k=1}^{m}\sum_{l=1}^{k} \zeta_m(v_{k-1}, v_{l-1})\big[B_H(v_k) - B_H(v_{k-1})\big]\big[B_H(v_l) - B_H(v_{l-1})\big]\Bigg|^2 \Bigg\} \\
= 2\sum_{i=1}^{n}\sum_{j=1}^{i}\sum_{k=1}^{n}\sum_{l=1}^{k} &\zeta_n(v_{i-1}, v_{j-1})\zeta_n(v_{k-1}, v_{l-1})E\Big\{ \big[B_H(v_i) - B_H(v_{i-1})\big] \\
&\times \big[B_H(v_j) - B_H(v_{j-1})\big]\big[B_H(v_k) - B_H(v_{k-1})\big]\big[B_H(v_l) - B_H(v_{l-1})\big] \Big\} \\
+ 2\sum_{i=1}^{m}\sum_{j=1}^{i}\sum_{k=1}^{m}\sum_{l=1}^{k} &\zeta_m(v_{i-1}, v_{j-1})\zeta_m(v_{k-1}, v_{l-1})E\Big\{ \big[B_H(v_i) - B_H(v_{i-1})\big] \\
&\times \big[B_H(v_j) - B_H(v_{j-1})\big]\big[B_H(v_k) - B_H(v_{k-1})\big]\big[B_H(v_l) - B_H(v_{l-1})\big] \Big\}.
\end{aligned}
$$
$$(5.4)$$

Both terms in the expression (5.4) are similar except for the m and n. Consider the first term

$$2\sum_{i=1}^{n}\sum_{j=1}^{i}\sum_{k=1}^{n}\sum_{l=1}^{k}\zeta_n(v_{i-1},v_{j-1})\zeta_n(v_{k-1},v_{l-1})E\Big\{\big[B_H(v_i)-B_H(v_{i-1})\big]$$

$$\times\big[B_H(v_j)-B_H(v_{j-1})\big]\big[B_H(v_k)-B_H(v_{k-1})\big]\big[B_H(v_l)-B_H(v_{l-1})\big]\Big\}$$

$$=2\sum_{i=1}^{n}\sum_{j=1}^{i}\sum_{k=1}^{n}\sum_{l=1}^{k}\zeta_n(v_{i-1},v_{j-1})\zeta_n(v_{k-1},v_{l-1})\Big\{E\big[B_H(v_i)-B_H(v_{i-1})\big]$$

$$\times\big[B_H(v_j)-B_H(v_{j-1})\big]E\big[B_H(v_k)-B_H(v_{k-1})\big]\big[B_H(v_l)-B_H(v_{l-1})\big]$$

$$+E\big[B_H(v_i)-B_H(v_{i-1})\big]\big[B_H(v_k)-B_H(v_{k-1})\big]$$

$$\times E\big[B_H(v_j)-B_H(v_{j-1})\big]\big[B_H(v_l)-B_H(v_{l-1})\big]$$

$$+E\big[B_H(v_i)-B_H(v_{i-1})\big]\big[B_H(v_l)-B_H(v_{l-1})\big]$$

$$\times E\big[B_H(v_j)-B_H(v_{j-1})\big]\big[B_H(v_k)-B_H(v_{k-1})\big]\Big\}.$$

There are three similar terms on the right-hand side. We consider the first. The others can be treated in a similar way. Let $\Delta v_r = v_r - v_{r-1}$. Then consider for every $v_i,\,v_{i-1},v_{j-1}$ and $v_j \in \pi_n \cup \pi_m$,

$$\lim_{\Delta v_i,\,\Delta v_j \to 0} E\left\{\frac{\big[B_H(v_i)-B_H(v_{i-1})\big]}{\Delta v_i}\frac{\big[B_H(v_j)-B_H(v_{j-1})\big]}{\Delta v_j}\right\}$$

$$\longrightarrow V_H H(2H-1)\big|v_i - v_j\big|^{2H-2}$$

from Barton and Poor (1988, Eqs. 2.5 & 2.13) and using Lemma 5.2. This expression is also bounded for $v_i,\,v_j \in [0,\,t]$. Hence for a sufficiently refined partition $\pi_n \cup \pi_m$, there must be an $\epsilon > 0$ independent of v_i and v_j such that,

$$V_H H(2H-1)|v_i - v_j|^{2H-2} - \epsilon$$

$$\leq E\left\{\frac{\big[B_H(v_i)-B_H(v_{i-1})\big]}{\Delta v_i}\frac{\big[B_H(v_j)-B_H(v_{j-1})\big]}{\Delta v_j}\right\}$$

$$\leq V_H H(2H-1)|v_i - v_j|^{2H-2} + \epsilon.$$

Thus we have for a sufficiently refined partition $\pi_n \cup \pi_m$,

$$\sum_{i=1}^{n}\sum_{j=1}^{i}\sum_{k=1}^{n}\sum_{l=1}^{k}\zeta_n(v_{i-1},v_{j-1})\zeta_n(v_{k-1},v_{l-1})\Big[V_H H(2H-1)|v_i-v_j|^{2H-2}-\epsilon\Big]$$

$$\times\Big[V_H H(2H-1)|v_k-v_l|^{2H-2}-\epsilon\Big]\Delta v_i\Delta v_j\Delta v_k\Delta v_l$$

$$\leq \zeta_n(v_{i-1},v_{j-1})\zeta_n(v_{k-1},v_{l-1})E\big[B_H(v_i)-B_H(v_{i-1})\big]$$

$$(5.5)\qquad\times\big[B_H(v_j)-B_H(v_{j-1})\big]E\big[B_H(v_k)-B_H(v_{k-1})\big]\big[B_H(v_l)-B_H(v_{l-1})\big]$$

$$\leq \sum_{i=1}^{n}\sum_{j=1}^{i}\sum_{k=1}^{n}\sum_{l=1}^{k}\zeta_n(v_{i-1},v_{j-1})\zeta_n(v_{k-1},v_{l-1})\Big[V_H H(2H-1)|v_i-v_j|^{2H-2}+\epsilon\Big]$$

$$\times\Big[V_H H(2H-1)|v_k-v_l|^{2H-2}+\epsilon\Big]\Delta v_i\Delta v_j\Delta v_k\Delta v_l.$$

Taking lim inf across the left-hand inequality in (5.5) and lim sup across the right-hand inequality and since the end expressions are Riemann integrable, we obtain

$$\int_0^t\int_0^s\int_0^t\int_0^r \zeta(s,\tau)\zeta(r,\alpha)[V_H H(2H-1)|s-\tau|^{2H-2}-\epsilon]$$

$$\times[V_H H(2H-1)|r-\alpha|^{2H-2}-\epsilon]dsd\tau drd\alpha$$

$$\leq \liminf \sum_{i=1}^{n}\sum_{j=1}^{i}\sum_{k=1}^{n}\sum_{l=1}^{k}\zeta_n(v_{i-1},v_{j-1})\zeta_n(v_{k-1},v_{l-1})E\big[B_H(v_i)-B_H(v_{i-1})\big]$$

$$\times\big[B_H(v_j)-B_H(v_{j-1})\big]E\big[B_H(v_k)-B_H(v_{k-1})\big]\big[B_H(v_l)-B_H(v_{l-1})\big]$$

$$\leq \limsup \sum_{i=1}^{n}\sum_{j=1}^{i}\sum_{k=1}^{n}\sum_{l=1}^{k}\zeta_n(v_{i-1},v_{j-1})\zeta_n(v_{k-1},v_{l-1})E\big[B_H(v_i)-B_H(v_{i-1})\big]$$

$$\times\big[B_H(v_j)-B_H(v_{j-1})\big]E\big[B_H(v_k)-B_H(v_{k-1})\big]\big[B_H(v_l)-B_H(v_{l-1})\big]$$

$$\leq \int_0^t\int_0^s\int_0^t\int_0^r \zeta(s,\tau)\zeta(r,\alpha)[V_H H(2H-1)|s-\tau|^{2H-2}+\epsilon]$$

$$\times[V_H H(2H-1)|r-\alpha|^{2H-2}+\epsilon]dsd\tau drd\alpha.$$

But $\epsilon>0$ was arbitrary so that

$$\lim_{n\to\infty}\sum_{i=1}^{n}\sum_{j=1}^{i}\sum_{k=1}^{n}\sum_{l=1}^{k}\zeta_n(v_{i-1},v_{j-1})\zeta_n(v_{k-1},v_{l-1})E\big[B_H(v_i)-B_H(v_{i-1})\big]$$

$$(5.6)\qquad\times\big[B_H(v_j)-B_H(v_{j-1})\big]E\big[B_H(v_k)-B_H(v_{k-1})\big]\big[B_H(v_l)-B_H(v_{l-1})\big]$$

$$=\int_0^t\int_0^s\int_0^t\int_0^r \zeta(s,\tau)\zeta(r,\alpha)[V_H H(2H-1)|s-\tau|^{2H-2}]$$

$$\times\big[V_H H(2H-1)|r-\alpha|^{2H-2}\big]dsd\tau drd\alpha.$$

Let us denote the integral in (5.6) by C for convenience. It follows immediately that

$$\lim_{n,m\to\infty} E\Bigg| \sum_{i=1}^{n}\sum_{j=1}^{i} \zeta_n(v_{i-1},v_{j-1})\big[B_H(v_i)-B_H(v_{i-1})\big]\big[B_H(v_j)-B_H(v_{j-1})\big]$$

$$-\sum_{k=1}^{m}\sum_{l=1}^{k} \zeta_m(v_{k-1},v_{l-1})\big[B_H(v_k)-B_H(v_{k-1})\big]\big[B_H(v_l)-B_H(v_{l-1})\big]\Bigg|^2 < \infty.$$

We expand this expression

$$E\Bigg| \sum_{i=1}^{n}\sum_{j=1}^{i} \zeta_n(v_{i-1},v_{j-1})\big[B_H(v_i)-B_H(v_{i-1})\big]\big[B_H(v_j)-B_H(v_{j-1})\big]$$

$$-\sum_{k=1}^{m}\sum_{l=1}^{k} \zeta_m(v_{k-1},v_{l-1})\big[B_H(v_k)-B_H(v_{k-1})\big]\big[B_H(v_l)-B_H(v_{l-1})\big]\Bigg|^2$$

$$= E\Bigg[\sum_{i=1}^{n}\sum_{j=1}^{i} \zeta_n(v_{i-1},v_{j-1})\big[B_H(v_i)-B_H(v_{i-1})\big]\big[B_H(v_j)-B_H(v_{j-1})\big]\Bigg]^2$$

$$- 2E\Bigg[\sum_{i=1}^{n}\sum_{j=1}^{i} \zeta_n(v_{i-1},v_{j-1})\big[B_H(v_i)-B_H(v_{i-1})\big]\big[B_H(v_j)-B_H(v_{j-1})\big]$$

$$\times \sum_{k=1}^{m}\sum_{l=1}^{k} \zeta_m(v_{k-1},v_{l-1})\big[B_H(v_k)-B_H(v_{k-1})\big]\big[B_H(v_l)-B_H(v_{l-1})\big]\Bigg]$$

$$+ E\Bigg[\sum_{k=1}^{m}\sum_{l=1}^{k} \zeta_m(v_{k-1},v_{l-1})\big[B_H(v_k)-B_H(v_{k-1})\big]\big[B_H(v_l)-B_H(v_{l-1})\big]\Bigg]^2.$$

Taking limits as n and m go to ∞, we have

$$\lim_{n,m\to\infty} E\Bigg| \sum_{i=1}^{n}\sum_{j=1}^{i} \zeta_n(v_{i-1},v_{j-1})\big[B_H(v_i)-B_H(v_{i-1})\big]\big[B_H(v_j)-B_H(v_{j-1})\big]$$

$$-\sum_{k=1}^{m}\sum_{l=1}^{k} \zeta_m(v_{k-1},v_{l-1})\big[B_H(v_k)-B_H(v_{k-1})\big]\big[B_H(v_l)-B_H(v_{l-1})\big]\Bigg|^2$$

$$= C - 2C + C = 0.$$

$\square$

DEFINITION 5.1. *The stochastic integral $\int_0^t X(s)dB_H(s)$ is defined as the quadratic mean limit of $\sum_{j=1}^{n}\sum_{k=1}^{j}\zeta_n(v_{j-1}, v_{k-1})[B_H(v_k) - B_H(v_{k-1})][B_H(v_j) - B_H(v_{j-1})]$ as $n \to \infty$.*

THEOREM 5.4. *The stochastic integral $\int_0^t X(s)dB_H(s)$ exists and is well defined. The usual properties of an integral hold.*

Proof. By Lemma 5.3, $\sum_{j=1}^{n}\sum_{k=1}^{j}\zeta_n(v_{j-1}, v_{k-1})[B_H(v_k) - B_H(v_{k-1})][B_H(v_j) - B_H(v_{j-1})]$ is a Cauchy sequence. Thus, we have

$\sum_{j=1}^{n} \sum_{k=1}^{j} \zeta_n(v_{j-1}, \ v_{k-1})[B_H(v_k) - B_H(v_{k-1})]$ converges in quadratic mean to a limit process since the space is complete. It is straightforward to show that the ordinary properties of an integral hold since the integral is approximated by the double sum. $\qquad\Box$

DEFINITION 5.2. *The stochastic integral,* $\int_0^t X(s)dX(s)$, *is defined as* $\int_0^t \theta(s)\Big[X(s)\Big]^2 ds + \int_0^t X(s)dB_H(s)$.

6. Summary. We have considered the stochastic differential equations, $dX(t) = \theta X(t)dt + dB_H(t); \ t > 0$, and $dX(t) = \theta(t)X(t)dt + dB_H(t); \ t > 0$ where $B_H(t)$ is fractional Brownian motion. We have found solutions for these differential equations and have shown the existence of the integrals related to these solutions. We then showed that $B_H(t)$ is not a martingale. This implies that several conventional methods for defining integrals on fractional Brownian motion are inadequate. We formally demonstrated the existence of an estimator for θ or $\theta(t)$ but that estimator depended on the existence of integrals which we did not know existence. We concluded by showing the existence and Riemann sum approximations for these integrals.

REFERENCES

BARTON R.J. AND POOR V.H. (1988), "Signal Detection in Fractional Gaussian Noise," IEEE Transactions on Information Theory, **34**: 943–959.

CHRISTOPEIT N. (1986), "Quasi-Least-Squares Estimation in Semimartingale Regression Models," Stochastics, **16**: 255–278.

CRAMER H. AND LEADBETTER M.R. (1967), Stationary and Related Stochastic Processes, John Wiley and Sons, Inc.: New York.

DOBRUSHIN R. (1979), "Gaussian and their subordinated generalized fields," Annals of Probability, **7**: 1–28.

GREGOTSKI M.E., JENSEN O., AND ARKANI-HAMED J. (1991), "Fractal Stochastic Modeling of Aeromagnetic Data," Geophysics, **56**(11): 1706–1715.

MAJOR P. (1981), Multiple Wiener-Ito Integrals, Lecture Notes in Mathematics, Springer-Verlag: New York.

MANDELBROT B.B. (1983), The Fractal Geometry of Nature, W.H. Freeman and Company: New York.

SHIRYAYEV A.N. (1984), Probability, Springer-Verlag: New York.

SOONG T.T. (1973), Random Differential Equations in Science and Engineering, Academic Press, Inc.: New York.

STEWART C.V., MOGHADDAM B., HINTZ K.J., AND NOVAK L.M. (1993), Fractional Brownian Motion Models for Synthetic Aperture Radar Imagery Scene Segmentation," Proceedings of the IEEE, **81**(10): 1511–1522.

WEGMAN E.J. AND HABIB M.K. (1992), "Stochastic Methods for Neural Systems," J. Statistical Planning and Inference, **33**: 5–26.

LIST OF WORKSHOP PARTICIPANTS

- Dale N. Anderson, Pacific Northwest National Laboratory
- Elizabeth M. Andrews, Department of Statistics, Colorado State University
- Ana Monica Costa Antunes, Department of Mathematics, University of Manchester Institute of Science and Technology (UMIST)
- Douglas N. Arnold, Institute for Mathematics and its Applications, University of Minnesota
- Santiago Betelu, Department of Mathematics, University of North Texas
- Christopher Binghamn, School of Statistics, University of Minnesota
- Jamylle Carter, School of Mathematics, University of Minnesota
- Christine Calynn T. Cheng, Department of Electrical Engineering and Computer Science, University of Wisconsin-Milwaukee
- Richard Davis, Department of Statistics, Colorado State University
- Doug Dokken, Department of Mathematics, University of St. Thomas
- Gregory S. Duane, NCAR
- Fabien Dubuffet, Minnesota Supercomputing Institute University of Minnesota
- William T.m. Dunsmuir Division of Biostatistics, School of Statistics, University of Minnesota
- Michael Efroimsky, AA Department, US Naval Observatory
- Selim Esedoglu, Department of Mathematics, University of California - Los Angeles
- Robert Gulliver, School of Mathematics, University of Minnesota
- Shaleen Jain, CIRES Climate Diagnostics Center, NOAA
- Daniel Kern Department of Mathematical Sciences, University of Nevada, Las Vegas
- Sung-Eun Kim Department of Mathematical Sciences University of Cincinnati
- Genshiro Kitagawa, The Institute of Statistical Mathematics, Mina-to-ku, Tokyo, Japan
- Yngvar Larsen, University of Minnesota
- Keh-Shin Lii, Department of Statistics, University of California - Riverside
- Catherine Majumder, Department of Geophysics, University of Minnesota

- Aurelia Minut, Institute for Mathematics and its Applications, University of Minnesota
- Gary W. Oehlert, School of Statistics, University of Minnesota
- Miao-Jung Yvonne Ou, University Central Florida
- Tohru Ozaki, The Institute of Statistical Mathematics, 4-6-7 Minami Azabu Minato-ku, Tokyo, Japan
- Donald B. Percival, Applied Physics Laboratory, University of Washington
- Jianliang Qian, Department of Mathematics, University of California - Los Angeles
- Gabriel A. Rodriguez-Yam, Department of Statistics, Colorado State University
- Murray Rosenblatt, Department of Mathematics, University of California - San Diego
- Fadil Santosa, Institute for Mathematics and its Applications, University of Minnesota
- Robert H. Shumway, Department of Statistics, University of California - Davis
- David S. Stoffer, Department of Statistics, University of Pittsburgh
- Tata Subba Rao, Department of Mathematics, University of Manchester Institute of Science and Technology (UMIST)
- Tze Chein Sun, Department of Mathematics, Wayne State University
- David J. Thomson, Bell Laboratories, Lucent Technologies
- Donald Turcotte, Department of Geological Sciences, Cornell University
- Edward J. Wegman, Center for Computational Statistics, George Mason University
- Wei Biao Wu, Department of Statistics, University of Chicago
- Zhongjie Xie, School of Mathematical Sciences, Peking University
- Toshio Yoshikawa, University of Minnesota
- David A. Yuen, Department of Geology and Geophysics, University of Minnesota

1999–2000 Reactive Flows and Transport Phenomena
2000–2001 Mathematics in Multimedia
2001–2002 Mathematics in the Geosciences
2002–2003 Optimization
2003–2004 Probability and Statistics in Complex Systems: Genomics,
           Networks, and Financial Engineering
2004–2005 Mathematics of Materials and Macromolecules: Multiple Scales,
           Disorder, and Singularities
2005-2006 Imaging
2006-2007 Applications of Algebraic Geometry

IMA SUMMER PROGRAMS

1987 Robotics
1988 Signal Processing
1989 Robust Statistics and Diagnostics
1990 Radar and Sonar (June 18–29)
      New Directions in Time Series Analysis (July 2–27)
1991 Semiconductors
1992 Environmental Studies: Mathematical, Computational, and
           Statistical Analysis
1993 Modeling, Mesh Generation, and Adaptive Numerical Methods
           for Partial Differential Equations
1994 Molecular Biology
1995 Large Scale Optimizations with Applications to Inverse Problems,
           Optimal Control and Design, and Molecular and Structural
           Optimization
1996 Emerging Applications of Number Theory (July 15–26)
      Theory of Random Sets (August 22–24)
1997 Statistics in the Health Sciences
1998 Coding and Cryptography (July 6–18)
      Mathematical Modeling in Industry (July 22–31)
1999 Codes, Systems, and Graphical Models (August 2–13, 1999)
2000 Mathematical Modeling in Industry: A Workshop for Graduate
           Students (July 19–28)
2001 Geometric Methods in Inverse Problems and PDE Control
           (July 16–27)
2002 Special Functions in the Digital Age (July 22–August 2)
2003 Probability and Partial Differential Equations in Modern
           Applied Mathematics (July 21–August 1)
2004 n-Categories: Foundations and Applications (June 7–18)

IMA "HOT TOPICS" WORKSHOPS

- Challenges and Opportunities in Genomics: Production, Storage,
  Mining and Use, April 24–27, 1999

- Decision Making Under Uncertainty: Energy and Environmental Models, July 20–24, 1999
- Analysis and Modeling of Optical Devices, September 9–10, 1999
- Decision Making under Uncertainty: Assessment of the Reliability of Mathematical Models, September 16–17, 1999
- Scaling Phenomena in Communication Networks, October 22–24, 1999
- Text Mining, April 17–18, 2000
- Mathematical Challenges in Global Positioning Systems (GPS), August 16–18, 2000
- Modeling and Analysis of Noise in Integrated Circuits and Systems, August 29–30, 2000
- Mathematics of the Internet: E-Auction and Markets, December 3–5, 2000
- Analysis and Modeling of Industrial Jetting Processes, January 10–13, 2001
- Special Workshop: Mathematical Opportunities in Large-Scale Network Dynamics, August 6–7, 2001
- Wireless Networks, August 8–10 2001
- Numerical Relativity, June 24–29, 2002
- Operational Modeling and Biodefense: Problems, Techniques, and Opportunities, September 28, 2002
- Data-driven Control and Optimization, December 4–6, 2002
- Agent Based Modeling and Simulation, November 3–6, 2003
- Enhancing the Search of Mathematics, April 26-27, 2004
- Compatible Spatial Discretizations for Partial Differential Equations, May 11-15, 2004

SPRINGER LECTURE NOTES FROM THE IMA:

The Mathematics and Physics of Disordered Media
    Editors: Barry Hughes and Barry Ninham
    (Lecture Notes in Math., Volume 1035, 1983)

Orienting Polymers
    Editor: J.L. Ericksen
    (Lecture Notes in Math., Volume 1063, 1984)

New Perspectives in Thermodynamics
    Editor: James Serrin
    (Springer-Verlag, 1986)

Models of Economic Dynamics
    Editor: Hugo Sonnenschein
    (Lecture Notes in Econ., Volume 264, 1986)

FSC
www.fsc.org
MIX
Papier aus verantwortungsvollen Quellen
Paper from responsible sources
FSC® C105338